AF879453

ASTROPHYSICAL JETS

ASTROPHYSICS AND SPACE SCIENCE LIBRARY

A SERIES OF BOOKS ON THE RECENT DEVELOPMENTS
OF SPACE SCIENCE AND OF GENERAL GEOPHYSICS AND ASTROPHYSICS
PUBLISHED IN CONNECTION WITH THE JOURNAL
SPACE SCIENCE REVIEWS

Editorial Board

J. E. BLAMONT, *Laboratoire d'Aeronomie, Verrières, France*
R. L. F. BOYD, *University College, London, England*
L. GOLDBERG, *Kitt Peak National Observatory, Tucson, Ariz., U.S.A.*
C. DE JAGER, *University of Utrecht, The Netherlands*
Z. KOPAL, *University of Manchester, England*
G. H. LUDWIG, *NOAA, Environmental Research Laboratories, Boulder, CO, U.S.A.*
R. LÜST, *President Max-Planck-Gesellschaft zur Förderung der Wissenschaften, München, F.R.G.*
B. M. McCORMAC, *Lockheed Palo Alto Research Laboratory, Palo Alto, Calif., U.S.A.*
H. E. NEWELL, *Alexandria, Va., U.S.A.*
L. I. SEDOV, *Academy of Sciences of the U.S.S.R., Moscow, U.S.S.R.*
Z. ŠVESTKA, *University of Utrecht, The Netherlands*

VOLUME 103
PROCEEDINGS

ASTROPHYSICAL JETS

PROCEEDINGS OF AN INTERNATIONAL WORKSHOP
HELD IN TORINO, ITALY, OCTOBER 7-9, 1982

Edited by

A. FERRARI
University of Torino, Italy

and

A. G. PACHOLCZYK
University of Arizona, Tucson, U.S.A.

D. REIDEL PUBLISHING COMPANY
A MEMBER OF THE KLUWER ACADEMIC PUBLISHERS GROUP
DORDRECHT / BOSTON / LANCASTER

Library of Congress Cataloging in Publication Data

CIP

Main entry under title:

Astrophysical jets.

(Astrophysics and space science library: v. 103)
Includes index.
1. Astrophysical jets–Congresses. 2. Radio sources (astronomy)–Congresses. I. Ferrari, Attilio, 1941– . II. Pacholczyk, A. G., 1935– . III. Università di Torino. Istituto di Fisica Generale. IV. Istituto di Cosmo-geofisica (Italy). V. Series.
QB466.J46.A78 1983 523 83–11116
ISBN-13: 978-94-009-7188-2 e-ISBN-13: 978-94-009-7186-8
DOI: 10.1007/978-94-009-7186-8

Published by D. Reidel Publishing Company,
P.O. Box 17, 3300 AA Dordrecht, Holland.

Sold and distributed in the U.S.A. and Canada
by Kluwer Academic Publishers,
190 Old Derby Street, Hingham, MA 02043, U.S.A.

In all other countries, sold and distributed
by Kluwer Academic Publishers Group,
P.O. Box 322, 3300 AH Dordrecht, Holland.

All Rights Reserved
Copyright © 1983 by D. Reidel Publishing Company, Dordrecht, Holland
Softcover reprint of the hardcover 1st edition 1983
No part of the material protected by this copyright notice may be reproduced or
utilized in any form or by any means, electronic or mechanical
including photocopying, recording or by any informational storage and
retrieval system, without written permission from the copyright owner

TABLE OF CONTENTS

INTRODUCTORY REMARKS

The idea of organizing a meeting on Extragalactic Jets originated at the time of the Albuquerque IAU Symposium on Extragalactic Radio Sources, when the presentation of the new high-resolution maps obtained at the Very Large Array made everybody confident that we were close to having a statistically significant sample of jets which would allow discussion of morphologies and physical parameters on a general, comprehensive basis. In Albuquerque most of the time was more inclined to discuss observations of jets and to test the validity of data reduction, rather than to fit those data into theoretical models. This was the right thing to do at the time, but a rich collection of possible interpretations were soon put forward so that theoretical predictions had to be discussed. We concluded, therefore, that a short interpretation-oriented meeting could be held.

Our small group of theorists here in Torino decided to promote such a meeting, with the additional aim of fostering the scientific activity of our university. We were glad to have the enthusiastic support of Andrzej Pacholczyk who worked at Torino many years ago prior to moving to the United States.

The response of the scientific community was very good, as witnessed by the long list of participants who came from all over the world. We were very glad to host them in our town in which is located the Istituto di Fisica, dedicated to Amedeo Avogadro, an illustrious representative of modern science, and professor at our University during the years 1820-1850. We are convinced that the discussions which took place during the two days of meetings complemented the scientific tradition of the old School.

There were 90 participants from 9 countries, and 35 papers were delivered (19 invited talks and 16 contributions). They are presented in these Proceedings; we are sure they will be useful to the astronomical community. In order to give the reader a feeling of the lively discussions that took place, we also tried to include the questions the speakers felt would clarify their talks.

In addition to the scientific efforts, Torino also sought to extend a warm welcome to the participants. In particular, a visit was organized to the Mount Blanc Laboratory of the Istituto di Cosmo-geofisica of the Consiglio Nazionale delle Ricerche; the scientific visit was preceded by an excursion to the top of the Mount

A. Ferrari and A. G. Pacholczyk (eds.), Astrophysical Jets, vii–viii.
Copyright © 1983 by D. Reidel Publishing Company.

Blanc by cableway. Those participants who could come enjoyed some cold but beautiful sightseeing, as illustrated in the group photograph.

In conclusion, we would like to personally express our deep appreciation to all who contributed to the demanding organization of the meeting. We thank the Local Committee, with Lorenzo Zaninetti, Lucia Bonafini, Silvano Massaglia, and above all, Mrs. Maria Luisa Agostini Marchese, who provided the most dedicated collaboration in looking after the accomodations for the visitors and preparing programs, lunches, transportation, etc. The Cassa di Risparmio di Torino continued its generous program in support of cultural events of the town, offering for our use the Centro Incontri, the theatre where the meeting was held. The Aeritalia, renowned for its participation in the space programs of the Shuttle, gladly sponsored our meeting in support of the scientific initiative. The Città di Torino offered hospitality as well as specific support for the editing of these Proceedings. Support was also provided by the Istituto Bancario San Paolo and by the Istituto di Fisica Nucleare, Sezione di Torino.

Finally, we once again extend our sincere thanks to all participants who contributed to the success of this initiative by their lively interest and discussions, and with their kind acceptance of any inconveniences which may have been related to organization. We also thank the members of the Scientific Committee whose advice was very helpful in selecting the topics to be discussed.

Attilio Ferrari
Andrzej G. Pacholczyk

SCIENTIFIC COMMITTEE

Attilio	FERRARI	Torino	(Chairman)
Francesco	BERTOLA	Padova	
Massimo	CALVANI	Padova	
Massimo	CAPACCIOLI	Padova	
Roberto	FANTI	Bologna	
Alberto	MASANI	Torino	
Luciano	NOBILI	Padova	
Giancarlo	SETTI	Bologna	
Edoardo	TRUSSONI	Torino	

LOCAL ORGANIZING COMMITTEE

Lorenzo Zaninetti (Secretary)
Maria Luisa Agostini Marchese
Lucia Bonafini
Silvano Massaglia

The Workshop was organized by:

Istituto di Fisica Generale, Università di Torino
Istituto di Cosmo-geofisica del Consiglio Nazionale delle Ricerche, Torino

Sponsoring organizations were:

Osservatorio Astronomico di Padova
Istituto di Radioastronomia del Consiglio Nazionale delle Ricerche, Bologna

Società Astronomica Italiana
Società Italiana di Fisica
Aeritalia, Società Aerospaziale Italiana
Cassa di Risparmio di Torino
Istituto Bancario San Paolo di Torino
Assessorato al Turismo della Città di Torino
Istituto Nazionale di Fisica Nucleare, Sezione di Torino

1. N. Dallaporta
2. G. Pelletier
3. M. Begelman
4. L. Nobili
5. Mrs. Dallaporta
6. Driver
7. J. Brodie
8. R. Davis
9. P. Barthel
10. Guest
11. G. Miley
12. S. Phinney
13. A. Ferrari
14. F. Fricke
15. P. Galeotti
16. M. Calvani
17. I. Browne
18. R. Porcas
19. E. Feigelson
20. S. Massaglia
21. Mrs. Fomalont
22. Mrs. Ferrari
23. M. Sikora
24. W. Jaegers
25. L. Zaninetti
26. M. Nepveu
27. E. Fomalont

The photograph was taken by R. Schilizzi, who kindly provided the Editors with a copy.

LIST OF PARTICIPANTS

Barthel, P.D.	Sterrewacht, Leiden, NL
Begelman, M.C.	University of Colorado, Boulder, USA
Benford, G.	University of California, Irvine, USA
Bertola, F.	Osservatorio Astronomico, Padova, Italy
Bianchi, L.	Osservatorio Astronomico, Pino T.se, Italy
Biermann, P.	MPIfRadioastronomie, Bonn, FRG
Bodo, G.	Osservatorio Astronomico, Pino T.se, Italy
Briel, U.	MPIfExtraterrestrische Physik, Garching, FRG
Brodie, J.	Institute of Astronomy, Cambridge, UK
Browne, I.A.	NRAL, Jodrell Bank, UK
Burns, J.O.	University of New Mexico, Albuquerque, USA
Calvani, M.	Istituto di Fisica, Padova, Italy
Capaccioli, M.	Osservatorio Astronomico, Padova, Italy
Castagnoli, C.	Istituto di Fisica Generale, Torino, Italy
Cazzola, P.	Istituto di Fisica, Padova, Italy
Chiuderi, C.	Osservatorio Astronomico, Arcetri, Italy
Cini Castagnoli, G.	Istituto di Cosmo-geofisica, Torino, Italy
Coppi, B.	MIT, Cambridge, USA
Dallaporta, M.	Int. School Advanced Studies, Trieste, Italy
Danziger, I.J.	ESO, Garching, FRG
Davis, R.J.	NRAL, Jodrell Bank, UK
Facondi Bonsignori S.	Istituto di Radioastronomia, Bologna, Italy
Fanti, C.	Istituto di Radioastronomia, Bologna, Italy
Fanti, R.	Istituto di Radioastronomia, Bologna, Italy
Feigelson, E.D.	MIT, Cambridge, USA
Ferrari, A.	Istituto di Fisica Generale, Torino, Italy
Feretti, L.	Istituto di Radioastronomia, Bologna, Italy
Fomalont, E.B.	NRAO, Socorro, USA
Fricke, K.J.	Universitäts Sternwarte, Göttingen, FRG
Gallino, R.	Istituto di Cosmo-geofisica, Torino, Italy
Gavazzi, G.	Istituto di Fisica Cosmica, Milano, Italy
Giovannini, G.	Istituto di Radioastronomia, Bologna, Italy
Gregorini, L.	Istituto di Radioastronomia, Bologna, Italy
Grewing, M.	Astronomical Institute, Tübingen, FRG
Jaegers, W.J.	Sterrewacht, Leiden, NL
Kotanyi, G.C.	ESO, Garching, FRG
Lalande, P.Q.	Institut d'Astrophysique, Paris, France
Leborgne, J.F.	Observatoire du Pc-du-Midi, France
Londrillo, P.	Istituto di Radioastronomia, Bologna, Italy
Maccagni, D.	Istituto di Fisica Cosmica, Milano, Italy
Macchetto, F.	ESA, Astronomy Division, Noordwjik, NL
Mantovani, F.	Istituto di Radioastronomia, Bologna, Italy
Maraschi, L.	Istituto di Fisica Cosmica, Milano, Italy
Masani, A.	Istituto di Fisica Generale, Torino, Italy
Massaglia, S.	Istituto di Fisica Generale, Torino, Italy
Messina, A.	Istituto di Astronomia, Bologna, Italy
Miley, G.	Huyghens Laboratorium, Leiden, NL
Nepveu, M.	Astronomical Institute, Bonn, FRG
Nieto, J.-L.	Observatoire du Pic-du-Midi, France

Nobili, L.	Istituto di Fisica, Padova, Italy
Norman, M.H.	MPIfAstrophysik, Garching, FRG
Pacholczyk, A.G.	Specola Vaticana, Città del Vaticano
Padrielli, L.	Istituto di Radioastronomia, Bologna, Italy
Palumbo, G.	Istituto TESRE, Bologna, Italy
Pelletier, G.	Groupe d'Astrophysique, Grenoble, France
Perley, R.	NRAO, Socorro, USA
Perola, G.C.	Istituto di Astronomia, Roma, Italy
Phinney, E.S.	Institute of Astronomy, Cambridge, UK
Piragino, G.	Istituto di Fisica Generale, Torino, Italy
Porcas, R.W.	MPIfRadioastronomie, Bonn, FRG
Poyet, J.P.	Observatoire de Toulouse, France
Preuss, E.	MPIfRadioastronomie, Bonn, FRG
Ray, T.P.	Astronomy Center, Sussex, UK
Ruffini, R.	Istituto di Fisica, Roma, Italy
Saggion, A.	Istituto di Fisica, Padova, Italy
Saikia, D.J.	Tata Institute, Bangalore, India
Schilizzi, R.T.	Radiosterrenwacht, Dwingeloo, NL
Sikora, M.	Astronomical Center, Warsaw, Poland
Silvestro, G.	Istituto di Fisica Generale, Torino, Italy
Sol, H.	Observatoire de Paris, France
Stoeger, W.R.	Specola Vaticana, Città del Vaticano
Tanzella Nitti, G.	Osservatorio Astronomico, Pino T.se, Italy
Torricelli Ciamponi G.	Osservatorio Astronomico, Arcetri, Italy
Treves, A.	Istituto di Fisica, Milano, Italy
Trussoni, E.	Istituto di Cosmo-geofisica, Torino, Italy
Turolla, R.	Int. School Advanced Studies, Trieste, Italy
Zamorani, G.	Istituto di Radioastronomia, Bologna, Italy
Zaninetti, L.	Istituto di Fisica Generale, Torino, Italy
Zieba, S.	Scuola Normale Superiore, Pisa, Italy

SCIENTIFIC FOREWORD

Attilio Ferrari

I want to recall here the basic points I raised at the beginning of the Workshop as the main targets of discussion (in the name of the Scientific Committee). I attempted to focus the attention of participants on the fact that, in many instances, we tend to discuss jets in terms of simple physics, more or less as one did at the time extragalactic radio sources were discovered: for instance, we still use equipartition arguments. However, we must realize that processes in jets, leading to their morphologies and energetics clearly depend on complex plasma phenomena. Therefore, the same standard arguments used to derive characteristic parameters should be questioned; some of the speakers were invited to attempt a critical analysis of this point, and in fact I believe that this "inquisitive attitude" was actually present all along the Workshop.

Observers were asked to choose the parameters to be used in a statistical sample of jets. For this they were urged, first of all, to distinguish between primary and secondary features. For instance, are knots and wiggles common to all jets? Are relativistic flow velocities expected in all active nuclei? Are jets denser or lighter than the external medium?

On the theoretical side I asked to discuss whether or not existing models are in accordance with the limited statistical sample that we have today. And which should be the lines of development to be pursued first, and to what extent.

In particular I listed the following highlights for a methodological confrontation:

1. Confidence level of parameters derived from model independent standard arguments applied to observations.

2. Definition of morphological and physical parameters of a "standard" jet, with a clear indication of their dependence on model assumptions.

A. Ferrari and A. G. Pacholczyk (eds.), Astrophysical Jets, xv–xvi.
Copyright © 1983 by D. Reidel Publishing Company.

3. Definition of the minimum number of assumptions required to develop a physical model.

4. Extent to which (theoretical) models must be worked out for being reasonably applicable to observations.

5. Elaboration of predictions useful to observers.

My personal feeling is that these points were actually discussed at the meeting as it is witnessed by these Proceedings; to judge whether valid answers were reached is a matter of taste, but I am sure that those discussions were deeply educating to all of us.

SMALL SCALE STRUCTURE OF NONTHERMAL RADIO SOURCES

E. Preuss
Max-Planck-Institut für Radioastronomie, Bonn, Germany

ABSTRACT

Observational facts from VLBI relevant to the discussion of "jets" are reviewed in the following sections: 1. Introduction, 2. VLBI and astrophysical jets, 3. Current instrumental limits to VLBI, 4. Observational material (1981/1982), 5. Radio nuclei of powerful radio sources, 6. Radio nuclei of weak radio sources (Seyfert galaxies and mildly active "normal" galaxies), 7. Concluding remarks. The tables attached include lists of objects and samples observed by VLBI (and reported about during the publication period 1981/1982) and of nearby galaxies detected with VLBI.

1. INTRODUCTION

In this review I will try to summarize observational results from VLBI on extragalactic radio sources which seem important for a discussion about "astrophysical jets". The radio structures seen by VLBI on angular scales $\lesssim$ 50 milliarcsec are obviously related, in a more or less direct way, to the larger scale jets which are thought to power the extended, sometimes giant radio lobes associated with active galactic nuclei (including quasars). Comparing the jet with a river we are not sure in a given case whether we are looking at the young stream itself, obstacles in the water, the fog in the valley or the immediate surroundings of the origin. Although the imaging capability of VLBI is modest when compared with that of local interferometry, VLBI is of particular interest as it is still the only technique which probes directly into spatial scales $\lesssim$ pc and the potential of the method is far from being exhausted.

I have subdivided this talk in the following way: after some general remarks on "VLBI and astrophysical jets" (section 2) I will outline the current limits to VLBI as an observing technique (section 3) and give a brief progress report on recent VLBI work based on papers from the past two years (section 4). The purpose of this section is to facilitate the access to the observational material which has recently

A. Ferrari and A. G. Pacholczyk (eds.), Astrophysical Jets, 1–25.
Copyright © 1983 by D. Reidel Publishing Company.

become available. Lists of objects and samples observed, intended to be complete, are included here. Section 5 presents in summary form observational results on powerful radio sources and section 6 does the same for weaker sources, i.e. those associated with Seyfert galaxies and mildly active "normal" galaxies, objects which are only now about to become amenable to high sensitivity high resolution observations. I will conclude with a few remarks on the bearing of the available evidence on jet models.

Compact radio sources in general have been reviewed by Kellermann and Pauliny-Toth (1981). Reviews on VLBI work have recently been written by Readhead and Pearson (1982) on "The milliarcsecond structure of radio galaxies and quasars", by Cohen and Unwin (1982) on "Superluminal radio sources", by Phillips and Mutel (1982) on "Symmetric structure in compact radio sources" and by Preuss (1981) on "VLBI observations of jets and active nuclei" (= paper I). This paper, together with paper I is intended to give complete lists of extragalactic objects and samples of objects observed by VLBI since 1975. See also the report by Cohen (1980) for a complete VLBI bibliography up to mid-1980.

2. VLBI AND ASTROPHYSICAL JETS

2.1 Topical concepts.

The general acceptance of the concept of a "beam" or "jet" model and the attraction of models involving relativistic jets in particular become obvious if one looks at recent papers. The discussion sections of current VLBI papers typically refer to one or more of the following concepts "jet", "relativistic beaming" or "unifying scheme" and it is probably the latter aspect which is the most fascinating one about the relativistic beam hypothesis: i.e. the explanation of the apparent variety (e.g. the difference between "compact" and "extended" sources on scales $\gtrsim$ 1") by a different perspective (viewing angle) of intrinsically similar objects.

The discernible features of radio images, or "radio components" are tentatively looked at as: optically thick "cores" of radiating jets, zones of localised energy redistribution or acceleration of particles (obstacles, "clouds", shocks, instabilities), cocoons, or even accretion disks in nearby objects. Relatively little mention is made of the "adiabatically expanding cloud of relativistic electrons".

2.2 Some basic VLBI results.

The continuous, collimated outflow (beams, jets) of energy, mass and momentum was first postulated about ten years ago as supply mechanism for the extended emission regions of the (large and strong) double radio sources (see, e.g., Longair, Ryle and Scheuer 1973 and

references therein). These jets, as efficient transport channels, had of course to be basically invisible.
The results from early VLBI observations, such as the common presence of very compact central components and their frequent characteristic linear structure, were considered as strong circumstantial evidence in favour of this hypothesis. The alignment of radio structure on different size scales spanning a range $>10^6:1$ in some cases indicated that the jet may have a well-defined and stable geometrical axis, established on a scale < 1 pc.

The following findings emerged fairly early and have since been substantiated by many observations:

(I) A substantial fraction (if not all) of the extended (and compact) extragalactic radio sources have a core of $\sim$ milliarcsec scale size. Sources with flat or inverted cm spectra invariably have milliarcsec structure.

(II) All very compact (<10 pc) radio sources coincide with the nucleus of the parent galaxy (if any) or the QSO, whereas "hot spots" or "knots" in the large scale emission regions are typically $\gtrsim$ 100 pc.

(III) Pc-scale radio components typically have linear structure and the source axis is closely related to the direction of larger scale features.

(IV) There is no compelling evidence for more than one active center in strong sources.

(V) VLBI observations have not furnished any evidence forcing us to invoke radiation mechanisms other than incoherent synchrotron emission from relativistic electrons.

All directly measured brightness temperatures so far are below the limit $\sim 10^{12}$ K imposed by the interplay of Synchrotron - and Synchro-Compton emission processes. But note: for weaker sources ($S_\nu \lesssim 1$ Jy) earth-based interferometers are not sufficient to measure brightness temperatures $\gtrsim 5 \times 10^{11}$ K (independent of wavelength, see section 3).

2.3 Scale sizes: "target zones"

In order to visualise the spatial scales that can be resolved at cm wave-lengths, let us look at the linear resolution of the baseline Effelsberg (Germany) - Owens Valley (California), length D $\sim$ 8200 km at λ = 1.3 cm. The angular resolution of this baseline $\sim 0.5\ \lambda/D$ = 0$''$00016 is about the highest currently achievable. The corresponding linear sizes for some objects typifying distance classes are given in column 3 and 4; in column 4 in units of the radius R_c of a "collapsed object" of $10^9\ M_\odot$. $R_c \sim 1.5 \times 10^7\ M/M_\odot$ cm = 50 R (Schwarzschild) is defined by $GM^2/R_c^2 \gtrsim 0.01\ Mc^2$ as the radius of a sphere in which for a given mass M the gravitational energy exceeds the maximum thermonuclear energy which can be released.

Table 1

Object	Distance[a)]	Size corresponding to 0".00016 (cm)	($R_c \times 10^9 M_\odot/M$)
M81	3.25 Mpc	7.9×10^{15}	0.53
M87	21.9 Mpc	5.3×10^{16}	3.5
NGC1275	110 Mpc	2.6×10^{17}	17.3
3C147	z=0.595	3.7×10^{18}	247

a) $H_o = 50$ km s^{-1} Mpc^{-1}

Table 1 shows that for nearby ($\lesssim$100 Mpc) objects VLBI is sensitive to scale sizes which may be characteristic of accretion disks or other fundamental regions. The volumes resolved by VLBI are in any case comparable to the Broad Line Emission Regions ($\lesssim 10^{18}$ cm).

3. CURRENT INSTRUMENTAL LIMITS TO VLBI

To get a better idea of what can be expected from VLBI it is necessary to look at the performance of the instrumentation available. This is briefly outlined in this section.

Observing wavelengths (λ): 2.8, 6, 18 and 21 cm are routinely, and 1.3, 3.8, 13, 50 and 90 cm are occasionally used. Successful pilot experiments have been made at 4 mm.

Angular resolution for a baseline of length D is $\sim 0.5\ \lambda/D$ = 0".001 (λ/cm)/(D/1000 km) and ranges from $\sim$ 0".1 to 0".0001

Maximum measurable brightness temperature is $\sim 4 \times 10^{11}$ (S_ν/Jy) x (D/10 000 km)^2

Positional accuracy is a few 0".01, achieved both by VLBI and local interferometry

Relative positional accuracy $\lesssim$0".001 for a separation of 0°.5 (Shapiro et al. 1979)

Sensitivity: with the Mark II (2 MHz-) system and combinations of the largest telescopes one achieves a rms noise in correlated flux density (fringe amplitude) of a few mJy at dm and cm wavelengths. A map requires a source strength $\gtrsim$ 0.5 Jy. The Mark III (50 MHz-) system which is currently implemented is about five times more sensitive.

The dynamic range for VLBI images is $\sim$ 10:1, in some cases $\sim$ 25:1. See, e.g. Wilkinson (1983) or Readhead & Wilkinson (1978).

Polarisation: only first steps towards VLBI mapping of polarisation parameters have been made; see Cotton (1982)

Network facilities: 7 stations in the US and 5 in Europe are by now regularly available for 1 to 2 weeks every 2 months, and normally 2 frequencies are scheduled in each period. VLBI processors are operational in Bonn at the Max-Planck-Institut für Radioastronomie (Mark II and Mark III), in Westford, Massachussetts, USA, at the Haystack Observatory (Mark III), in Charlottesville, Virginia, at the National Radio Astronomy Observatory (Mark II) and in Pasadena, California, at the California Institute of Technology (Mark II, Mark III soon).

The monitoring capability of current networks for observations at regular intervals and at several frequencies is severely limited. VLBI at present relies mainly on a random collection of multi-purpose telescopes, some of which are equipped for only some of the observing frequencies. The requirements for rigorous monitoring can only be met by dedicated arrays possibly backed by a satellite station; see, e.g. NRAO-report (May 1982).

4. OBSERVATIONAL MATERIAL (1981/82)

This section is intended to be a short progress report covering the publication period 1981-1982. All relevant publications and preprints from this period which have come to my knowledge have been used to compile the lists presented in tables 2, 3 and 4. The lists of objects and samples are intended to be complete according to the defining criteria given below.

Some aspects of the observing strategies noticeable in current programs are:

- detailed studies of well-conditioned objects, "key" objects, if possible on different angular scales and at several frequencies;
- observations of statistically well-defined samples;
- advancements in the technique: towards higher frequencies, higher sensitivity and dynamic range, measurement of polarisation, and so forth.

Table 2 lists 77 extragalactic objects which have been detected by VLBI and at least individually discussed (if not mapped) in one of the references given. The latter criterion naturally makes some arbitrariness unavoidable.
6 objects from papers published earlier than 1981 have been added so that the list includes all 52 objects known to have been mapped by VLBI at the time of writing.

The objects mapped so far are typically powerful radio sources, at large distances and bright at cm wavelengths. They include:

- 16 of 82 sources listed by Bridle (1981) as having large scale (> 1")

Table 2

Selection of extragalactic objects observed with VLBI (publication period covered: 1981-1982)

object		z or distance	type[a]	code[b]	references
0046+316	Mk348	0.0148	SO		33
0055+300	NGC315	0.0167	E	J *	23, 52
0133+476	OC457	0.86?	Lac	*	37
0212+735			Lac	*	6, 40, 12
0219+428	3C66A		Lac		2
0235+164			Lac		2
0316+162	CTA21		U		71
0316+413	3C84	0.0176	E	*	35, 37, 53, 56, 66, 75
0333+321	NRAO 140	1.258	Q	*	25, 26, 27
0355+508	NRAO 150		U	*	30
0415+379	3C111	0.0485	N	J *	23, 52
0429+415	3C119	0.408	Q	*	36
0430+052	3C120	0.033	N	*	35, 68
0428+205		0.219	G	*	42
0454+844			Lac		6, 12
0518+165	3C138	0.76	Q		15
0538+498	3C147	0.545	Q	J *	50, 52, 62
0552+398	DA193	2.365	Q		57
0609+710	Mk3	0.014	S		33
0710+439	OI417		G	*	40
0711+356	OI318	1.620	Q	*	40
0716+714			Lac		6, 12
0723+67	3C179	0.843	Q	J	45, 46
0735+178		0.424	Lac		2
0804+499	OJ508		Q	*	40
0814+425	OJ425		Q	*	40
0836+714		2.17	Q	*	19, 40
0850+581	4C58.17	1.322	Q	*	40
0851+203	OJ287		Lac	*	35
0859+470	4C47.29	1.462	Q	*	37
0906+430	3C216	0.670	Q	*	40
0923+392	4C39.25	0.699	Q	*	35, 37
0945+408	4C40.24	1.252	Q	*	40
0951+693	M81	3.25 Mpc	S		3, 77
0957+561		1.405	Q		17, 44
1003+351	3C236	0.098	E	*	4, 58
1101+384	Mk421	0.03	Lac		2
1146+597	NGC3894	0.011	E		5
1208+397	NGC4151	0.0033	S	J	51
1226+023	3C273	0.158	Q	J *	35, 38, 39

Table 2 continued

object		z or distance	type[a]	code[b]	references
1228+126	3C274	20 Mpc	E	J *	10, 35, 54, 55
1253-055	3C279	0.536	Q	J *	35
1254+571	Mk231	0.041			51
1322-427	Cen A	9 Mpc	E	J	18, 49
1323+321	DA344		G	*	32
1328+307	3C286	0.846	Q	J *	36, 71
1458+718	3C309.1	0.904	Q	J *	22
1518+047			U	*	42
1607+268	CTD93		G?	*	41
1624+416	4C41.32		U	*	40
1633+382	4C38.41	1.814	Q	*	40
1637+574	OS 552	0.745	Q	*	40
1637+826	NGC6251	0.023	E	J *	9
1641+399	3C345	0.595	Q	J *	1, 53, 61, 63, 64, 65, 78
1642+690	4C69.21		Q	J *	40
1652+398	4C39.49	0.0337	G	*	40
1717+490	ARP102B	0.025	E		5
1730-130	NRAO530	0.901	Q		25
1749+701			Lac		2
1753+183	NGC6500	0.011	S		5
1803+784			Lac		6, 12
1807+698	3C371	0.05	N	J *	37
1823+568	4C56.27		Q?	*	40
1828+487	3C380	0.692	Q	J *	37
1845+797	3C390.3	0.0561	N	J *	23, 24, 52
1901+319	3C395	0.635	Q	*	19
1928+738	4C73.18	0.36	Q	*	6, 40, 12
1954+513	OV591	1.230	Q	*	40
1957+406	CYG A	0.056	E	J *	21, 23
2007+777			Lac		6, 12
2021+614			Q	*	40, 76
2050+364	DA529		U	*	42
2134+004	PKS...	1.94	Q	*	35
2200+420	BL Lac	0.0695	Lac	*	2, 31, 35, 37, 43
2223-052	3C446		Q		8
2251+158	3C454.3	0.859	Q	J *	11, 35
2351+456	4C45.51		G	*	40

a) "S, SO, E, N" have the usual meaning; "Lac" means Bl-Lac type object, "Q" quasar, "G" galaxy, and "U" unidentified object.

b) "J" means: object has large scale ($\gtrsim$1") radio jet, "*": object has been mapped by VLBI.

radio jets, see table 2;
- 7 double radio galaxies: NGC315, 3C111, 3C236, 3C274, NGC6251, 3C390.3 and Cyg A;
- 3 galaxies with broad emission line regions: 3C111, 3C120 and 3C390.3.

The three closest objects are 3C274, NGC315 and 3C84. The sources most frequently observed are 3C84, 3C345 and BL Lac. First reasonably detailed maps at 1.3 cm have recently been obtained for 3C84 and 3C345 by Readhead et al. (1983).

Table 3 lists 14 "samples" of sources observed by VLBI to which two more should be added:
1) 33 compact radio sources observed at 18 cm by Matveyenko et al. (1981) and
2) 10 extragalactic sources "with distinctive spectra" observed at 6 and 18 cm by Spangler et al. (1981).

"Sample" is used here in a broad sense. Seven of them are indeed statistically well-defined.

5. RADIO NUCLEI OF POWERFUL RADIO SOURCES

In this section I shall summarize observational results about the very compact ($\lesssim$ 10 pc) radio components in the centers of strong sources associated with radio galaxies and quasars. I call a source "strong" if P(6 cm) $\gtrsim 10^{31}$ erg s^{-1} Hz^{-1} or S(6cm)/Jy $\gtrsim$ 0.8 (distance/100 Mpc). The radio nuclei of these objects which include both compact (< 1") and extended ($\gtrsim$ 1") sources, are often also strong and therefore most amenable to VLB interferometry. On the assumption of unbeamed emission their radio luminosities $\lesssim 10^{46}$ erg s^{-1} and minimum energy requirements $\lesssim 10^{60}$ erg are in a range comparable to that found for extended emission regions. The magnetic field strengths derived from arguments based on synchrotron self-absorption are typically stronger than those in extended emission regions, and range from $\sim 10^{-4}$ to 1 Gauss.

5.1 Compactness

On the milliarcsec scale, all the stronger sources with flux densities, say, $\gtrsim$ 100 mJy at cm wavelengths, are at least partially resolved. Yet of 50 objects selected and mapped at 6 cm about 1/3 are only slightly resolved on the milliarcsec scale (Readhead & Pearson 1982). BL Lac-type objects appear to be the most compact ones. In 0454+844, for example, the total 6 cm flux density $\sim$ 1.4 Jy arises from a region $\lesssim$ 0".003 within the calibration accuracy of a few % (Eckart et al. 1982).

Table 3

Samples observed with VLBI (publication period: Jan. 1981 - Sep. 1982)

First author[c)]	a)	Characteristic feature of sources in sample	λ (cm)	No. of objects	Refs[b)]
v. Breugel		Mixed selection of extragal. objects	6	62	7
Eckart	*	S(6 cm) > 1 Jy, α(6,2.8)>-0.5 δ ≥ 70°	6	13	12
Graham		Selected galaxies	6	14	16
Hummel	*	Core sources of spiral galaxies, S(21 cm) > 100 mJy	6	18	73
Jones		Optically bright galaxies with nuclear radio sources	18	58	20
Morabito		Extragalactic radio sources	13	546	28
Morabito	*	S(6 cm) > 0.5 Jy, α(6,2.8)>-0.6	13	48	29
Pearson&Readhead	*	S(6 cm) ≥ 1.3 Jy	6	51	37,40
Preston	*	Radio quasars, S(11cm)≥0.5 Jy	13	31	47
Preston		Extragalactic radio sources	3.8,13	>800	48
Preuss & Fosbury		Galaxies with broad line nuclei	6	21	51
Waltman		Compact radio sources at δ>67°	6	15	69
Wehrle	*	Bright galaxies with B2 radio sources, S(75cm)>0.25 Jy, m_V<16.5,	13	26	70
Zensus	*	S(6cm)>1 Jy, α(6,2.8)>-0.5	6	59	72

a) "*" means: sample is statistically well-defined
b) see references to tables 2, 3 and 4
c) see text for additional two samples observed by Matveyenko et al. (1981) and Spangler et al. (1981)

Table 4

VLBI observations of compact continuum sources (possibly) associated with galactic objects (publication period: Jan. 1981 - Sep. 1982)

Object		Refs.[a)]	Remarks
0125+62	G127.11+0.54	13, 14	in centre of SNR NGC559
1617-15	SCO X-1	74	triple radio structure 2'; not detected by VLBI
1909+04	SS433	34,59,60,67	in SNR W50
2048+31	CL4	13	in Cygnus Loop
2001+43	G84	13	not detected by VLBI

a) see references to tables 2, 3 and 4

References to tables 2, 3, and 4

1 Baath et al. (1981a)
2 Baath et al. (1981b)
3 Bartel et al. (1982a)
4 Barthel et al. (1983)
5 Biermann et al. (1981a)
6 Biermann et al. (1981b)
7 van Breugel et al. (1981)
8 Brown et al. (1981)
9 Cohen et al. (1979)
10 Cotton et al. (1981)
11 Cotton et al. (1982)
12 Eckart et al. (1982)
13 Geldzahler & Shaffer (1981a)
14 Geldzahler & Shaffer (1982)
15 Geldzahler et al. (1981b)
16 Graham et al. (1981)
17 Haschik et al. (1981)
18 Jauncey et al. (1982)
19 Johnston et al. (1981)
20 Jones et al. (1981)
21 Kellermann et al. (1981)
22 Kus et al. (1981)
23 Linfield (1981a)
24 Linfield (1982)
25 Marscher & Broderick (1981a)
26 Marscher & Broderick (1982)
27 Marscher & Broderick (1981b)
28 Morabito et al. (1982)
29 Morabito et al. (1981)
30 Mutel & Phillips (1980)
31 Mutel et al. (1981a)
32 Mutel et al. (1981b)
33 Neff & de Bruyn (1983)
34 Niell et al. (1982)
35 Pauliny-Toth et al. (1981)
36 Pearson et al. (1980)
37 Pearson & Readhead (1981a)
38 Pearson et al. 1981b)
39 Pearson et al. (1982a)
40 Pearson et al. (1982b)
41 Phillips & Mutel (1980)
42 Phillips & Mutel (1981)
43 Phillips & Mutel (1982a)
44 Porcas et al. (1981)
45 Porcas (1981)
46 Porcas (1982)
47 Preston et al. (1981)
48 Preston et al. (1982a)
49 Preston et al. (1982b)
50 Preuss et al. (1982)
51 Preuss & Fosbury (1983)
52 Preuss et al. (1983)
53 Readhead et al. (1983)
54 Reid et al. (1982a)
55 Reid et al. (1982b)
56 Romney et al. (1982)
57 Schilizzi et al. (1981a)
58 Schilizzi et al. 1981b)
59 Schilizzi et al. (1981c)
60 Schilizzi et al. (1982)
61 Schraml et al. (1981)
62 Simon et al. (1981)
63 Spencer et al. (1981)
64 Unwin et al. (1982a)
65 Unwin et al. (1982b)
66 Unwin et al. (1982c)
67 Walker et al. (1981)
68 Walker et al. (1982)
69 Waltman et al. (1981)
70 Wehrle et al. (1981)
71 Wilkinson et al. (1979)
72 Zensus et al. (1982)
73 Hummel et al. (1982)
74 Geldzahler et al. (1981c)
75 Romney et al. (1983)
76 Wittels et al. (1982)
77 Bartel et al. (1982b)
78 Cohen et al. (1981)

5.2 Main structural characteristics in brief

The structure of the very compact radio components is frequently

- elongated or linear with a well-defined position angle (p.a.) of the source axis,
- aligned with larger scale features, in particular in objects with symmetric large scale structure,
- curved, i.e. showing a systematic change of p.a. on successively larger scales through angles > 20°, generally in those objects which have dominant central components,
- asymmetric or one-sided with a bright "core" at the end of a diffuse elongated feature,
- variable on time scales $\gtrsim$ months; in 6 sources "superluminal" motion has been found with separation velocities of subcomponents (i.e. peaks in the radio brightness) > c. In a further 4 sources, such motions have been surmised.

Table 5 shows a breakdown of structural types for 33 properly resolved sources which have been mapped by VLBI.

Table 5

Type of small scale structure for 33 resolved sources (numbers from Readhead & Pearson (1982)).

structure	quasar or Lacertid	galaxy	unidentified	total
a) asymmetric ("core-jet")	11	9	1	21
b) equal double	1	1	3	5
c) uncertain ((a) or (b))	4	1	0	5
d) complex	0	2	0	2
total	16	13	4	33

5.3 On Symmetry/Asymmetry

The degree of asymmetry of the brightness distribution may well be frequency dependent for a given source. In a physically more meaningful definition an object is said to have a one-sided "core-jet" structure (e.g. Readhead & Pearson 1982) if it has a bright flat spectrum component at the end of an elongated steep-spectrum feature. The terminology is strongly interpretative, but in widespread use. The "core" found at a particular frequency may resolve into a "core-jet" structure again when observed at a higher frequency (e.g. 3C273). The core of a given frequency appears to be the region where the jet becomes optically thick at that frequency, and it seems plausible to assume that it surrounds the center of activity.

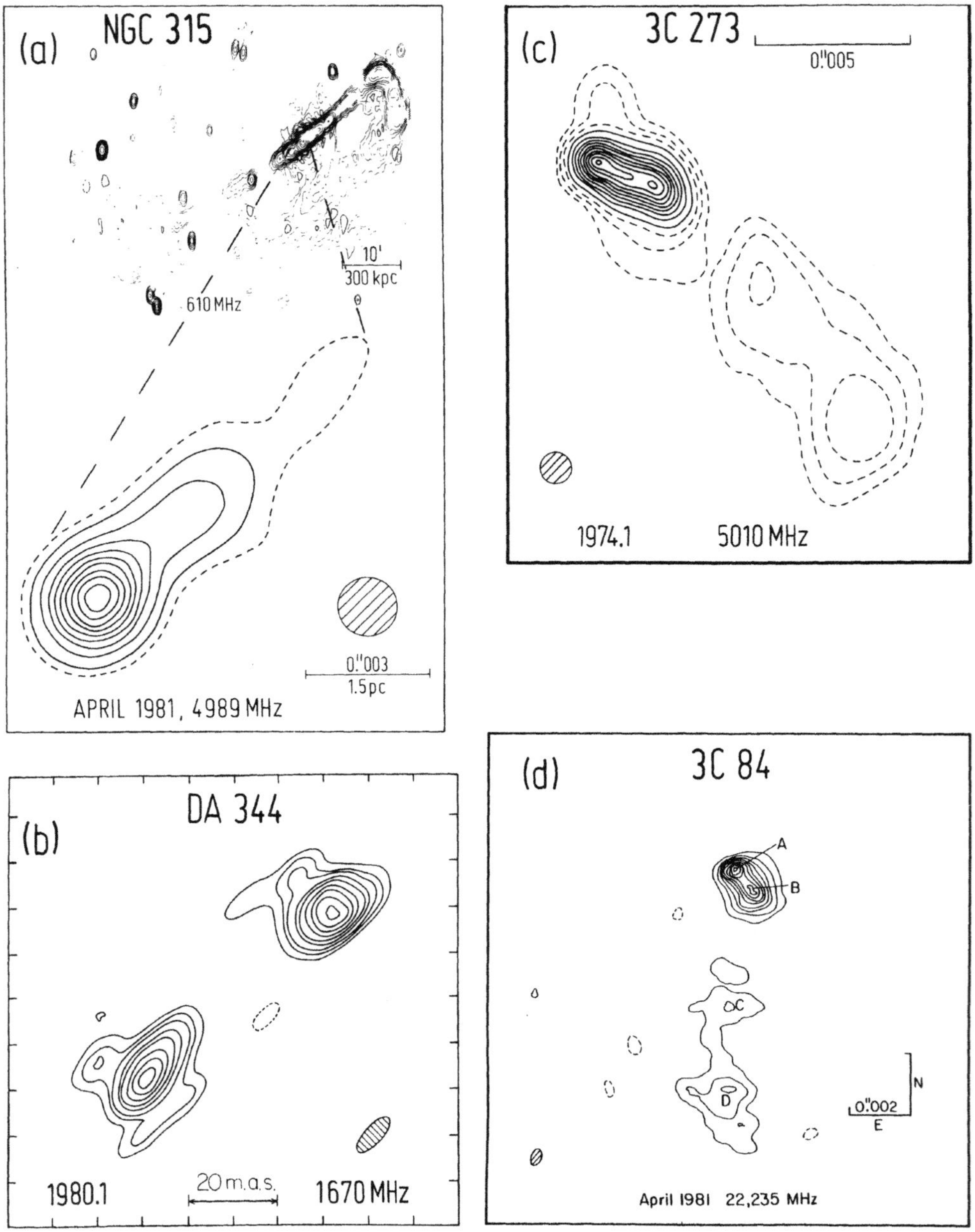

Fig. 1: VLBI maps of (a) NGC315 (Preuss et al. 1983) shown in comparison with a map of the large scale structure from Bridle et al. (1979); (b) DA344 (Mutel, Phillips and Skuppin 1981); (c) 3C273 (Pauliny-Toth et al. 1981); (d) 3C84 (Readhead et al. 1983).

Higher dynamic range than currently available is required to find out about the details of the "jets". The evidence so far available indicates that they have an inhomogeneous, complex structure.

Small scale asymmetry occurs in objects with all types of large scale symmetry. It has been found in more than 20 sources, i.e. > 80% of all sources which are properly resolved and have been mapped on scales $\lesssim$ 10 pc. 6 of the 7 symmetric classical double sources mapped by VLBI are one-sided on the milliarcsec scale with a "jet/counterjet" ratio R $\gtrsim$ 10. The lower limit is due to the limited dynamic range of VLBI imaging. 3C236 at 6 cm is two-sided (Barthel et al. 1983) and there is an indication for "counterjets" in 3C390.3 (Linfield 1982).

Really symmetric, "equal double" sources without any bright core have only been found on scales $\gtrsim$100 pc (Phillips & Mutel 1982b): 1518+047, CTD93, 3C395, 2050+364 and DA344. The spectra of these sources look like those of homogeneous synchrotron sources with a self absorption turnover near 1 GHz.

"Complex" sources may well show one-sided, core-dominated structure, when observed at sufficiently high frequency, as for example, recent 1.3 cm observations have shown for 3C84 (Readhead et al. 1983).

5.4 On alignment, curvature and orientation

As mentioned before there is in general a strong relation between the directions of small and larger scale features. In some cases (3C345, NGC6251) the milli-arcsec jet is continuous with the larger scale jet. There is good alignment in large scale double sources, but sometimes (Cyg A, 3C236) the small jets point some (<20°) degrees away from the outer lobes. In objects with a dominating central component, so called "core-objects" (3C147, 3C273, 3C309.1, e.g.) one frequently finds a systematic change in p.a. between the very compact core and the extended features, i.e. the jet is bent through angles >20°, in some cases >100°, up to 180° (3C454.3) with most of the bending occurring near the core (see e.g. Readhead et al. 1983).

With regard to orientation:

- in all 4 classical doubles with large scale jets (3C274, N315, NGC6251, Cyg A) the milliarcsec jet points in the same direction as the large scale jets;
- milliarcsec jets in double sources tend to point toward the more compact outer lobe (Linfield 1982a);
- all 6 superluminal sources with clearly defined VLBI structures have asymmetries on the milliarcsec and the arcsec scale in the same sense (Browne et al. 1982).

There is a good correlation between the magnetic field direction, inferred from linear polarisation measurements and the direction of the source axis, e.g. in 3C273, 3C345, 3C454.3 (Davis et al. 1978, Browne et al. 1982).

5.5 On structural variability

Only a fraction of all objects mapped by VLBI have been properly examined for structural variability. About 10 sources have definitely been shown to have variable structure, e.g., 3C84 and the "superluminal sources". Not all bright compact sources are variable; e.g. 4C39.25 appears to have a stable structure. 3C84 underwent drastic changes of its brightness distribution at 2.8 cm on a timescale $\gtrsim$ 1/2 year (Preuss et al. 1979, Romney et al. 1982) but measured separation velocities are clearly "subluminal". Many (if not all) compact objects with flat cm spectra tend to vary in their total flux density. But the relation between changes in spatial structure and total flux density is not clear as yet. There is still fairly little known on the relatively weak central components in extended double sources. There are indications for structural variability in 3C390.3 (Preuss et al. 1980, Linfield 1981a) and in the cores of extended quasars (Barthel et al., work in progress). "Superluminal" flux variation is reported for 3C111 (Hine & Scheuer 1980), that is, the brightness calculated from the variability diameter is higher than allowed by the inverse Compton cooling operating in incoherent Synchrotron emitters.

The properties and problems of superluminal radio sources have been reviewed by Cohen and Unwin (1982), see also the article by Browne et al. (1982). Roughly speaking, $\sim$ 50% of the compact sources properly looked at show the phenomenon. Table 6 lists some properties of the 6 sources with confirmed superluminal motion.

Table 6

Superluminal radio sources (numbers from Cohen & Unwin (1982))

Object	z	S(3 cm)[a)] (Jy)	separation velocity (0.″001/Y)	v/c $(\frac{H_o}{100})$
3C120	0.033	4	1.35	2.1
3C273	0.158	35	0.76	5.3
3C279	0.538	10	(0.5)	(10)
3C345	0.595	12	0.36	8.2
3C179	0.846	0.5	0.14	4.2
NRAO140	1.258	2	0.13	5.4

a) Nominal flux density

In addition to the objects listed in table 6, BL Lac and 3C446 (references in table 2) have to be mentioned as likely candidates. All objects in table 6 have large scale structure in addition to their compact cores. 3C179 is a large scale double source with a relatively strong central component (see Porcas, this workshop). Cohen and Unwin (1982) give as essential characteristics of superluminal radio sources: core-jet structure with several moving components; spectral gradient; evolution and decay of outer components; outer one-sided jet; spatial curvature, mainly near core; superluminal motion and spacing independent of wavelength (2-6 cm).

So far it has not been possible to tell which of the separating components is moving and which one is stationary (if any). It can be hoped that phase referencing with respect to neighbouring sources will allow the measurement of "relative absolute position" of the subcomponents in superluminal sources.

6. RADIO NUCLEI OF WEAK RADIO SOURCES

In this section I will briefly describe the status of VLBI observations of (optically selected) Seyfert galaxies and mildly active "normal" galaxies. The total radio emission of these objects is typically weak (P(6 cm) $\lesssim 10^{31}$ erg s^{-1} Hz^{-1}) and so are their compact radio nuclei (if any). Detectable sources of this kind are naturally relatively nearby. The highest available sensitivity is required for their study and the current instrumental performance is just sufficient to tackle a few of them in the hope of obtaining maps.

Note: the total radio luminosities of these objects, typically in the range $\sim 10^{39}$-10^{41} erg s^{-1}, are high when compared with "radio normal" galaxies but low when compared with powerful radio galaxies. The theoretical interest in these objects - in the context of this workshop - lies in questions such as these: Is the radio phenomenon and the "central engine" in these objects qualitatively the same as in powerful sources, with only certain characteristic properties (such as power) scaled down? Are "jets" the energetic backbone for the larger scale emission regions or are there other feeding mechanisms at work? From the observational point of view, one hopes that the strongest of the nearby nuclei will allow us to really probe into the small volumes where the radio phenomenon originates.

Table 7 lists 41 galaxies of the category described, here simply called "nearby galaxies". The list is intended to include all objects from VLBI papers which have come to my attention and which meet the following 2 criteria:

a) they are detected with VLBI at least at one wavelength on an angular scale $\lesssim$ 50 milliarcsec and
b) the total radio emission is weak as defined before or the object is closer than 100 Mpc ($50/H_o$).

The numbers in table 7 are taken from the references given or references therein.

Table 7

Nearby Galaxies detected with VLBI (Sep. 1982)

Name (1)	Name (2)	z or dist. (3)	Optical type (4)	(5)	L.A.S. (6)	S.A.S. (0".001) (7)	References VLBI (8)	other (9)	Remarks (10)
0017+296	N0076		Compact			5	1		
0046+316	N0262	0.014	S0	SEY2		1	2, 3, 4		Mk348; variable radio source
0055+300	N0315	0.0167	E		30' D	1	5, 6		Giant radio galaxy
0238-084	N1052	28 Mpc	E	EM	20"	5	7, 8		HI detected
0240-002	N1068	22 Mpc	S	SEY2	14" D	50	1, 4, 9	13	M77, 3C71
0305+039	N1218	172 Mpc	S0	EM	2".5	5	1	10	3C78
0359+229	N1497		S0			5	1		
0609+710	Mk3	0.0137	S	SEY2	1".5 D	30	2, 4	11	4C70.05
0645+744	Mk6	0.0176	S0	SEY1	1" D	50	2, 4	12	IC450
0840+504	N2639	67 Mpc	S		0".7	10	14		
0931+103	N2911	60 Mpc	S0	EM	<1"	5	1, 3, 7		
0951+693	N3031	3.25Mpc	S	SEY1		1	1,9,15,20		M81
0951+699	N3034	3.25Mpc	Irr	EM	15"	1	1,2,8,16	17	M82, 3C231
1122+39	N3665	40 Mpc	S0		30" D	20	7	18	
1139+267	N3826		E			5	1, 3		
1146+596	N3894	0.011	E			1	19		in galaxy pair
1155+557	N3998	24 Mpc	S0		4' D	10	14		
1208+396	N4151	19 Mpc	S	SEY1	10" D	20	2, 7	13 21 22	in galaxy pair
1216+061	N4261	44 Mpc	E		9' D	1	1, 9		3C270
1217+29	N4278	21 Mpc	E	EM	$\lesssim$1"	5	1,7,8	12	HI detected
1222+131	N4374	22 Mpc	E		2'.4 D	1	1, 9	10	M84, 3C272.1
1228+126	N4486	22 Mpc	E	EM	50" D	1	23,24,25	26 27	M87, 3C274, VIR A 10' radio halo
1233+128	N4552	22 Mpc	E		<1"	5	1,7,9	10	M89, variable radio source
1237-113	N4594	18.6Mpc	S			5	8, 28		M104, "Sombrero"
1254+571	Mk231	0.0410		SEY1	10"	1	2, 4	12	variable radio source
1317-12	N5077	50 Mpc	E	EM		20	7		
1322-427	N5128	9 Mpc	E		10° D	1	29		CENA
1348+339	N5318		S0?			5	1, 3		brightest in a group
1351+405	N5353	46 Mpc	S0		<0".2	10	14		
1353+054	N5363	22 Mpc	Irr		5"	5	1, 3,14		

table 7 continued

Name (1)	Name (2)	z or dist. (3)	Opt. type (4)	(5)	L.A.S (6)	S.A.S (0".001) (7)	References VLBI (8)	Other (9)	Remarks (10)
1353+186	Mk463	0.05		SEY2	1".3	50	4	12	galaxy with double nucleus
1426+276	N5635	78 Mpc	S		<1"	5	1	10	variable radio source
1430+365	N5675	86 Mpc	S		2".5	5	1, 3	10	
1553+246		260 Mpc	E			20	3, 19		in galaxy pair
1717+490	ARP102B	150Mpc	E			1	19		in galaxy pair
1742-289	SAG A	10 kpc	S			1	30,31, 32	33	Galactic Centre
1753+183	N6500	64 Mpc	S		5".3	1	1,3,7, 19	10	in galaxy pair
2116+262	N7052	0.0164	E			5	1, 3		
2303+338	N7485		SO			5	3		
2322+282			SO			5	1, 3		
2337+268	N7728	189 Mpc	E		4"	5	1, 3	10	NRAO 716

Meaning of symbols and abbreviations (where not self-explanatory):
column (5): SEY1 and SEY2: Seyfert type 1 and 2, EM = Emission lines present
column (6): L.A.S. = largest radio angular size, D = double source
column (7): S.A.S. = smallest angular scale on which radio structure has been detected by VLBI

REFERENCES

1 Jones et al. (1981)
2 Preuss & Fosbury (1983)
3 Crane (1979)
4 Neff & de Bruyn (1983)
5 Linfield (1981)
6 Preuss et al. (1983)
7 van Breugel et al. (1981)
8 Shaffer et al. (1979)
9 Preuss et al. (1977)
10 Jones et al. (1981)
11 Wilson et al. (1980)
12 Ulvestad et al. (1981)
13 Wilson & Ulvestad (1982)
14 Hummel et al. (1982)
15 Kellermann et al. (1976)
16 Geldzahler et al. (1977)
17 Kronberg et al. (1981)
18 Kotanyi (1979)
19 Biermann et al. (1981)
20 Bartel et al. (1982)
21 Johnston et al. (1982)
22 Booler et al. (1982)
23 Pauliny-Toth et al. (1981)
24 Cotton et al. (1981)
25 Reid et al. (1982)
26 Owen et al. (1980)
27 Charlesworth & Spencer (1982)
28 Graham et al. (1981)
29 Preston et al. (1982)
30 Geldzahler & Kellermann (1979)
31 Lo et al. (1981)
32 Backer (1978)
33 Brown (1981b)

The majority (> 70%) of the objects listed in table 7 are at distances < 100 Mpc. 21 (51%) objects including 6 3C-sources have extended ($\gtrsim$ 1") radio structure and 11 of these are double (or triple) sources. The list includes 4 of the 6 optically selected Seyfert galaxies (NGC1068, Mk3, Mk6, NGC4151) which were shown to have double structure by VLA measurements (see references in table 7). VLBI observations by Neff and de Bruyn (1983) have revealed a double structure in Mk3 on a scale $\sim$ 30 milliarcsec which is aligned with the arcsec scale structure. Jones, Sramek & Terzian (1981) also report close alignment of pc-scale and larger scale radio structure found with the VLA in NGC1218, 4374, 5675, 6500 and 7728.

7. CONCLUDING REMARKS

In the previous sections I have presented observational facts obtained from VLBI observations which I think are relevant to a discussion of jets. Before I close, let me briefly discuss the significance of the evidence produced so far by VLBI for beam models, in particular for the relativistic beaming hypothesis. I will do this by listing some of the relevant conclusions which have been reached in recent work.

But let me first remind you of the general situation of beam or jet models (see, e.g., Rees 1982) which are suggested by and applicable to evidence coming from observations at widely different wavelengths and angular resolutions. The main facts here are: a) the concept of collimated continuous outflow as an energy supply mechanism for the stronger radio sources has generally been accepted, to the extent that the term "jet" has come to mean not only the underlying hydrodynamic process but also elongated regions of emission on angular scales ranging from a few pc to hundreds of kpc, but b) at the same time there is a high degree of controversy (ignorance) about almost all fundamental physical characteristics such as beam speed, material, density, the role of the magnetic field, the transverse confinement and about important questions such as the origin, collimation, structure and dynamics, and the energy conversion which makes the jet visible. In fact, it may well turn out that a variety of models may be needed. Unfortunately, the beam speed cannot be measured directly. Taken together, the estimates and assumptions made cover the whole range between a few 100 km/s to highly relativistic velocities with Lorentz factors $\gamma \sim 5$ or even $\gamma \gtrsim 100$ (Kundt & Gopal-Krishna 1980). In the current discussion one notices an interesting contrast: in the case of extended radio sources the arguments seem to favour non-relativistic speeds whereas arguments based on observations of very compact radio sources seem to favour relativistic speeds. So it is of great interest to investigate more thoroughly the possibility of beam models which involve relativistic speeds near the nucleus and non-relativistic speeds further out, in other words, the possibility of strong deceleration without disruption.

The relativistic beam hypothesis (see Scheuer & Readhead 1979, Readhead et al. 1978, Rees 1978, Blandford & Königl 1979) is suggested by, or can explain effects such as, superluminal motion, asymmetry, curvature (see section 5) or rapid flux density variations, for which the brightness temperature derived from variability diameters exceeds 10^{12} K. The standard picture involves relativistic "twin beams", the observable characteristics of which are strongly influenced by Doppler enhancement and light-travel-time effects (aberration). These effects depend strongly on the viewing angle between the line-of-sight and the beam axis. If $\theta < 1/\gamma$ the receding jet (counterjet) may become invisible in a map with a limited dynamic range, and an intrinsic bend of the jet may be strongly amplified by projection. Some more general attractive features of the relativistic models are: a) they only require conventional incoherent synchrotron emission from electrons, b) they provide an efficient means of energy transport, c) they ease excessive energy requirements for instance in certain BL Lac type objects, and d) they explain a number of seemingly different phenomena in a "unifying scheme".

In the following I will list some conclusions which have recently been reached in the discussion of observational results (described in the previous sections) in terms of the relativistic beam model. This may roughly outline the current status of acceptance and/or applicability of such theoretical models.

a) Superluminal motion. According to Cohen and Unwin (1982) the relativistic jet model best explains the phenomenology of superluminal sources at present. The values required for γ and θ are in the range 3 to 10 and 6°-20° respectively (H_o = 100 km s^{-1} Mpc^{-1}). Superluminal motion should occur in only a few of the classical double sources supposed to be seen at large (see Porcas, this workshop). While superluminal motion certainly suggests bulk relativistic motion, it is at present unknown what fraction of the sources which are presumed to be relativistically beamed and pointing toward us should show superluminal motion, because this also requires a certain structure of the jet. Therefore superluminal motion cannot so far be predicted. The only hint one may use here is the current number of roughly 50% of all the strong compact sources (properly mapped) that show the effect.

b) Symmetry/asymmetry. The frequently observed asymmetry of the small scale structure may either be caused by the Doppler effect rendering the receding side of a relativistic twin jet invisible or may be intrinsic, i.e. the jet is one-sided, but alternates sides (beam switching, flip-flop behaviour) and thus accounts for any large scale symmetry.

In the case of superluminal sources, there is so far no way to decide (distinguish) between these possibilities (Cohen & Unwin 1982, Browne et al. 1982). For some classical double radio galaxies, Linfield (1982c) reaches the conclusion that the small scale asymmetry is probably due to an intrinsic asymmetry, rather than geometrical or radiative effects. This is supported by indications that in some of the double sources the extended lobes on the same side as the jet are systematically different from those on the opposite side (see section 5.4). Physical arguments based on synchrotron theory show that the jet velocities are at least weakly relativistic. But note that VLBI observations by Barthel et al. (1983) reveal a two-sided structure for the radio nucleus of 3C236.

Phillips and Mutel (1982b) point out that "equal double sources" (Section 5.3) are examples of powerful compact sources which are certainly not highly beamed emitters. For the observed flux ratios $S_1/S_2 \lesssim 1.5$ the required viewing angles are large ($\gtrsim 85°$), even if the motion is only mildly relativistic.

c) Curvature. Readhead et al. (1983) find that the distribution of all observed differences in position angle between the small-scale (<< 1 kpc) and large-scale (>> 1 kpc) features is roughly consistent with a simple relativistic model if one assumes a typical intrinsic bend of $\sim 10°$. The available sample of about 20 sources is not, however, statistically complete.

In summary:

(1) At present the relativistic beaming hypothesis cannot be ruled out. On the contrary it seems hard to avoid the conclusion that relativistic motion plays an important role in the physics of many small scale phenomena.

(2) Only rigorous statistical investigations of various classes of objects will provide the crucial test of models involving relativistic beams and their range of applicability, i.e. their potential to provide "unifying schemes" for different phenomena. Orr & Browne (1982) recently found that the statistical properties of flat and steep spectrum quasars are entirely consistent with the predictions of a simple relativistic-beam model if an average bulk Lorentz factor $\gamma \sim 5$ and $H_0 = 100$ km s^{-1} Mpc^{-1} is assumed.

(3) So far only simple models have been employed. It may well be possible to overcome remaining statistical difficulties by using minimally modified theoretical models.

ACKNOWLEDGEMENT

I thank Dr. I. Pauliny-Toth for critically reading the manuscript.

REFERENCES

Bääth, L.B., Rönnäng, B.O., Pauliny-Toth, I.I.K., Kellermann, K.I., Preuss, E., Witzel, A., Matveyenko, L.I., Kogan, L.R., Kostenko, V.I., Moiseev, I.G., Shaffer, D.B. 1981a. Ap. J. Lett. 243:L123

Bääth, L.B., Eldered, G., Lundqvist, G., Graham, D., Weiler, K.W., Seielstad, G.A., Tallqvist, S., Schilizzi, R.T. 1981b. Astron. Astrophys. 96:316

Backer, D.C. 1978. Ap. J. Lett. 222:L9
Bartel, N., Corey, B.E., Shapiro, I.I., Rogers, A.E.E., Whitney, A.R., Graham, D.A., Romney, J.D., Preston, R.A. 1982a. Proc. IAU Symp. 97, Extragalactic Radio Sources, p. 387
Bartel, N., Shapiro, I.I., Corey, B.E., Marcaide, J.M., Rogers, A.E.E., Whitney, A.R., Capallo, R.J., Graham, D.A., Romney, J.D., Preston, R.A. 1982b. Ap. J. 262:556
Barthel, P., Schilizzi, R.T., Miley, G.K. 1983. In preparation
Biermann, P., Kronberg, P.P., Preuss, E., Schilizzi, R.T., Shaffer, D.B. 1981a. Ap. J. Lett. 250:L49
Biermann, P., Duerbeck, H., Eckart, A., Fricke, K., Johnston, K.J., Kühr, H., Liebert, J., Pauliny-Toth, I.I.K., Schleicher, H., Stockman, H., Strittmatter, P.A., Witzel, A. 1981b. Ap. J. Lett. 247:L53
Blandford, R.D., Königl, A. 1979. Ap. J. 232:34
Booler, R.V., Pedlar, A., Davies, R.D. 1982. MNRAS 199:229
Breugel van, W.J.M., Schilizzi, R.T., Hummel, E., Kapahi, V.K. 1981. Astron. Astrophys. 96:310
Bridle, A.H., Davis, M.M., Fomalont, E.B., Willis, A.G., Strom, R.G. 1979. Ap. J. Lett. 228:L9
Bridle, A.H. 1981. Private list circulated at IAU Symp. 97
Brown, R.L., Johnston, K.J., Briggs, E.H., Wolfe, A.M., Neff, S.G., Walker, R.C. 1981a. Astrophys. Lett. 21:105
Brown, R.L. Johnston, K.J., Lo, K.Y. 1981b. Ap. J. 250:155
Browne, I.W.A., Clark, R.R., Moore, P.K., Muxlow, T.W.B., Wilkinson, P.N., Cohen, M.H., Porcas, R.W. 1982. Nature 299:788
Charlesworth, M., Spencer, R.E. 1982. MNRAS 200:933
Cohen, M.H., Readhead, A.C.S. 1979. Ap. J. Lett. 233:L10
Cohen, M.H. (ed.) 1980. A transcontinental radio telescope, report, California Institute of Technology
Cohen, M.H., Unwin, S.C., Simon, R.S., Seielstad, G.A., Pearson, T.J., Linfield, R.P., Walker, R.C. 1981. Ap. J. 247:774
Cohen, M.H., Unwin, S.C. 1982. Proc. IAU Symp. 97, Extragalactic Radio Sources, p. 345
Cotton, W.D., Shapiro, I.I., Wittels, J.J. 1981. Ap. J. Lett. 244:L57
Cotton, W.D., Geldzahler, B.J., Shapiro, I.I. 1982. Proc. IAU Symp. 97, Extragalactic Radio Sources, p. 301
Crane, P.C. 1979. Astron. J. 84:281
Davis, R.J., Stannard, D., Conway, R.G. 1978. MNRAS 185:435
Eckart, A., Hill, P., Johnston, K.J., Pauliny-Toth, I.I.K., Spencer, J.H., Witzel, A. 1982. Astron. Astrophys. 108:157
Geldzahler, B.J., Kellermann, K.I., Shaffer, D.B., Clark, B.G. 1977. Ap. J. Lett. 215:L5
Geldzahler, B.J., Kellermann, K.I., Shaffer, D.B. 1979. Astron. J. 84:186
Geldzahler, B.J., Shaffer, D.B. 1981a. Ap. J. 248:132
Geldzahler, B.J., Fanti, C., Fanti, R., Schilizzi, R.T., Weiler, K.W., Kellermann, K.I., Shaffer, D.B. 1981b. Preprint
Geldzahler, B.J., Fomalont, E.B., Hilldrup, K., Corey, B.E. 1981c. Astron. J. 86:1036
Geldzahler, B.J., Shaffer, D.B. 1982. Ap. J. Lett. 260:L69

Graham, D.A., Weiler, K.W., Wielebinski, R. 1981. Astron. Astrophys. 97:388
Haschick, A.D., Moran, J.M., Reid, M.J., Davis, M., Lilley, A.E. 1981. Ap. J. Lett. 243:L57
Hine, R.G., Scheuer, P.A.G. 1980. MNRAS 193:285
Hummel, E., Fanti, C., Parma, P., Schilizzi, R.T. 1982. Astron. Astrophys. 114:400
Jauncey, D.L., Preston, R.A., Kellermann, K.I., Shaffer, D.B. 1982. Ap. J. Lett. submitted
Johnston, K.J., Spencer, J.H., Witzel, A., Fomalont, E.B. 1981. Preprint
Johnston, K.J., Elvis, M., Kjer, D., Shen, B.S.P. 1982. Ap. J. 262:61
Jones, D.L., Sramek, R.A., Terzian, Y. 1981a. Ap. J. 246:28
Jones, D.L., Sramek, R.A., Terzian, Y. 1981b. Ap. J. Lett. 247:L57
Kellermann, K.I., Shaffer, D.B., Pauliny-Toth, I.I.K., Preuss, E., Witzel, A. 1976. Ap. J. Lett. 210:L121
Kellermann, K.I., Downes, A.J.B., Pauliny-Toth, I.I.K., Preuss, E., Shaffer, D.B., Witzel, A. 1980. Astron. Astrophys. 97:L1
Kellermann, K.I., Pauliny-Toth, I.I.K., 1981. Ann. Rev. Astron. Astrophys. 19:373
Kotanyi, C.G. 1979. Astron. Astrophys. 74:156
Kronberg, P.P., Biermann, P., Schwab, F.R. 1981. Ap. J. 246:751
Kundt, W., Gopal-Krishna 1980. Nature 288:149
Kus, A.J., Wilkinson, P.N., Booth, R.S. 1981. MNRAS 194:527
Linfield, R.P. 1981a. Ap. J. 244:436
Linfield, R.P. 1981b. Ap. J. 250:464
Linfield, R.P. 1982a. Ap. J. 254:465
Linfield, R.P. 1982b. preprint
Lo, K.Y., Cohen, M.H., Readhead, A.C.S., Backer, D.C. 1981. Ap. J. 249:504
Longair, M.S., Ryle, M., Scheuer, P.A.G. 1973. MNRAS 164:243
Marscher, A.P., Broderick, J.J. 1981a. Ap. J. 249:406
Marscher, A.P., Broderick, J.J. 1981b. Ap. J. Lett. 255:L11
Marscher, A.P., Broderick, J.J. 1981c. Ap. J. Lett. 247:L49
Marscher, A.P., Broderick, J.J. 1982. Proc. IAU Symp. 97, Extragalactic Radio Sources, p. 359
Matveyenko, L.I., Kostenko, V.I., Papatsenko, A. Kh., Bartel, N., Massi, M., Romney, J.D., Weiler, K.W., Ficarra, A., Mantovani, F., Padrielli, L., Moiseev, I.G., Bääth, L., Nicolson, G. 1981. Soviet Astron. Lett. 7:259
Morabito, D.D., Preston, R.A., Faulkner, J. 1981, NASA TDA Progress Report 42-66
Morabito, D.D., Preston, R.A., Slade, M.A., Jauncey, D.L. 1982. Astron. J. 87:517
Mutel, R.L., Phillips, R.B. 1980. Ap. J. Lett. 241:L73
Mutel, R.L., Aller, D.D., Phillips, R.B. 1981a. Nature 294:236
Mutel, R.L., Phillips, R.B., Skuppin, R. 1981b. Astron. J. 86:1600
National Radio Astronomy Observatory 1982. report: A Program for the Very Long Baseline Array Radio Telescope
Neff, S.G., de Bruyn, A.G. 1983. Proc. International Conference on VLBI Techniques held in Toulouse, France, Aug./Sep. 1982, in press

Niell, A..E, Lockhart, T.G., Preston, R.A. 1982. Preprint
Orr, M.J.L., Browne, I.W.A. 1982. MNRAS 200:1067
Owen, F.N., Hardee, P.E., Bignell, R.C. 1980. Ap. J. Lett. 239:L11
Pauliny-Toth, I.I.K., Preuss, E., Witzel, A., Graham, D., Kellermann, K.I., Rönnäng, B. 1981. Astron. J. 86:371
Pearson, T.J., Readhead, A.C.S., Wilkinson, P.N. 1980. Ap. J. 236:714
Pearson, T.J., Readhead, A.C.S. 1981a. Ap. J. 248:61
Pearson, T.J., Unwin, S.C., Cohen, M.H., Linfield, R.P., Readhead, A.C.S., Seielstad, G.A., Simon, R.S., Walker, R.C. 1981b. Nature 290:365
Pearson , T.J., Unwin, S.C., Cohen, M.H., Linfield, R.P., Readhead, A.C.S., Seielstad, G.A., Simon, R.S., Walker, R.C. 1982a. Proc. IAU Symp. 97, Extragalactic Radio Sources, p. 355
Pearson, T.J., Readhead, A.C.S. 1982b. in preparation
Phillips, R.B., Mutel, R.L. 1980. Ap. J. 236:89
Phillips, R.B., Mutel, R.L. 1981. Ap. J. 244:19
Phillips, R.B., Mutel, R.L. 1982a. Ap. J. Lett. 257:L19
Phillips, R.B., Mutel, R.L. 1982b. Astron. Astrophys. 106:21
Porcas, R.W., Booth, R.S., Browne, I.W.A., Walsh, D., Wilkinson, P.N. 1981. Nature 289:758
Porcas, R.W. 1981. Nature 294:47
Porcas, R.W. 1982. Proc. IAU Symp. 97, Extragalactic Radio Sources, p. 361
Preston, R.A., Morabito, D.D., Jauncey, D.L. 1981. BAAS 13:898
Preston, R.A., Morabito, D.D., Jauncey, D.L., Williams, J.G., Slade, M.A., Harris, A.W. 1982a. List available at request, JPL, Pasadena, California
Preston, R.A., Wehrle, A.E., Morabito, D.D., Jauncey, D.L., Batty, M., Haunes, R.F., Wright, A.E., Nicolson, G.D. 1982. Proc. IAU Symp. 97, Extragalactic Radio Sources, p. 119
Preuss, E., Pauliny-Toth, I.I.K., Witzel, A., Kellermann, K.I., Shaffer, D.B. 1977. Astron. Astrophys. 54:297
Preuss, E., Kellermann, K.I., Pauliny-Toth, I.I.K., Witzel, A., Shaffer, D.B. 1979. Astron. Astrophys. 79:268
Preuss, E., Kellermann, K.I., Pauliny-Toth, I.I.K., Shaffer, D.B. 1980. Ap. J. Lett. 240:L7
Preuss, E. 1981. Proc. ESO/ESA Workshop Optical Jets in Galaxies, ESA SP-162, p. 97
Preuss, E., Alef, W., Pauliny-Toth, I.I.K., Kellermann, K.I. 1982. Proc. IAU Symp. 97, Extragalactic Radio Sources, p. 289
Preuss, E., Fosbury, R.A.E. 1983. MNRAS in press
Preuss, E., Alef, W., Pauliny-Toth, I.I.K., Kellermann, K.I. 1983. in preparation
Readhead, A.C.S., Wilkinson, P.N. 1978. Ap. J. 223:25
Readhead, A.C.S., Cohen, M.H., Pearson, T.J., Wilkinson, P.N. 1978. Nature 276:768
Readhead, A.C.S., Pearson, T.J. 1982. Proc. IAU Symp. 97, Extragalactic Radio Sources, p. 279
Readhead, A.C.S., Hough, D.H., Ewing, M.S., Walker, R.C., Romney, J.D. 1983. Ap. J. in press
Rees, M.J. 1978. Nature 275:516

Rees, M.J. 1982. Proc. IAU Symp. 97, Extragalactic Radio Sources, p. 211
Reid, M.J., Schmitt, J.H.M.M., Owen, F.N., Booth, R.S., Wilkinson, P.N., Shaffer, D.B., Johnston, K.J., Hardee, P.E. 1982a. preprint submitted to Ap. J.
Reid, M.J., Schmitt, J.H.M.M., Owen, F.N., Booth, R.S., Wilkinson, P.N., Shaffer, D.B., Johnston, K.J., Hardee, P.E. 1982b. Proc. IAU Symp. 97, Extragalactic Radio Sources, p. 293
Romney, J.D., Alef, W., Pauliny-Toth, I.I.K., Preuss, E., Kellermann, K.I. 1982. Proc. IAU Symp. 97, Extragalactic Radio Sources, p. 291
Romney, J.D., Alef, W., Pauliny-Toth, I.I.K., Preuss, E., Kellermann, K.I. 1982. In preparation
Scheuer, P.A.G., Readhead, A.C.S. 1979. Nature 277:182
Schilizzi, R.T., Shaver, P.A. 1981a. Astron. Astrophys. 96:365
Schilizzi, R.T., Miley, G.K., Janssen, F.L.J., Wilkinson, P.N., Cornwell, T.J., Fomalont, E.B. 1981b. Proc. ESO/ESA Workshop Optical Jets in Galaxies, ESA SP-162, p. 107
Schilizzi, R.T., Miley, G.K., Romney, J.D., Spencer, R.E. 1981c. Nature 290:318
Schilizzi, R.T., Fejes, I., Romney, J.D., Miley, G.K., Spencer, R.E., Johnston, K.J. 1982. Proc. IAU Symp. 97, Extragalactic Radio Sources, p. 205
Schramal, J., Pauliny-Toth, I.I.K., Witzel, A., Kellermann, K.I., Johnston, K.J., Spencer, J.H. 1981. Ap. J. Lett. 251:L57
Shaffer, D.B., Marscher, A.P. 1979. Ap. J. Lett. 233:L105
Shapiro, I.I., Wittels, J.J., Counselman, C.C., Robertson, D.S., Whitney, A.R., Hinteregger, H.F., Knight, C.A., Rogers, A.E.E., Clark, T.A., Hutton, L.K., Niell, A.E. 1979. Astron. J. 84:1459
Simon et al. 1981. BAAS 13:898
Spangler, S.R., Benson, J.M., Cordes, J.M., Hall, R.B., Jones, T.W., Johnston, K.J. 1981. Astron. J. 86:1155
Spencer, J.H., Johnston, K.J., Pauliny-Toth, I.I.K., Witzel, A. 1981. Ap. J. Lett. 251:L61
Ulvestad, J.S., Wilson, A.S., Sramek, R.A. 1981. Ap. J. 247:419
Unwin, S.C. 1982a. Proc. IAU Symp. 97, Extragalactic Radio Sources, p. 357
Unwin et al. 1982b. In preparation
Unwin, S.C., Mutel, R.L., Phillips, R.B., Linfield, R.B. 1982c. Ap. J. 256:83
Walker, R.C., Readhead, A.C.S., Seielstad, G.A., Preston, R.A., Niell, A.E., Resch, G.M., Crane, P.C., Shaffer, D.B., Geldzahler, B.J., Neff, S.G., Shapiro, I.I., Jauncey, D.L., Nicolson, G.D. 1981. Ap. J. 243:589
Walker, R.C., Seielstad, G.A., Simon, R.S., Unwin, S.C., Cohen, M.H., Pearson, T.J., Linfield, R.P. 1982. Ap. J. 257:56
Waltman, E., Johnston, K.J., Spencer, J.H., Pauliny-Toth, I.I.K., Schraml, J., Witzel, A. 1981. Astron. Astrophys. 101:49
Wehrle, A.E., Preston, R.A., Meier, D.L., Gorenstein, M.V., Rogers, A.E.E., Shapiro, I.I., Rius, A. 1981. BAAS 13:898
Wilkinson, P.N., Readhead, A.C.S, Anderson, B., Purcell, G.H. 1979. Ap. J. 232:365

Wilkinson, P.N. 1983. Proc. International Conference on VLBI Techniques held in Toulouse, France, Aug./Sep. 1982, in press
Wilson, A.S., Pooley, G.G., Willis, A.G., Clements, E.D. 1980. Ap. J. Lett. 237:L61
Wilson, A.S., Ulvestad, J.S. 1982. Ap. J.
Wittels, J.J., Shapiro, I.I., Cotton, W.D. 1982. Ap. J. Lett. 262:L27
Zensus, A., Porcas, R.W., Pauliny-Toth, I.I.K., Kellermann, K.I. 1982. In preparation

ARC SECOND STRUCTURE OF COMPACT RADIO SOURCES

I.W.A. Browne, M. Charlesworth, T.W.B. Muxlow, A. Tzanetakis and P.N. Wilkinson
University of Manchester
Nuffield Radio Astronomy Laboratories
Jodrell Bank, Macclesfield, Cheshire, SK119DL, England

I. INTRODUCTION

Radio sources which have the majority of their emission arising from regions $\lesssim$ 2 arc sec in diameter are relatively common, forming approximately $\sim$30% of all strong sources in surveys made at $\sim$ 1 GHz. Compact sources usually have high radio luminosities and are found at high redshifts. At such distances the linear scales being discussed are $\lesssim$ 20 kpc so the majority of the radio emission arises within the parent galaxy. Some of the sources have prominent jets. These are invariably one-sided and often show quite large changes in position angle of elongation at different angular scales. We are interested in the causes of the one-sidedness (whether Doppler favouritism or not) and of the bends (whether due to winds or ballistic). In the ballistic case the apparent bends are due to precession or rotation of the central massive object and information of fundamental importance about the object may be deduced. Alternatively, if the bending is due to an interaction of the jet with its surroundings the bendings becomes a useful probe of the galactic environment.

Amongst the compact sources some have overall flat spectra and others have steep power law spectra. Over 50% of the flat spectrum (or core-dominated) sources, including most of the superluminal sources, have detectable arc second emission (Moore et al. 1981; Perley et al. 1982). Just as numerous as the core-dominated sources are the steep spectrum compact sources. Despite the marked difference in spectral shape some of these sources have structures very similar to the core-dominated sources; e.g. compact cores and asymmetric jets.

In this paper we present MERLIN maps which illustrate the wide variety of structures found amongst compact sources. Some statistical properties are summarized and evidence discussed for the relationship between compact sources and the "normal" extended doubles. Special attention will be focussed on the misalignment of the milliarcsecond structure and the core second structure.

A. Ferrari and A. G. Pacholczyk (eds.), Astrophysical Jets, 27–36.
Copyright © 1983 by D. Reidel Publishing Company.

II. MERLIN MAPS OF COMPACT SOURCES

1. Steep spectrum objects

An interesting question is whether steep spectrum compact sources are simply scaled down versions of the normal classical doubles or separate class of object. A study of a sample of 19 compact 3CR sources with the Jodrell Bank-Defford interferometer at frequencies of 408 and 1666 MHz shows 8 asymmetrical double sources ($\sim$ 5% of all 3CR sources), 8 asymmetrical sources and 3 compact ones (<0.1). Many of the classifications have been confirmed by detailed 5 GHz MERLIN maps. Fig. 1a and Fig. 1b show two small, relatively symmetrical doubles 3C237 and 3C277.1. 3C237 has no optical identification while 3C277.1 is a $17^{m}.9$ quasar with redshift 0.320. By contrast there are other sources like 3C147 and 3C309.1 (Fig. 2a, Fig. 2b, Fig. 2c) which show much more complex and asymmetric structure. Both have unresolved cores, jets, and a region of much more extended emission. These latter sources are obviously not scaled down classical doubles, although it is possible that they could be such doubles seen in projection, with the core and arc sec scale jet emission enhanced by Doppler beaming effects. No superluminal motion has yet been reported in such sources.

2. Core-dominated sources

The arc second structure of core-dominated sources have been extensively studied (Browne et al. 1982a, Perley, 1982). The two well known quasars 3C345 and 3C454.3 show structures typical of many of the sources (Fig. 3a, Fig. 3b). They have compact cores, and a hotspot on one side only joined to the core by a weak jet. It is interesting that only when very high dynamic range maps are available, does the arc second structure resemble a jet. In addition to the jet-like emission many of these sources including most of the superluminal ones, have more diffuse emission (see Schilizzi, this conference), and by virtue of this emission alone these sources become strong low frequency objects. Such diffuse emission assumes great significance if the brightness of the cores and superluminal effects are interpreted in terms of the bulk relativistic motion of the emitting material in the core. The emission would be visible in those objects not aligned with the observer, and this prompts the hypothesis that core-dominated sources are normal doubles seen end-on (Browne et al. 1982a).

III. JETS CONTINUITY IN CORE-DOMINATED SOURCES

Jets are commonly found in core-dominated sources. On the arc second scale they are invariably asymmetric and the same is true on the milliarcsecond scale. Is there a continuity between the milliarcsecond and the arc sec structure? The answer is almost certainly yes, since the sense of jet asymmetry on the two scales always seems to be the same. In particular this holds for the superluminals NRAO140, 3C120, 3C273, 3C279, 3C345 and 3C454.3 (Browne et al. 1982b). This continuity

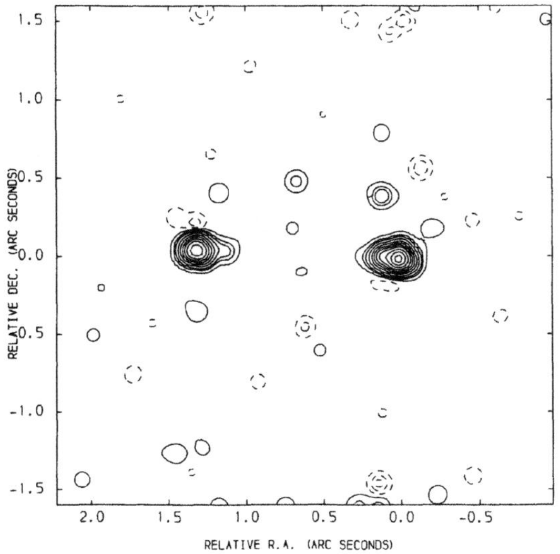

Figure 1a: 5 GHz MERLIN map of 3C237 showing classical double structure.

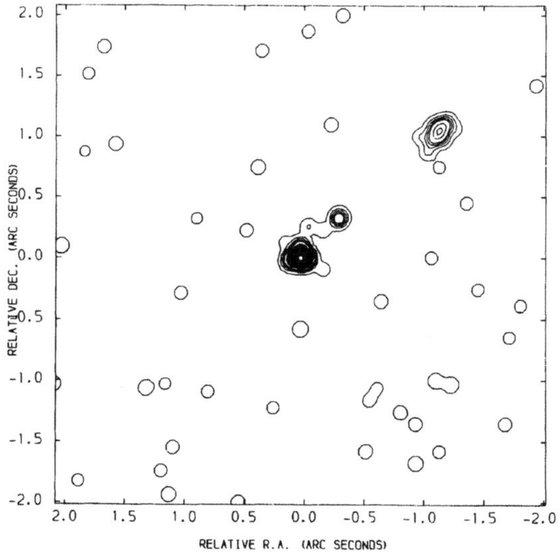

Figure 1b: 5 GHz MERLIN map of 3C277.1 showing classical double structure with a central component.

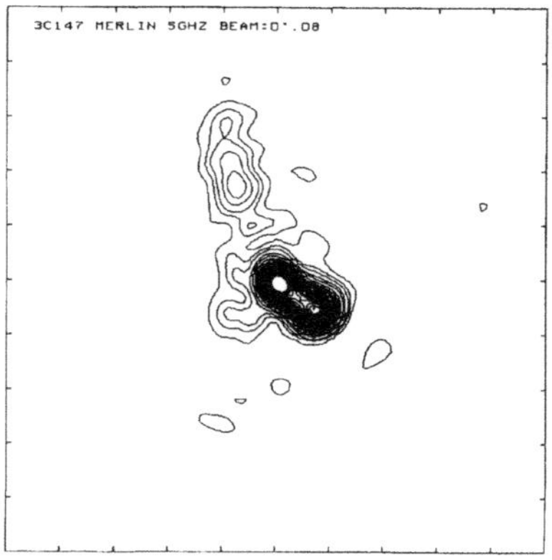

Figure 2a: 5 GHz MERLIN map of 3C147 showing the strong core and a jet to the SW. Note the extended emission to the north.

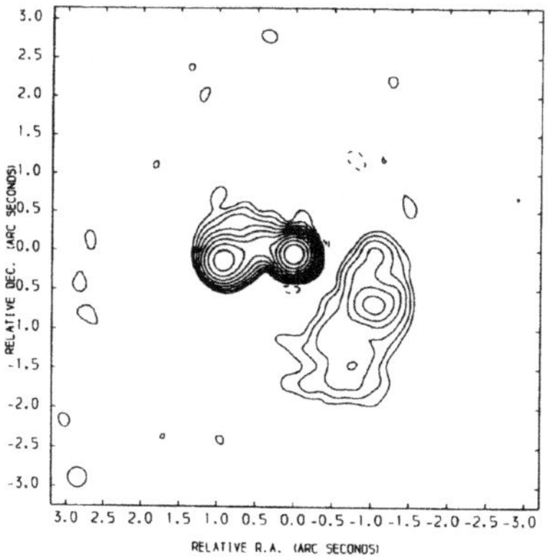

Figure 2b: 1666 MHz MERLIN map of 3C309.1

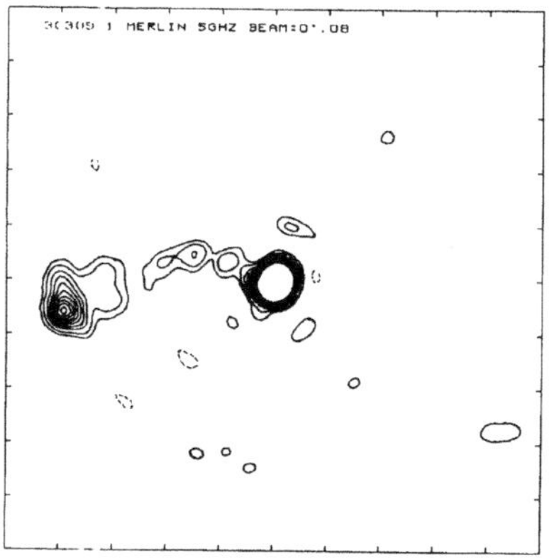

Figure 2c: 5 GHz MERLIN map of 3C309.1 showing jet pointing east.

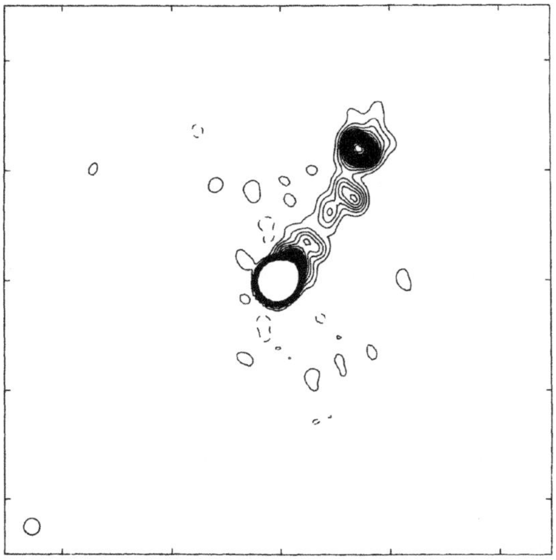

Figure 3a: 1666 MHz MERLIN map of 3C345.

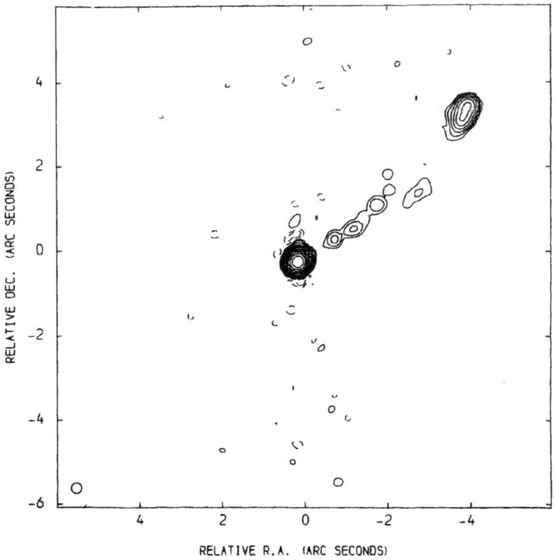

Figure 3b: 1666 MHz MERLIN map of 3C454.3

indicates a common cause for the asymmetry on all scales. As Doppler beaming is capable of making an intrinsically symmetrical twin beam source look one sided, and the required velocities seem to exist in superluminals, the observations point towards the arc second jets being relativistic too.

IV. ARE BENT JETS BALLISTIC?

It is possible to model the curvature of some jets in terms of precession of the central object and subsequent ballistic motion of the jet material (Linfield, 1981; Gower et al. 1982). If this is the true explanation of bent jets, then very useful information about the behaviour of the central massive object can be obtained such as, for instance, its precession period.

3C418 has a remarkable curved jet whose shape strongly suggests precession (Fig. 4). A ballistic model can be obtained which gives a good fit to the jet trajectory, but it does require the half angle of the precession cone to increase with time (Muxlow et al. in preparation). The periods which give an acceptable fit are in the range 10^4 to 10^5 years. Such short periods can only arise if there exists a close binary black hole system in the centre of the quasar. In addition, the increase in the precession cone angle with time, suggests that the binary system is evolving rapidly (Begelman, Blandford & Rees 1980).

The apparent bending in the cores of superluminal sources may be another example of ballistic motion. The large bend in 3C345 (Fig. 5) illustrates an extreme case where 60° of bending occurs in the first 4 m.a.s. of the jet. Such a large amount of bending in the presence of superluminal motion enables further tests of the ballistic hypothesis to be made on reasonable timescales. Observations of the different relative velocities along the advancing front of ejecta should provide enough information to determine not only the precession period of the central object but also the true speed of the jet material. This latter possibility is particularly important because it suggest a new way to determine Hubble's constant by comparing the observed velocities with the model values.

V. THE DISTRIBUTION OF MISALIGNMENT ANGLES BETWEEN VLBI CORES AND ARCSECOND JETS

As the observations of 3C418 and 3C345 indicate, even for a continuous jet the position angle of the structure in the VLBI core of a source may be very different to that of the arc second structure. In relativistic beam models these large apparent bends can be explained in terms of intrinsically small bends amplified by projection effects (Readhead et al. 1978). For a sample of sources selected on the basis of the strength of their core emission alone it is possible to predict

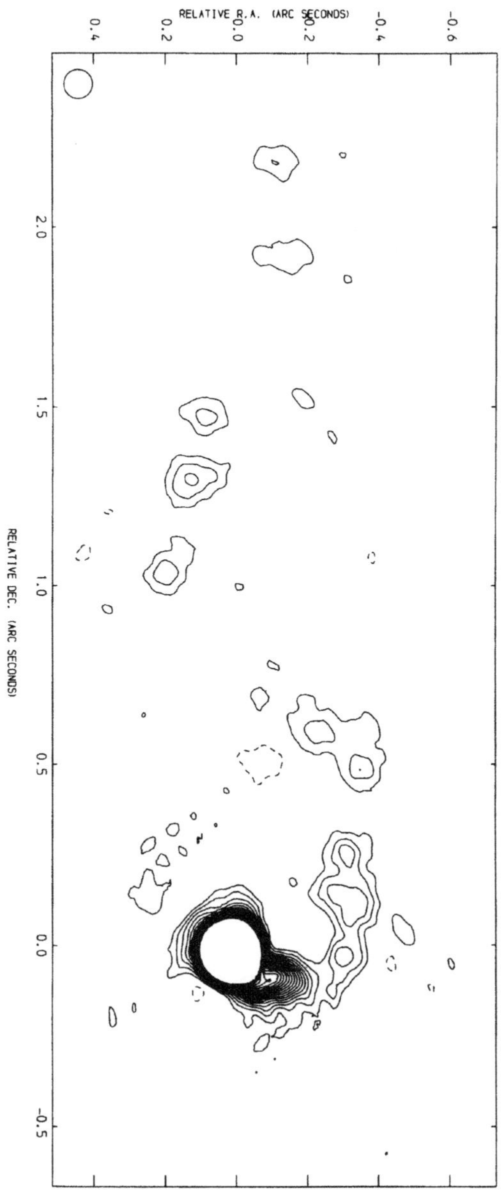

Figure 4: 1666 MHz MERLIN map of 3C418 made by combining MERLIN and European VLBI measurements.

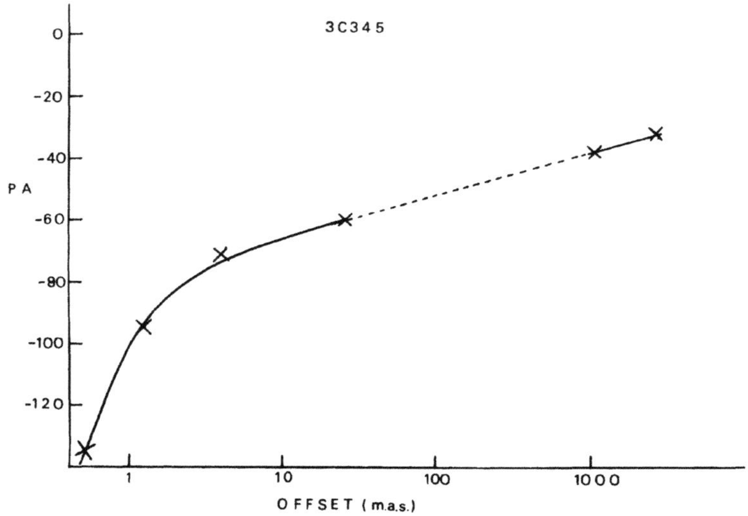

Figure 5: The bend in 3C345. The position angle of a point in the jet as seen from the core is plotted against the distance of that point from the core.

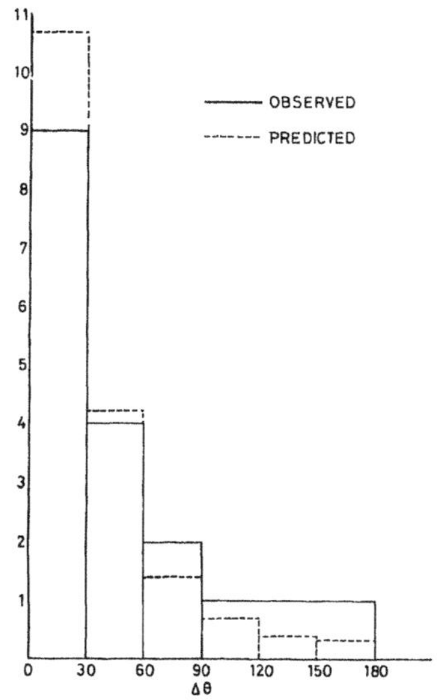

Figure 6: Histogram of the position angle difference (Δθ) between the milliarcsecond and arcsecond structure of eighteen core-dominated sources (continuous line). The dotted histogram is that predicted for a mean intrinsic bend of 5° and a mean Lorentz factor of 5.

the expected distribution of misalignment angles knowing a) the average intrinsic bend and b) the mean Lorentz factor in the cores (Moore et al. 1981; Browne et al. 1982; Readhead et al. 1983). If a unified scheme is adopted in which core-dominated sources are regarded as normal doubles seen end-on, the average size of the intrinsic bends in sources can be estimated from observations of steep spectrum sources. Macklin (1981) shows that the intrinsic misalignment between core and lobe amongst 3CR double sources in $\sim 5°$. Quite separately Orr & Browne (1982) find that in order to match the relative numbers of flat and steep spectrum quasars within a unified scheme, a mean Lorentz factor of 5 is required in quasar cores. Using $\overline{\theta} = 5°$ and $\overline{\gamma} = 5$ enables a prediction of the expected distribution to be made.

Information for complete samples of core-dominated sources is not available but we have found data on 18 objects and, since they have not been selected with any bias towards large bends, they should form a representative group. Fig. 6 compares the misalignment histogram for these 18 sources with the theoretical distribution. There is a remarkable good fit.

VI. THE LINEAR SIZES OF CORE-DOMINATED SOURCES

The relative numbers of flat and steep spectrum sources, and the distribution of bend angles are consistent with a unified scheme. Another prediction of such schemes is that the linear sizes of sources should be related to the fraction of source flux density in the core (strong cores imply a small angle to the line of sight). Amongst a well defined sample of 36 core-dominated sources Moore (1982) finds a median linear size of 30 kpc compared to 350 kpc for steep spectrum double sources. Also Kapahi and Saikia (1982) find that in a sample of 78 well observed double quasars the fraction of flux density in the core is anticorrelated with the observed linear size. Likewise they find that sources of smaller linear sizes appear more misaligned, and that the degree of misalignment is correlated with the ratio of the separations of the outer hot spots. Although other explanations are possible both Moore's, and Kapahi and Saikia's results are just what are expected from relativistic beaming type unified schemes.

REFERENCES

Belgelman, M.C., Blandford, R.D. & Rees, M. 1980, Nature 287, 307
Browne, I.W.A., Orr, M.J.L., Davis, R.J., Foley, A., Muxlow, T.W.B. & Thomasson, P. 1982a, Mon. Not. R. astr. Soc., 198, 673
Browne, I.W.A., Clark, R.R., Moore, P.K., Muxlow, T.W.B., Wilkinson, P.N., Cohen, M.H. & Porcas, R.W., 1982b, Nature 299, 788
Gower, A.C., Gregory, P.C., Hutchings, J.B. & Unrah, W., 1982, Ap. J. 262, 478
Linfield, R., 1981, Ap. J., 250, 464
Macklin, J.T., 1981, Mon. Not. R. astr. Soc., 196, 967
Moore, P.K., Browne, I.W.A., Daintree, E.J., Noble, R.G. & Walsh, D.,

1981, Mon. Not. R. astr. Soc., 197, 325
Moore, P.K., 1982, Mon. Not. R. astr. Soc., submitted
Orr, M.J.L. & Browne, I.W.A., 1982, Mon. Not. R. astr. Soc., 200, 1067
Perley, R.A., 1982. Astr. J., 87, 859
Readhead, A.C.S., Cohen, M.H., Pearson, T.J. & Wilkinson, P.N., 1978. Nature, 276, 768
Readhead, A.C.S., Hough, D.H., Ewing, M.S., Walker, R.C. & Romney, J.D. 1982, preprint

DISCUSSION

Mike Norman Would you interpret the large bending angle in 3C418 as an angle-of-viewing effect on an intrinsically nearly-straight jet?

Ian Browne Yes. I think the bend in 3C418 is probably magnified by projection effects. A small angle of viewing follows naturally if one believes the reason for the bright core is Doppler Boosting.

Gregory Benford Turning jets through $\sim 180°$ is very destabilizing for hose-like motions. Wouldn't you rather describe 3C418 as a lesser turning, amplified by projection effects?

Ian Browne Yes.

Richard Porcas Eugen Preuss drew attention to a correlation between the side of the source showing an arc-second scale jet and the side with the most compact hot spot in the lobe. You in turn have spoken of a similar correlation between the side of the m.a.s. and arcsecond scale jets, and suggest this indicates a common cause. Together one must conclude that the reason for the VLBI asymmetry is the same as that for the hot spots. Since the hot spots are not moving relativistically, does this not imply that beaming does not cause the VLBI asymmetry?

Ian Browne The correlation referred to by Eugen Preuss may well be real. If it is, then either hot spot emission has to be beamed in some way or, more likely, the arc second jets are intrinsically asymmetric. Certainly in the latter case you are correct to point out that the correlation between the side of the milliarcsecond and the arcsecond jets indicates that Doppler beaming is not the sole cause of the observed VLBI asymmetry. This of course does not necessarily mean that the VLBI jets are slow.

A SUMMARY OF PROPERTIES OF RADIO JETS

Edward B. Fomalont
National Radio Astronomy Observatory
Socorro, NM 87801, USA

It is now commonly accepted that enormous amounts of energy are generated in the nucleus of some giant elliptical galaxies and quasars. This energy in some form is transported non-isotropically, often well beyond the boundaries of the galaxy. The energy is then disrupted in radio lobes which are often dominated by several intense regions of radio emission.

Although this overall scenario is in good standing at the present time, the physical mechanisms associated with most of the phenomena in luminous radio sources are poorly, if at all, understood. The wealth of details in the radio maps now being produced at the VLA and elsewhere are not only a theoretician's dilemma because he cannot explain them, but also an experimentalist's dilemma because he cannot easily extract the important properties from much of the clutter of individual differences.

With this in mind the following measurable radio properties, some of which may be important in theoretical models, have been listed. The remainder of the paper will expound on this list in some detail.

1. The incidence of radio jets
2. Jet asymmetries
3. Intensity distribution in jets
4. Jet collimation properties
5. Bends and wiggles in jets
6. The radio spectrum in jets
7. Magnetic field alignment in jets
8. Results from linear-polarization studies
9. Equipartition pressures
10. Jet velocities

1. The Incidence of Radio Jets:

Radio jets were first discovered in low radio luminosity, relatively close galaxies. With better sensitivity and higher

A. Ferrari and A. G. Pacholczyk (eds.), Astrophysical Jets, 37–46.
Copyright © 1983 by D. Reidel Publishing Company.

resolution now available even a significant proportion of high luminosity sources shows some evidence of jet structure (Wardle and Potash 1982). About 50 percent of the flat-spectrum quasars also show extended structure which is often composed of a jet with one hot spot (Perley et al. 1982).

Thus, there is reason to believe that most if not all bifurcated radio sources will contain a radio jet when sufficiently sensitive observations are made. This is not surprising since the presence of a radio lobe and a hot spot suggest energy flow into this region, presumably from the radio core.

2. Jet Asymmetries:

The generally asymmetric nature between a jet and the oppositely directed counter jet is very common. Only about 20 percent of the low luminosity radio sources have a jet/counter-jet intensity ratio within a factor of two. Some of these sources when observed at sufficiently high resolution become more asymmetric or even one-sided. A few exceptions include 3C449 (Cornwell and Perley 1982) or IC4296 (Killeen and Ekers 1982).

No luminous radio source greater than 10^{26} W/Hz at 2600 MHz shows any sign of a counter jet. In some cases the limits are better than about 100 to 1. It is still unclear whether the jet asymmetry is intrinsic or is caused by Doppler favoritism in one side of the jet. This will be discussed in more detail later.

Many jets are broken into several pieces along the radio core to hot spot trajectory. Rudnick (1982) has recently suggested that ejection from the nucleus occurs only in one direction at a time and that gaps in the jet on one side of the core fit with pieces of jet on the other side. Confirmation of this asymmetry would strengthen the argument that jet asymmetries are intrinsic. However, the evidence is still sketchy.

3. The Radiation Distribution in Jets:

The intensity profiles transverse to the jet axis are center-brightened in all well-documented cases. Hence a model in which most of the radiation comes from sheath or shell around the jet, is unlikely.

The intensity profiles along the jet axis show the presence of two components; a somewhat uniform brightness which often decrease with distance from the galaxy, and knots of enhanced radiation which are often more than ten times the brightness of the background jet emission. For some sources only the brightest knots can be detected. Perhaps, the best example is the source M87 (Biretta et al. 1982), seen in Figure 1, which shows strong knots on top of a uniform background.

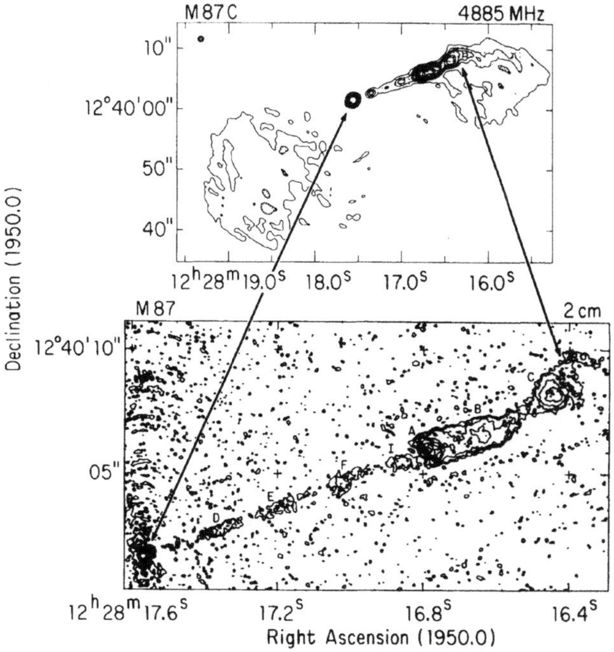

Figure 1a. Contour map of M87 at 5 GHz with 1.0 arcsec resolution.

Figure 1b. Contour map of the jet in M87 at 15 GHz with 0.2 arcsec resolution.

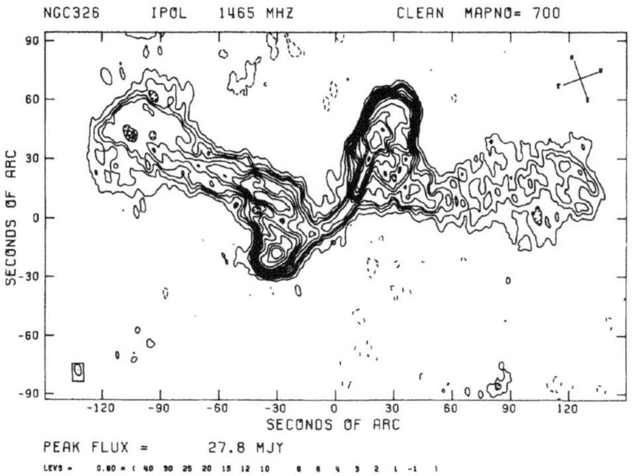

Figure 2. Contour map of NGC326 at 1.5 GHz with 4 arcsec resolution.

The ratio of knotiness to smooth component does not seem to be a strong function of the source luminosity.

The decrease in the intensity of the smooth component along the jet varies as $I \propto R^{-1.5}$ (Bridle 1982) where I is the surface brightness and R is the radius of the jet. This decrease is much slower than that for for flux conserving flow of unaccelerated electrons.

4. Jet Collimation Properties:

The average opening angle for jets with low luminosity is about 5 to 15 degrees; for high luminosity sources the angle is about three degree or less. In the well-studied sources, all of which are low luminosity sources, the rate or expansion of the jet is not uniform. There are alternating regions of rapid expansion and relatively slow or no expansion. There are no cases yet where the width of the jet has significantly decreased. The regions where the collimation properties change are not obviously connected with any other radio or optical property.

For the high luminosity sources the jets have not been resolved in sufficient detail transverse to their axis to know much about their collimation properties. Because of the predominance of a parallel magnetic field in these sources, the collimation properties may differ from that of the lower luminosity sources which have a predominantly perpendicular field.

5. Jet Wiggles and Bends:

Jets are never perfectly straight. The most obvious examples are head-tail sources, like 3C83.1, where the overall bending is caused by pressure of the external medium produced by the motion of the galaxy. Similar bending should occur to a lesser extent. The source NGC326, Fig. 2, is an example (Fomalont 1981). Some luminous quasars show significant curvature in their radio jets (Perley et al. 1982). This curvature is unlikely to be caused by the effects of an external medium.

More intriguing are the wiggles and oscillations which occur for many sources when sufficiently detailed maps are available. Examples are 3C273 (Perley et al. 1982) and 4C32.69 (Wardle and Potash 1982), see Figures 3 and 4. The wiggles are generally small in amplitude and show some periodicity. In a few cases there is symmetry between wiggles on the jet and counter-jet (Bridle et al. 1979). Finally, there are sources which have wiggles or bends of over 90 degrees (Fomalont 1981), as in 3C293 (Fig. 4b).

Several alternatives have been suggested for the cause of the wiggles or oscillations; kink instabilities in the flow, precession of the nuclear engine, Kelvin-Helmholtz instabilities or refraction from the galactic medium. Support and problems for each of these mechanisms

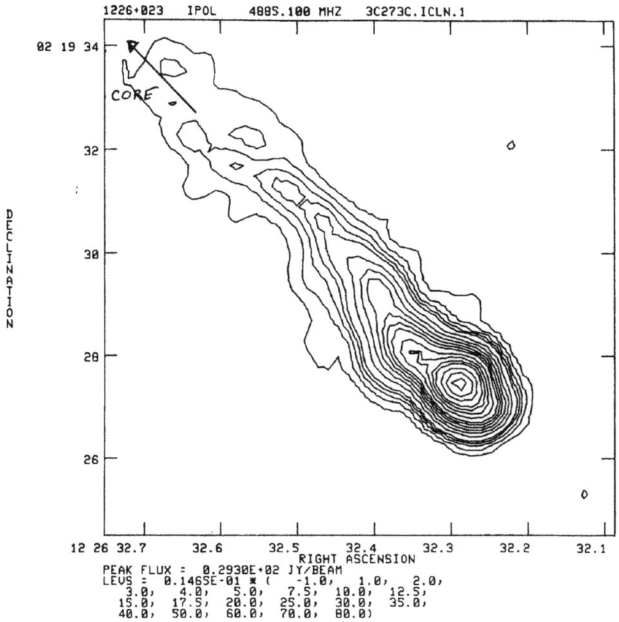

Figure 3. Contour map of 3C273 at 5 GHz with 0.6 arcsec resolution.

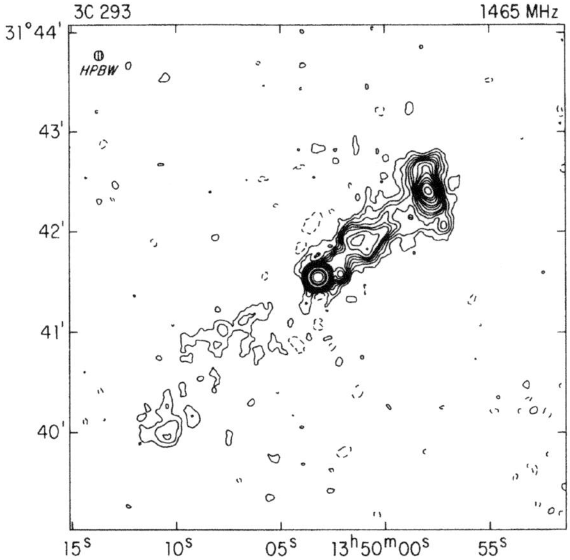

Figure 4. Contour map of 3C293 at 1.5 GHz with 1.8 arcsec resolution.

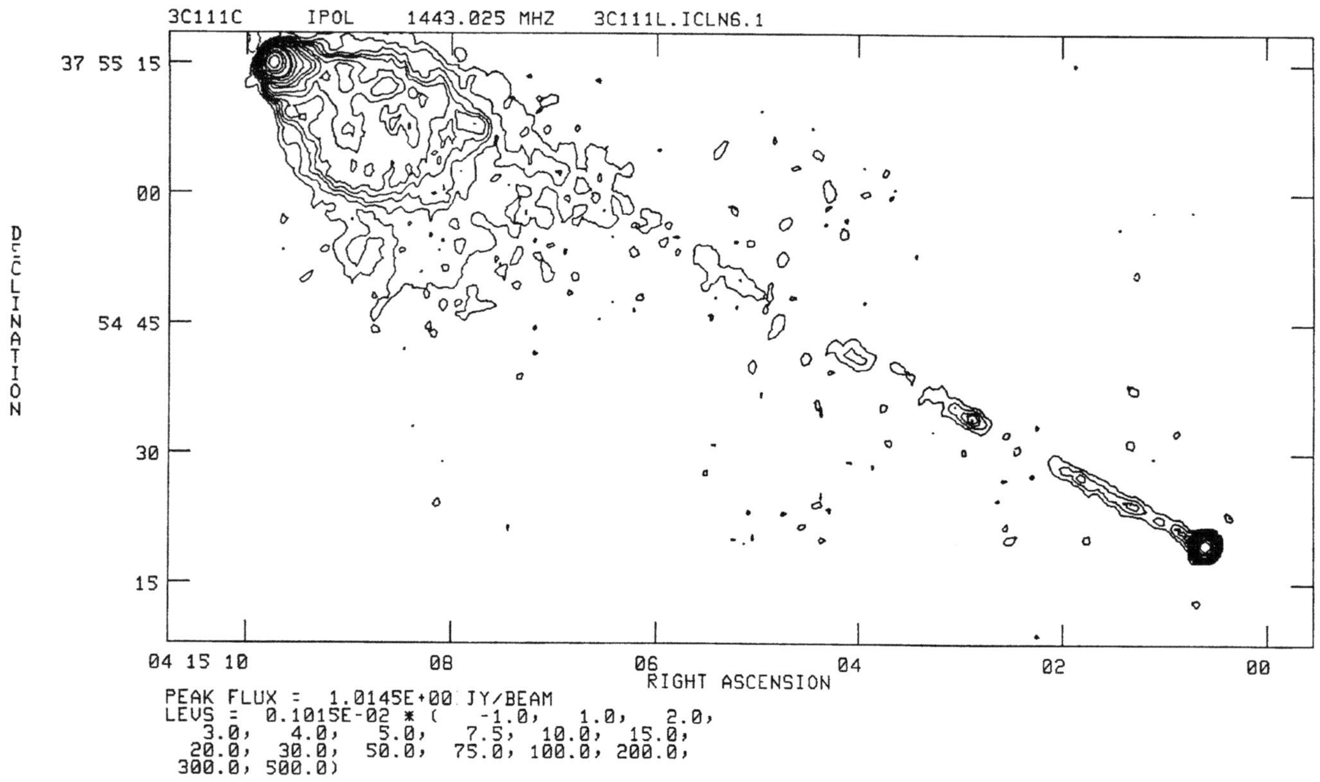

Figure 5. Contour map of 3C111 at 1.4 GHz with 1.5 arcsec resolution.

can be found in the set of jet data. For example; before many jets terminate into a radio lobe, there is a large amount of curvature. This may support the instability theories.

6. Spectrum of the Jets:

The spectral index of an overwhelming majority of jets lies between -0.5 and -0.7. In the well-studied cases the distribution of spectral index is remarkably constant over the entire jet emission region. In a few cases the jet is embodied in a steeper spectrum envelope (eg. NGC326). The jet in Fornax A and some other very weak sources appears to steepen and then disappear (Fomalont 1981).

7. Magnetic Field Alignment:

New data has not materially changed the earlier results on the alignment of the magnetic field in radio jets (Bridle 1982). The field directions in the most luminous sources are parallel to the jet axis. The field direction even appears to follow wiggles in the jet directions.

For the lower luminosity sources the field in the dominant jet is parallel to the axis near the galaxy but then flips to perpendicular. Near the edge of the jet, however, the field remains parallel to the jet axis. This configuration can be explained by a helical field or by a circumferential field with field shearing at the edges. The magnetic field direction for the weaker jet is perpendicular to the axis. There is no parallel field region. There are two known exceptions for this rule, 3C449 and IC4296. Both jets show regions near the core with parallel field.

8. Linear Polarization Studies:

Multiple-frequency observations of several radio sources show that there is little Faraday depolarization in the jets. This places a limit of about 10^{-3} cm^{-3} for the density of thermal matter in them. One exception is the source 3C449 which is depolarized near the core. However, this may be caused by rapid changes of polarization angle.

This limit is consistent with the lack of internal Faraday rotation in jets. Large variations in Faraday rotation have been seen in NGC6251 and M84; however there is good evidence to suggest that all of the variation are occurring outside of the jet and in the parent galaxy. Whether the medium causing the Faraday rotation is produced by random clouds in the galaxy or is in some way related to the jet, is unknown.

The degree of polarization in jets is about 25 to 50 percent. For some sources the region near the radio core may have less polarization but this may be caused by beam depolarization. Many field configurations and field randomness properties can produce such polarization degrees and these have been discussed elsewhere (e.g. Laing 1981).

9. Equipartition Pressures:

Typical equipartition values in radio jets are in the range 10^{-8} to 10^{-10} dyne cm^{-2} . These values are, of course, uncertain since the equipartition assumption may be far from valid. However, the internal pressures calculated are not likely to be much lower unless the beams are relativistically brightened. In this case the internal density is overestimated and the possible ram pressure confinement is underestimated.

The internal pressures now obtained from the jets in high luminosity quasars, as well as in the knots of many jets (M87 for example), are too high to be balanced by a reasonable external medium. Limits to the density of thermal material in a hot gas around several jet sources can be obtained by the limits or detection of X-rays from the galaxies using reasonable values of the temperature. The derived densities are too low by an order of magnitude to supply the necessary confining pressure.

There are several alternatives to external pressure confinement. Magnetic confinement requires circumferential fields, but the most luminous sources have fields parallel to the jet axis. The possible circumferential field may, of course, be outside of the radiating plasma. The strength of the current and the closing of the parallel field components are problems. Another possibility is that the jet alternates as a free jet and a confined jet (Sanders 1982). This model is also supported by the alternating regions of significant expansion and little expansion seen in many jets.

10. Jet Velocities:

Knowledge of the velocity of the flow in jets is extremely important in understanding many facets of radio sources. This velocity can be inferred by several lines are argument:

1. The energy flux needed in the jet to power the source.
2. Using the opening of the jet to deduce a Mach number in the flow. The sound speed is obtained from equipartition measurements.
3. The pressure flux needed to support the hot spot.
4. The momentum flux needed to move the hot spot.

These calculations lead to results of about 5000 km/s for the low luminosity sources and about 15000 km/s for the high luminosity sources. Much lower velocities necessitate large mass flow to supply the necessary energy and much higher velocities are not compatible with some of the properties of the low luminosity sources; such as the significant curvature in jets, the lack of enhancement of radiation on the outer edge of the jet.

The only direct measurements of velocities are:

1. VLB observations which are consistent with velocities of >0.8c within about 5 to 10 degrees of the line of sight. These velocities are only observed for luminous quasars on pc-size scale.
2. Emission lines are associated with some jets (eg. 3C277.3, Miley et al. 1981) and the derived velocities are about 300 km/s. However, this velocity need not be associated with the flow in the jet.
3. Observations of the galactic source SS433 show that material is flowing at about 0.25c from the star. The flow appears to be jet-like and the source of the flow seems to be precessing.
4. Observations of ScoX-1, a year apart, show that the hot spots are moving at less than about 20 km/s. The only way to supply a large amount of energy to the hot spot, but a relatively small momentum, is to have an efficient, relativistic beam impinging on the hot spot.

The question whether the flow in jets is relativistic is devilishly difficult to answer. The arguments for and against relativistic flow have been summarized by Rees (1982). Some new observational evidence appear to hurt the case for relativistic flow in kpc-sized jets.

Many luminous, core-dominated radio sources display a kpc-sized jet linked to a hot spot. The sources, some of which are superluminal, are very asymmetric and often show no emission on one side of the core (Perley et al. 1982) to a limit of over 100 to 1 in some cases. If the lobe and jet asymmetry is due solely to Doppler boosting, then the lobe velocity must be in excess of 0.7c. Conway (1982) has shown that it would take an absurdly low external density to keep the lobe from decreasing in velocity much below the level where Doppler boosting is important. Hence, the faintness of the other lobe must in part be because it is intrinsically much weaker. If the lobes are intrinsically different, then it is not unreasonable to assume that the difference in the kpc-sized jet and counter jet are in part intrinsic as well.

Long, one-sided jets have recently been detected in 3C111 and 3C405 (Perley 1982), Figure 5. They have prominent lobes and hot spots on both sides of the radio core. From the VLB jet, through the VLA detected jet to the hot spots, both sources are remarkably straight to within a few degrees. If the VLB jet, which is one sided, is aligned nearly along the line of sight, then one is forced to concede that the entire source is as well. This makes Cygnus A over 3 Mpc in size and 3C111 only slightly smaller. In addition, the core in Cygnus A is exceptionally weak relative to the rest of the emission. Thus, it is unlikely that these sources are Doppler boosted in their kpc structure.

REFERENCES

Biretta, J.A., Owen, F.N., and Hardee, P.E., 1982, submitted to Ap. J.
Bridle, A.H., 1982, I.A.U. Symposium No. 97 on Extragalactic Radio Sources, edited by D. Heeschen and C. Wade, (D. Reidel), p. 121
Bridle, A.H., Davis, M.M., Fomalont, E.B., Willis, A.G., and Strom, R.G., 1979, Ap. J., 228, L9
Conway, R.G., 1982, I.A.U. Symposium No. 97 on Extragalactic Radio Sources, edited by D. Heeschen and C. Wade, (D. Reidel), p. 167
Cornwell, T.J. and Perley, R.A., 1982, I.A.U. Symposium No. 97 on Extragalactic Radio Sources, edited by D. Heeschen and C. Wade, (D. Reidel), p. 139
Fomalont, E.B., 1982, I.A.U. Symposium No. 94 on Origin of Cosmic Rays, edited by G. Setti, G. Spado, and A. Wolfendale, (D. Reidel), p. 111
Killeen, N. and Ekers, R.D., 1982, in preparation
Laing, R.A., 1981, Ap. J., 248, 87
Miley, G.K., Heckman, T.M., Butcher, H.R., and van Breugel, W.J.M., 1981, Ap. J. (Lett.), 247, L5
Perley, R.A., 1982, I.A.U. Symposium No. 97 on Extragalactic Radio Sources, edited by D. Heeschen and C. Wade, (D. Reidel), p. 175
Perley, R.A., Johnston, K.J., and Fomalont, E.B., 1982, in preparation
Rees, M., 1982, I.A.U. Symposium No. 97 on Extragalactic Radio Sources, edited by D. Heeschen and C. Wade, (D. Reidel), p. 211
Rudnick, L., 1982, I.A.U. Symposium No. 97 on Extragalactic Radio Sources, edited by D. Heeschen and C. Wade, (D. Reidel), p. 47
Sanders, R.H., 1982, in preparation
Wardle, J.F.C. and Potash, R.I., 1982, I.A.U. Symposium No. 97 on Extragalactic Radio Sources, edited by D. Heeschen and C. Wade, (D. Reidel), p. 129

DISCUSSION

M. NEPVEU: Is there any well-defined mean separation of "radio lumps" inside jets?

E. FOMALONT: No. Lumps tend to occur with separations about 5 times the jet width, but no well-defined mean separation is seen.

RECENT OBSERVATIONS OF SUPERLUMINAL SOURCES

Richard W. Porcas
Max-Planck-Institut für Radioastronomie, Bonn, F.R.G.

ABSTRACT

One of the most fascinating phenomena associated with the topic of astrophysical jets is the observation of apparent faster-than-light ("superluminal") motions in the cores of some extragalactic objects. A thorough review of the subject was given at the I.A.U. Symposium No. 97 in Albuquerque last year (Cohen and Unwin, 1982). Since then a number of papers have been published, reporting in detail the most recent observational results on individual objects (Porcas, 1981; Marscher and Broderick, 1982; Phillips and Mutel, 1982; Walker et al., 1982). Here I wish to give a progress report on one of the objects - 3C179 - and discuss the significance of the fact that most of its flux density at around 1 GHz comes from extended, double radio lobes.

The initial report of superluminal motion in the core of the quasar 3C179 (Porcas, 1982) was based on transatlantic VLBI observations at only two epochs. Since then data has been obtained at a further two epochs using the same telescope array and observing frequency (10.7 GHz). The angular separation of the two milli-arc-second components (obtained from fitting models to the data) is plotted against epoch in Figure 1. Despite the poorer quality of the more recent data (due to less than perfect weather and equipment and a decrease in the flux density of the core) it is readily apparent that the expansion has continued, with separations even slightly larger than those predicted. Thus 3C179 can join the ranks of the other superluminal sources, with a consistent, steady rapid motion between components, confirmed at a number of epochs.

Also plotted in Figure 1 is a preliminary value for the separation obtained from observations in July this year, at an observing frequency of 5.0 GHz. This falls a little below the prediction from the 10.7 GHz expansion; this may indicate that at this lower frequency, with corresponding lower angular resolution, there is an extended component which affects the estimate of the separation of the two compact components seen at 10.7 GHz.

A. Ferrari and A. G. Pacholczyk (eds.), Astrophysical Jets, 47–49.
Copyright © 1983 by D. Reidel Publishing Company.

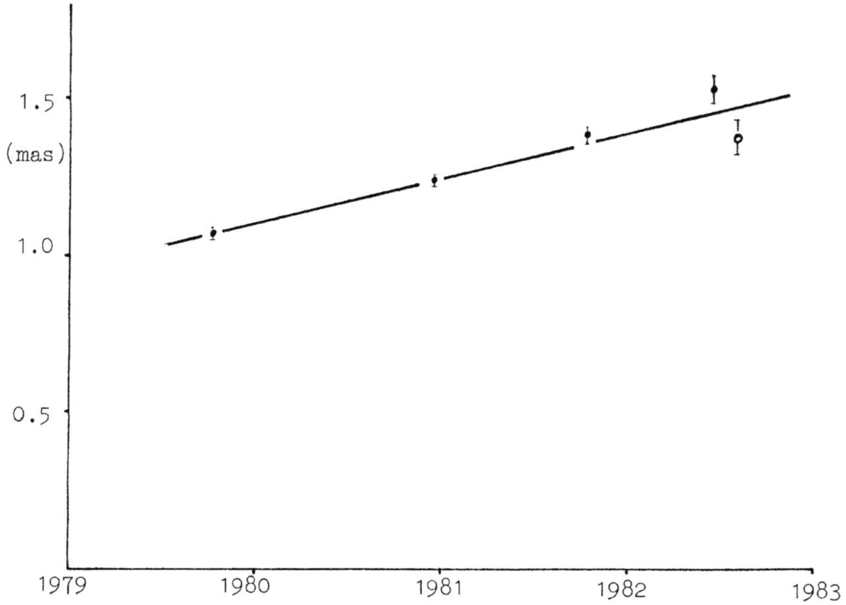

Fig. 1: Angular separation v. time for 3C179.
Filled circles 10.7 GHz; open circles 5.0 GHz

Now I would like to discuss the significance of the prominent, double-lobed structure seen on the arc-second scale in 3C179. Although many different explanations of the superluminal "illusion" have been proposed, including non-cosmological redshifts and gross re-evaluation of the Hubble constant, the most resilient explanation seems to be that of relativistic component motion (high Lorentz factor, γ) at a small angle ($\theta \sim \gamma^{-1}$) to the line of sight. This model has the advantage that it is economical of hypotheses, embraces the current astrophysical ideas of relativistic plasma beams, and makes specific observational predictions understandable to observers! It attributes the necessary high incidence of small θs in the superluminal sources (despite the low probability $\sim\frac{1}{2}\theta^2$) to a selection bias caused by the associated relativistic beaming at small θ. 3C179 certainly has a stronger than usual radio core for sources of its morphological class, which is the reason why it is possible to observe it.

The model also requires, however, that for every one superluminal source seen, there must exist another $\sim\gamma^2$ sources which are intrinsically similar but in which the superluminal effect and core flux boosting are absent because of an unfavourable alignment. Where are they? When the arc-second structure of the known superluminal sources was thought to be asymmetric (and hence itself possibly relativistically beamed) it was suggested that they were the radio quiet quasars (Scheuer and Readhead, 1979). The significance of arc second scale double structure is that the unaligned counterparts of superluminal sources exhibiting it cannot be "hidden" from radio telescopes in this way because such symmetric structure cannot have relativistic velocities (Longair and Riley, 1979). They must, then, exist amongst the known population of radio sources.

A number of the superluminal sources are now known to have some extended structure on both sides of the core (see e.g. Browne et al., 1982; Schilizzi and de Bruyn, 1983) but 3C179 is the only source where it produces the dominant flux at its survey selection frequency. As this is a workshop I will proceed to do statistics with N = 1 source and ask where the ∿50 unaligned 3C179s are. They must be "amongst" the other 30 quasars from the Jodrell Bank 966 MHz survey which are in the sample from which 3C179 was selected. Thus, if one accepts the model one must conclude that essentially all such quasars are intrinsically similar to 3C179 ($\gamma \sim 5$ with $H_0 \sim 100$ km $s^{-1}Mpc^{-1}$) but simply have unfavourable alignments. Orr and Browne (1982) have considered the consequences for quasar statistics of a "unified scheme" for quasars based on such an idea.

Finally I point out that the model, therefore, cannot allow any differences in the intrinsic properties of superluminal and non-superluminal quasars. A difference in the widths of the FeII emission lines has been suggested. If such claims are substantiated we may need a modified relativistic model (e.g. the "spray" model of Rees (1981), revived by Perola at this Workshop) or we can resort to explanations involving illuminated screens (Miley, this Workshop).

REFERENCES

Browne, I.W.A. et al.: 1982, Nature (in press)
Cohen, M.H., and Unwin, S.C.: 1982, I.A.U. Symposium No. 97, p. 345.
Longair, M.S., and Riley, J.M.: 1979, Monthly Notices Roy. Astron. Soc. 188, p. 625.
Marscher, A.P., and Broderick, J.J.: 1982, Astrophys. J. 255, L11
Orr, M.J.L., and Browne, I.W.A.: 1982, Monthly Notices Roy. Astron. Soc. 200, p. 1067.
Phillips, R.B., and Mutel, R.L.: 1982, Astrophys. J. 257, L19
Porcas, R.W.: 1981, Nature 294, p. 47.
Porcas, R.W.: 1982, I.A.U. Symposium No. 97, p. 361.
Rees, M.J.: 1981, I.A.U. Symposium No. 94, p. 139.
Scheuer, P.A.G., and Readhead, A.C.S.: 1979, Nature 277, p. 182.
Walker, R.C., Seielstad, G.A., Simon, R.S., Unwin, S.C., Cohen, M.H., Pearson, T.J., and Linfield, R.P.: 1982, Astrophys. J. 257, p. 56.

THE JET IN THE QUASAR 3C273

R.J. Davis
University of Manchester
Nuffield Radio Astronomy Laboratories
Jodrell Bank, Macclesfield, Cheshire, SK119DL

The quasar 3C273 has now been mapped with the six telescope array, MERLIN, at both 408 MHz and 1666 MHz, yielding maps with resolutions of 0.9 and 0.35 arc s. Fig. 1 shows the jet only.

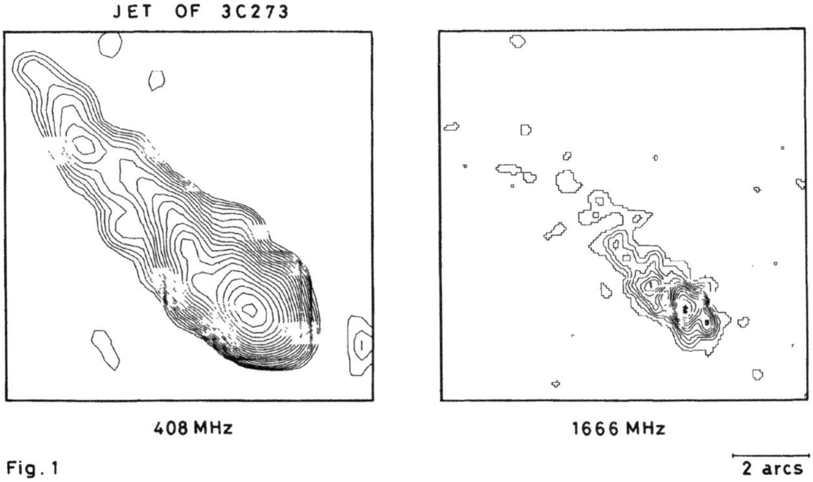

Fig. 1

The apparently smooth distribution in the low resolution map is seen to break up into complex subcomponents in the brightest regions. Such subcomponents may exist in the weakest regions also, but dynamic range limitations make them impossible to detect. Both maps show evidence for quasi-sinusoidal wiggles in the ridge line of emission.

Polarization measurements (Davis, Stannard and Conway, in preparation) yield an unusually low value of 11 m^{-3} for the density of thermal electrons in the jet. Making the usual equipartition assumptions, we calculate a sound speed for the plasma of 0.3c.

A. Ferrari and A. G. Pacholczyk (eds.), Astrophysical Jets, 51–52.
Copyright © 1983 by D. Reidel Publishing Company.

Two models for the wiggles have been investigated. In the first model nuclear precession proposed by Begelman, Blandford and Rees (1980), based on measured bending in the VLBI position angle, is assumed as the cause of the wiggle. Since the wavelength of the wiggle ∿ 5 kpc and the precession period is ∿ 1200 yrs, relativistic flow is inferred. In the second model, helical modes of the Kelvin Helholtz instability are proposed as the cause of the wiggle. Since high Mach numbers ($3 \leq m \leq 10$) (Ray 1981 and others) are required for this mode, and since the sound speed = 0.3c, again relativistic flow is inferred. The apparent decrease in the wavelength observed in the bright head of the jet on both models could be caused by a reduction in the Lorentz factor of the bulk flow.

The very high sound speed together with the small opening angle ∿ 1° implies that some form of sideways confinement is required. However, X-ray limits from the Einstein satellite give a maximum allowed density for the surrounding medium of 5000 particles m^{-3} which is insufficient to confine even the weakest regions of the jet. If the jet flow is relativistic with Lorentz factor ∿ 5, then sideways confinement is just possible with this limit on gas density. If either this limit is reduced or it is demonstrated that the jet is slow, then some other confinement mechanism must be found. One suggestion is magnetic confinement as mentioned by G. Benford. Another possibility is that this jet contains an invisible low pressure beam which carries energy in a very uniform manner, leaking energy into the visible radio emitting region. With this model, this outer region would not need confining, the narrow opening angle being provided by the invisible low pressure beam.

Finally, comparison of the maps convolved to the same resolution shows that the spectral index is remarkably constant along the length of the jet at + 0.75 ± .10. This spectral information, together with the overall spectrum of the source, enables us to rule out an inverse Compton model for the optical emission of the jet, and thus we favour the synchrotron process for this emission. We are able also to predict the presence of a steep spectrum ($\alpha \geq 1.4$) halo at low frequency with angular size $\geq$ 50 arc s^2.

REFERENCES

Begelman, M.C., Blandford, R.D. & Rees, M.J., 1980. Nature 287, 307
Ray, T.P., 1981. Mon. Not. R. astr. Soc., 196, 195.

THREE ARCHETYPAL RADIO JETS

D.J. Saikia
Radio Astronomy Group,
Tata Institute of Fundamental Research,
Post Box 1234, Bangalore 560 012, India

T.J. Cornwell
National Radio Astronomy Observatory, VLA Program
Post Box 0, Socorro, New Mexico 87801, USA

Radio jets, which provide the most direct evidence for supply of energy from the nuclei of active galaxies to the outer extended lobes, are believed to arise largely from the inefficiencies in the energy transport processes. These jets exhibit a wide variety of structures (Browne 1983; Fomalont 1983), possibly providing clues to the different physical processes responsible for their observed morphology. In this paper, we present maps of three interesting radio jets, two of them, 1033+003 and 1759+211, associated with galaxies and one, 1857+566, with a high redshift quasar. These sources were observed at the VLA as part of programs to study the radio properties of a large number of galaxies and quasars. They will be discussed in detail elsewhere (Saikia, Cornwell and Kapahi 1982; Saikia, Shastri, Cornwell and Banhatti 1982).

The radio source 1033+003 is associated with a 15^m galaxy whose optical minor axis (PA $\sim$ 50°) is inclined at about 20° to the inner region of the jet (PA $\sim$ 70°). Although the jet seems to wiggle close to the core, its overall structure when observed with coarser resolution appears to curve towards the optical minor axis but continues to bend in the same direction till it lies close to the position angle (PA$\sim$ 8°) defined by the outer radio lobes (Palimaka, Bridle, Fomalont and Brandie 1979). This observed morphology for the jet could arise through buoyant refraction of a light relativistic beam by pressure gradients in a dense atmosphere (Henriksen, Vallée and Bridle 1981). This mechanism is essentially the continuous jet version of the buoyancy model proposed by Gull and Northover (1973). Future, sensitive, X-ray observations would help clarify the situation.

The second interesting jet we present here is associated with a 16^m galaxy, 1759+211, which is the brightest member of a faint cluster (Goodson, Palimaka and Bridle 1979). The jet appears to oscillate, the amplitude being more pronounced at larger distances from the core. These oscillations can be understood if the energy-carrying beam is

A. Ferrari and A. G. Pacholczyk (eds.), Astrophysical Jets, 53–56.
Copyright © 1983 by D. Reidel Publishing Company.

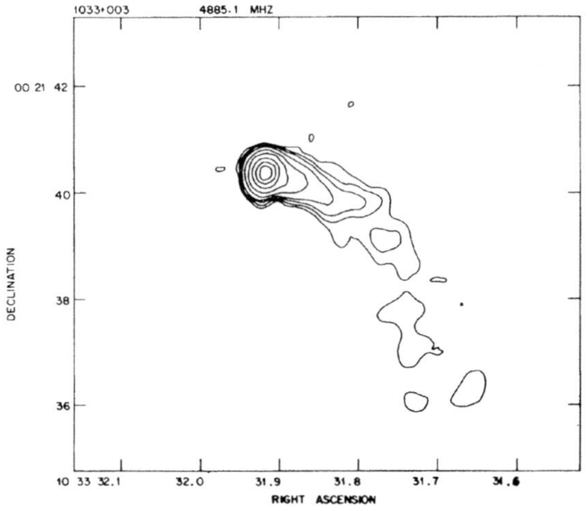

Figure 1: The total intensity distribution of the jet in 1033+003 at 5 GHz. The peak flux density is 42.5 mJy/beam. The contours are -1.2,1.2, 2, 3, 5, 10, 20, 40, 60 and 80 percent of the peak flux. The half-power width of the fitted beam is 0.46 x 0.39 arcsec along PA -3°.

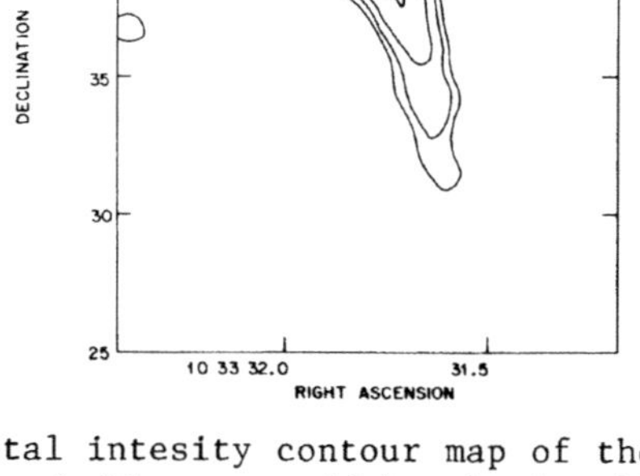

Figure 2: Total intesity contour map of the jet in 1033+003 at 20 cm. The half-power width of the fitted beam is 1.43 x 1.31 arcsec along PA 11°. The countours represent -2, 3, 5, 10, 20, 30, 40, 60 and 80 percent of the peak flux density which is 55 mJy/beam.

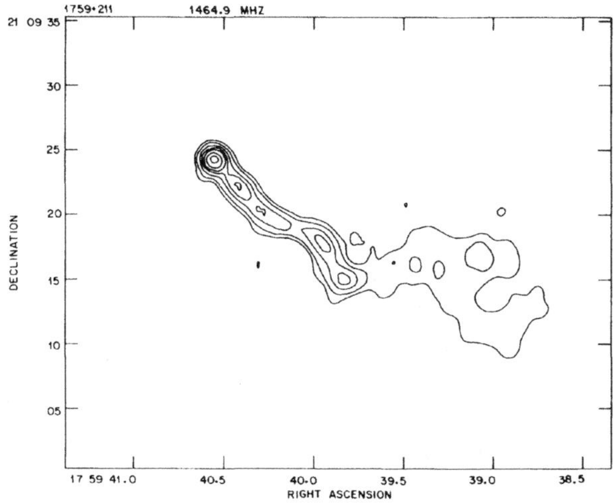

Figure 3: Radio map of the jet in 1759+211 at 20 cm. The half-power beam width is 1.37 x 1.33 arcsec along PA -58°. The peak flux density is 19 mJy. The contours are -5, 5, 10, 20, 30, 40, 60 and 80 percent of the peak flux density.

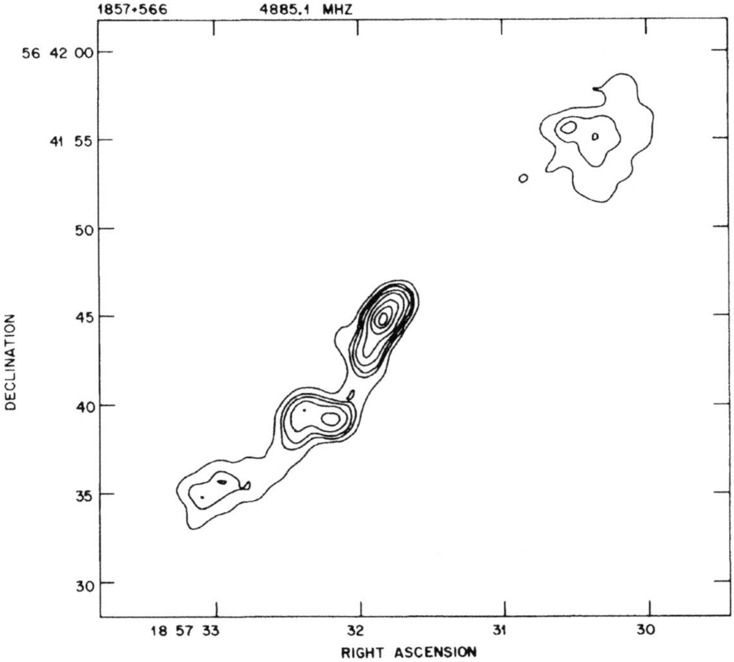

Figure 4: Total intensity distribution of the high redshift quasar 1857+566. The half-power width of the fitted beam is 1.04 x 1.04 arcsec along PA -44°.5. The peak flux density in 76 mJy/beam. The contours are -1, 1, 3, 5, 10, 20, 60 and 80 percent of the peak flux density.

Kelvin-Helmholtz unstable to long-wavelength helical waves which can affect its overall appearance (Ferrari, Trussoni and Zaninetti 1981, and references therein; Benford 1981; Hardee 1981). Earlier, within this framework, attempts have been made to explain the apparent helical pattern of the jets in nearby galaxies such as M87 (Benford 1981) and 3C449 (Hardee 1981), as well as in the luminous quasar 3C280.1 (Swarup, Sinha and Saikia 1982).

The jet in the quasar 1857+566 (z = 1.595) exhibits two sharp bends along its path and has no hot-spot at its end. On the opposite side of the nucleus there is a normal radio lobe containing a warm-spot. There is also a suggestion of a counter-jet near the nucleus. Although relativistic precessing jets can give rise to such apparently distorted and complex morphologies (cf. Gower, Gregory, Hutchings and Unruh 1982), features such as the absence of a hot-spot at the end of the jet, its apparent asymmetry and the possibility of the source being inclined at a large angle to the line of sight (it lies close to the upper envelope of the angular size-redshift diagram) suggest that the appearance of the jet is likely to be due to an asymmetric dissipation of energy by the beams. Possibly, the sharp bends arise due to interaction of the jet with inhomogeneities in the intergalactic medium. Recent observations of this source with somewhat better resolution by Owen and Puschell (1982) and Barthel, Miley and Schilizzi (1982) show both the bends in the jet and the counter-jet near the core in more detail. Deep optical and X-ray observations would be extremely useful in understanding this jet.

REFERENCES

Barthel, P., Miley, G.K. and Schilizzi, R.T., 1982. In preparation

Benford, G., 1981. Astrophys. J., 247, 792

Browne, I.W.A., 1983, this volume.

Ferrari, A., Trussoni, E. and Zaninetti, L., 1981. Mon. Not. R. astr. Soc., 196, 1051

Fomalont, E.B., 1983, this volume.

Goodson, R.E., Palimaka, J.J. and Bridle, A.H., 1979. Astr.J., 84, 1111

Gull, S.F. and Northover, K.J.E., 1973. Nature, 244, 80

Gower, A.C., Gregory, P.C., Hutchings, J.B. and Unruh, W.G., 1982, Astrophys. J., in press

Hardee, P.E., 1981, Astrophys. J., 250, L9

Henriksen, R.N., Vallée, J.P. and Bridle A.H., 1981, Astrophys. J., 249, 40

Owen, F.N. and Puschell, J.J., 1982. In preparation

Palimaka, J.J., Bridle, A.H., Fomalont, E.B. and Brandie, G.W., 1979. Astrophys. J., 231, L7

Saikia, D.J., Cornwell, T.J. and Kapahi, V.K., 1982. In preparation

Saikia, D.J., Shastri, P., Cornwell, T.J. and Banhatti, D.G., 1982. In preparation

Swarup, G., Sinha, R.P. and Saikia, D.J., 1982. Mon. Not. R. astr. Soc., 201, 393

PHYSICAL PROPERTIES OF THE JET IN NGC6251

A.H. Bridle and R.A. Perley
National Radio Astronomy Observatory
P.O. Box 0, Socorro, New Mexico 87801, USA

The jet in the radio galaxy NGC6251 (Waggett et al. 1977) has been observed with the VLA using resolutions from 1.15" to 15" at 1370, 1480, 1662 and 4885 MHz, and a resolution of 50" at 1446 and 1662 MHz. The resulting maps (e.g. Figure 1) give exceptionally detailed information on the distributions of the total and polarized intensities over the jet. The following physical properties can be deduced from these observations:

1. The jet expands laterally at a variable rate, exhibiting several stages of confinement and re-expansion (Figure 2). A regime of rapid expansion in the first 20" (8 kpc) of the jet is followed by alternating regimes of lateral confinement and slower expansion. We conclude that the jet cannot be free (unconfined) over its entire length; also that the Mach number of the flow is <12.5 over the first 8 kpc. The lateral confinement observed beyond 8 kpc from the core could be produced by thermal pressure of an external medium whose density variation is similar in form to that of the X-ray halo of M87; the required medium would have only about 1/10th the 0.5-4.5 keV luminosity of the M87 halo however (assuming a temperature of 3×10^7 K). The inner 8 kpc of the jet must either be free, or be confined by a medium at a temperature much below, or much above, 3×10^7 K, to avoid implying an excessive X-ray luminosity. The longer wavelength changes in the expansion rate shown in Figure 2 can be ascribed to adjustments of the jet's internal pressure gradients that are induced when it first encounters the confining "halo" pressure, as discussed by Bridle et al. (1981) and by Chan and Henriksen (1980).

2. The brightness of the jet generally decreases much more slowly with increasing radius than would be expected if the jet contained adiabatically expanding conserved fluxes of relativistic electrons and magnetic field. Relativistic particle reacceleration and/or magnetic field amplification must occur many tens of kpc from the core if the jet is a steady flow of material ejected from the core. The average exponent of the brightness-radius relation changes from -1.5 within 120" (50 kpc) of the core to -4.5 beyond 220" (94 kpc) from the core.

A. Ferrari and A. G. Pacholczyk (eds.), Astrophysical Jets, 57–66.
Copyright © 1983 by D. Reidel Publishing Company.

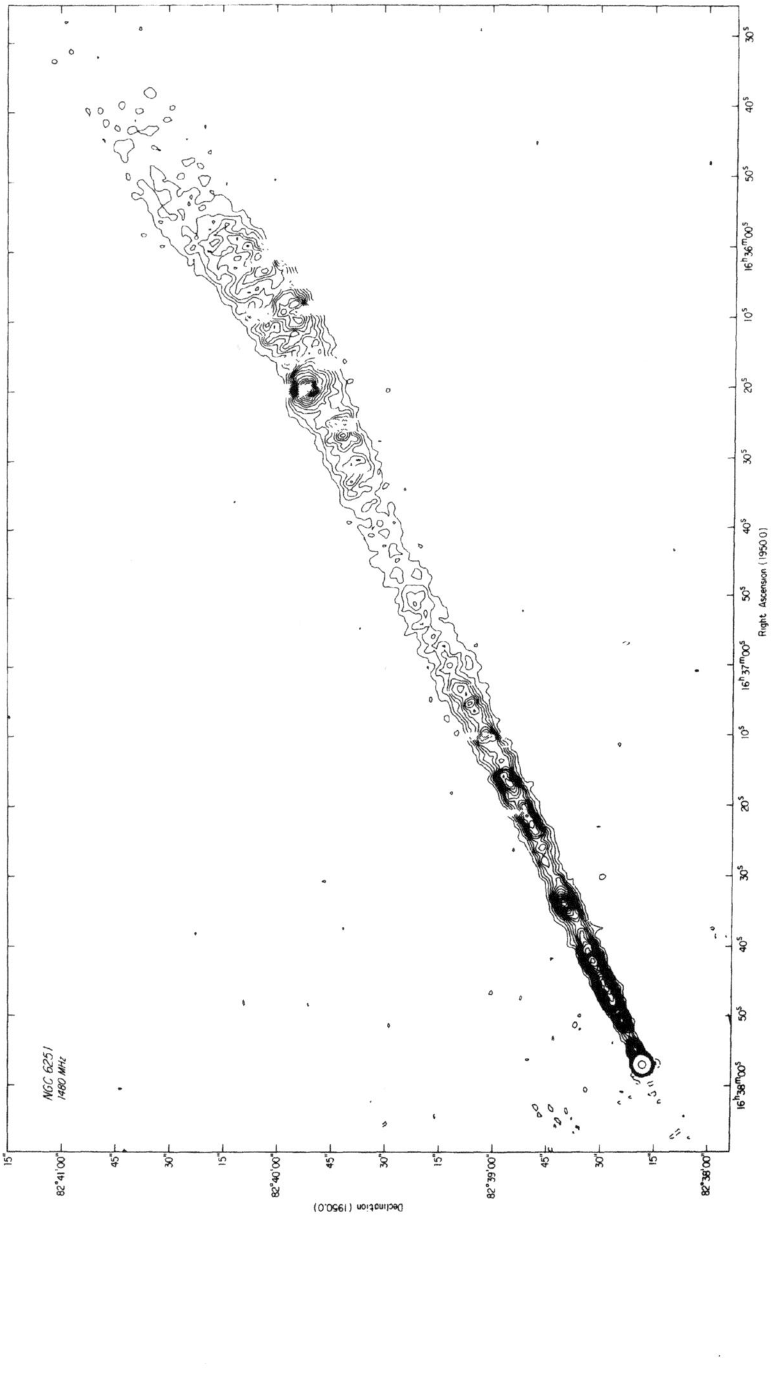

Figure 1: VLA map of the first 4.4' of the jet at 1480 MHz with 2.11" resolution. Contours are drawn at -0.448 (dashed), and +0.448 mJy/beam, then at equal intervals of 0.448 mJy/beam until 6.27 mJy/beam, then at 7.17, 8.06, 8.96 and 224 mJy/beam. The highest contour delineates the FWHM of the beam around the peak of the core source.

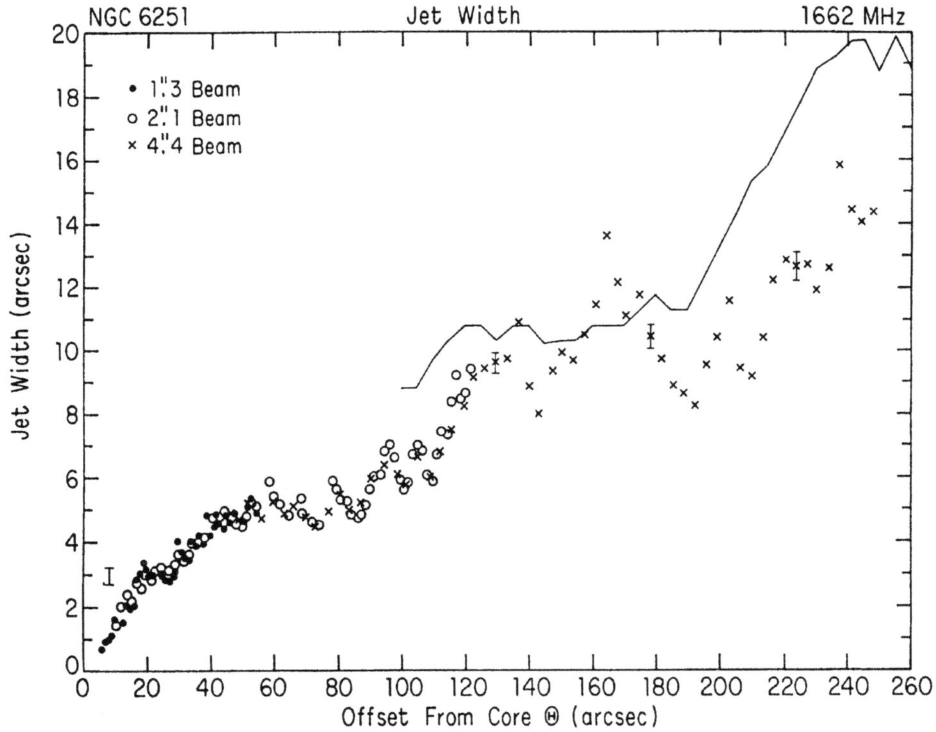

Figure 2: Collimation (expansion) properties of the first 250" of the jet. Individual points show the deconvolved FWHMs of Gaussian profiles fitted to transverse sections through the jet at 1662 MHz, at the resolutions indicated. The solid line shows the full width between the 0.45 mJy/beam contours at 1662 MHz, minus twice the FWHM of the instrumental beam. The isophotal width varies much more smoothly than the Gaussian FWHM beyond 100" from the core, due to the knotty substructure of the outer jet.

The spectral index of the jet between 1480 and 4885 MHz is everywhere within errors of -0.64.

3. There is a weak counterjet (Figure 3), whose brightness is about 1/40th that of the main jet for the first 130" (56 kpc) from the core, but less than 1/300 that of the main jet farther from the core. The counterjet is therefore not a fainter replica of the main jet, but has a different variation of brightness with projected distance from the core.

4. The only limb-brightened region of the jet is a discrete knot about 18" (7.7 kpc) from the core, immediately on the coreward side of the first slowdown of the lateral expansion. Except for this one localised knot, there is no evidence for any concentration of the synchrotron emissivity towards the outer boundaries of the jet.

5. Beyond about 100" (43 kpc) from the core, the Faraday rotation measure (RM) in front of the jet is constant at -50 rad/m^2 to within the errors of measurement (Figure 4). The mean of the RMs at 123 locations in this region of the jet where the signal to noise ratio of the polarized intensity exceeds 12:1 at all four frequencies is -48.9 ± 0.2 rad/m^2. The limits to the RM gradients across the outer jet, combined with the equipartition estimates of the magnetic field strengths, imply that the mean density within the jet at these distances from the core is $<1-2 \times 10^{-4}$/cc. (The numerical value of this limit depends on the assumptions made about the geometry of the ordered field components).

6. Closer to the core (within the stellar envelope of the galaxy) there are large and complex gradients in the RM both along and across the jet (Figure 4). The RM varies from 0 rad/m^2 to -130 rad/m^2 over the inner 50" (22 kpc) of the jet. We interpret these variations as due to a magnetoionic medium outside the jet but within NGC6251. The medium cannot lie within the jet because such large deviations in the RM would then be incompatible with the observed lack of significant depolarization between 21cm and 6cm. The medium produces RM gradients of order 50 rad/m^2 over angular scales of order 5" (2 kpc); these RM fluctuations cause a false depolarization in low-resolution maps made at frequencies below 2 GHz.

7. The projected magnetic field structure of the jet (Figure 5) is dominated by components parallel to the jet axis over most of the first 90" (40 kpc) both at the center and the edges of the jet. Adjacent vectors in the field maps are well ordered, but there are localised deviations from strictly longitudinal field that are well above the uncertainties due to noise in the correction for the RM. Further from the core, the magnetic field is predominantly perpendicular to the jet axis at the center of the jet and parallel to the axis at its edges. The outer jet contains several knots in which the projected magnetic field deviates substantially from the center-perpendicular, edge-parallel configuration however.

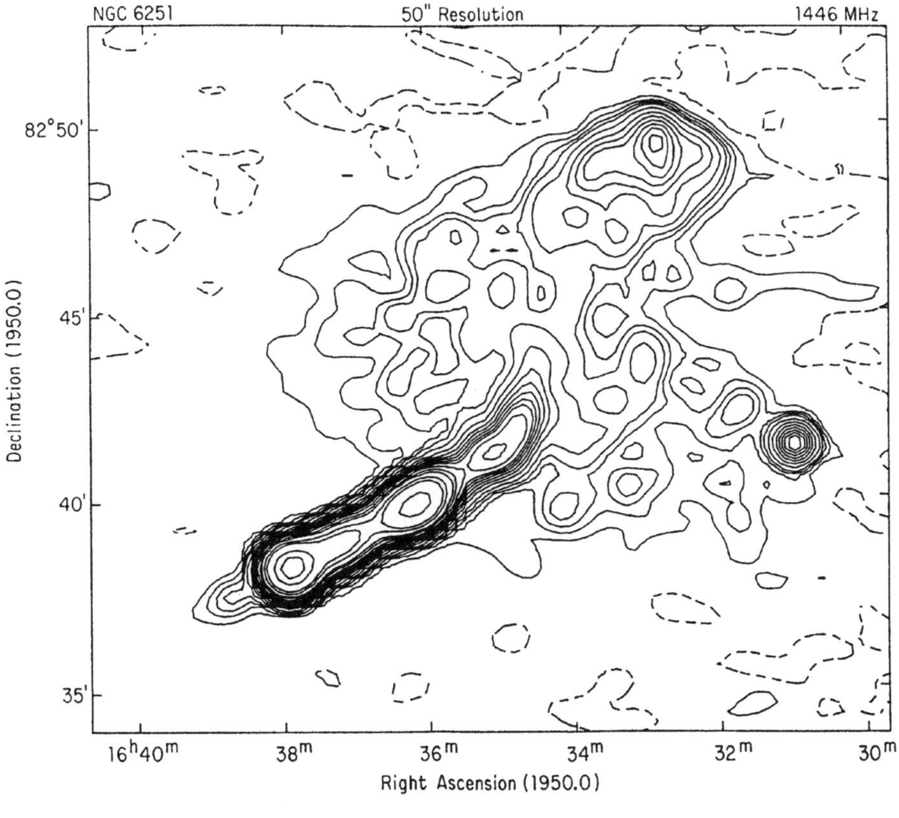

Figure 3: VLA "snapshot" map of the jet and northwest lobe with 50" resolution at 1446 MHz. Contours are drawn at -1.33 (dashed) and +1.33 mJy/beam, then at 2.66, 3.99, 5.32, 7.98, 10.6, 13.3, 16.6, 20.0, 23.3, 26.6, 30.0, 33.3, 40, 66.5, 100, 133, 266 and 400 mJy/beam. Note the evidence for the weak counterjet to the southeast of the core source.

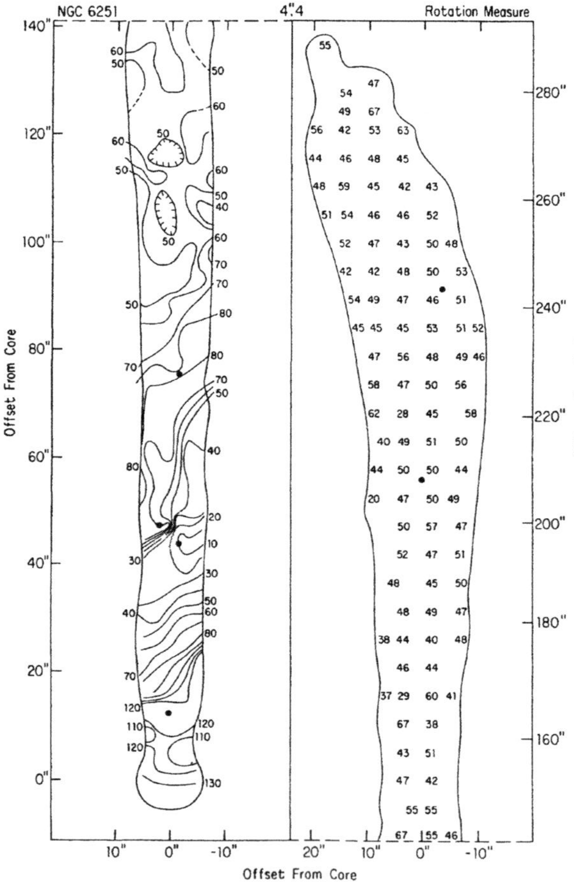

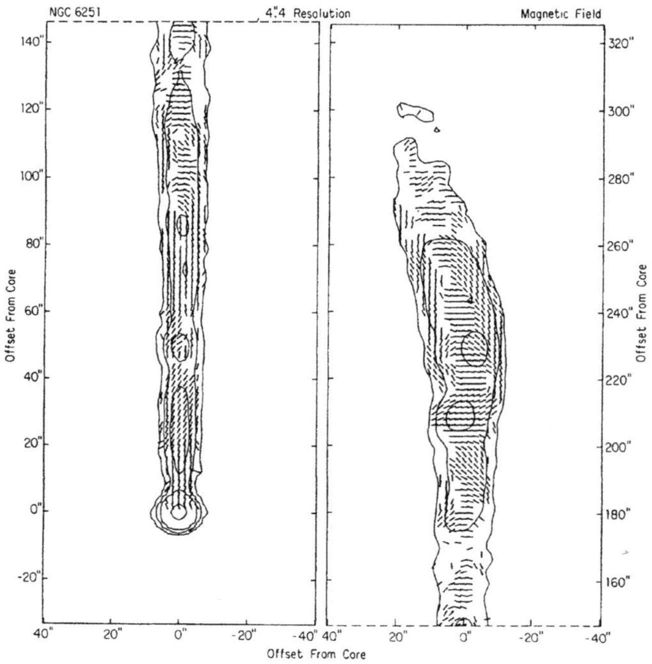

Figure 5: Map of the projected magnetic field configuration over the first 4.4' of the jet at 4.4" resolution, derived by extrapolating the observed E vector position angles at 1370, 1480, 1662 and 4885 MHz to infinite frequency, then adding 90 degrees.

Figure 4: Distribution of the Faraday rotation measure RM (negative sign omitted) over the entire jet at 4.4" resolution. Left panel: contour map of RM fluctuations over the inner jet; dotted contours show regions where the RM variation is poorly determined. Right panel: individual RM values in the outer jet, where there are no systematic RM gradients.

8. The variations of the projected magnetic field components and of the degree of linear polarization both along and across the jet can be fitted by several models of the three-dimensional field structure which contain mixtures of organised and random field components. A partially randomized version of the self-similar helical proposed by Chan and Henriksen (1980) provides a good description of most of the data if the jet is within 10 degrees of the plane of the sky within a few tens of kpc from the core, but is inclined by up to 40 degrees from the plane of the sky further out. Departures from axial symmetry of the field and relativistic particle distributions must however be assumed to explain the local regions of oblique projected field in terms of this model. The partially ordered field configuration proposed by Laing (1980, 1981) can also be fitted to the data if longitudinal field components such as those produced by a thick boundary shear layer are added to his proposed configuration. The added shear must lack axial symmetry almost everywhere along the jet to give acceptable fits in terms of this model. In either the "randomized CH" or "sheared Laing" fits, the ratio of organized to random components of the field must vary with distance along the jet.

9. The limits to the true depolarization of the jet about 30" (13 kpc) from the core imply that the thermal densities within it at these distances from the core are $<1-4 \times 10^{-3}$ /cc. The lower values apply if the organised fields are uniform throughout the jet; the higher if they have the "randomized CH" configuration which best fits the intensity and polarization data.

10. The ridge line of the jet is not straight. There is a rich spectrum of lateral oscillations (Figure 6) in which various wavelengths dominate at different distances from the core. The projected wavelengths of the most significant lateral oscillations are 143", 31", 17.5", 12" and 9" (61, 13, 7.5, 5 and 4 kpc). There is a trend for the dominant wavelength to lengthen with increasing jet radius.

11. If the large-scale confinement of the jet is produced by an external medium, then unless its temperature greatly exceeds the 3×10^7 K of the M87 halo, results (1), (5) and (9) imply that the thermal density within the jet is <1-2 times that in the external medium between 10 and 100 kpc from the core. This limit to the density contrast between the jet and the external medium constrains interpretation of the complex lateral oscillation spectrum (Figure 6) in terms of Kelvin-Helmholtz instabilities. The more rapid oscillations could not be identified with the most rapidly growing helical mode unless the Mach number of the jet is of order 2-3. If the jet is almost free in the first 8 kpc, so that its Mach number is close to the upper limit of 12.5 estimated in (1), then only the 143" oscillation could be a Kelvin-Helmholtz mode (according to linear perturbation analyses such as those of Hardee 1982 and Ferrari et al. 1982). The oscillation with 143" wavelength has an amplitude in the outer jet that is comparable to the jet radius there however, so it is not clear that the linear

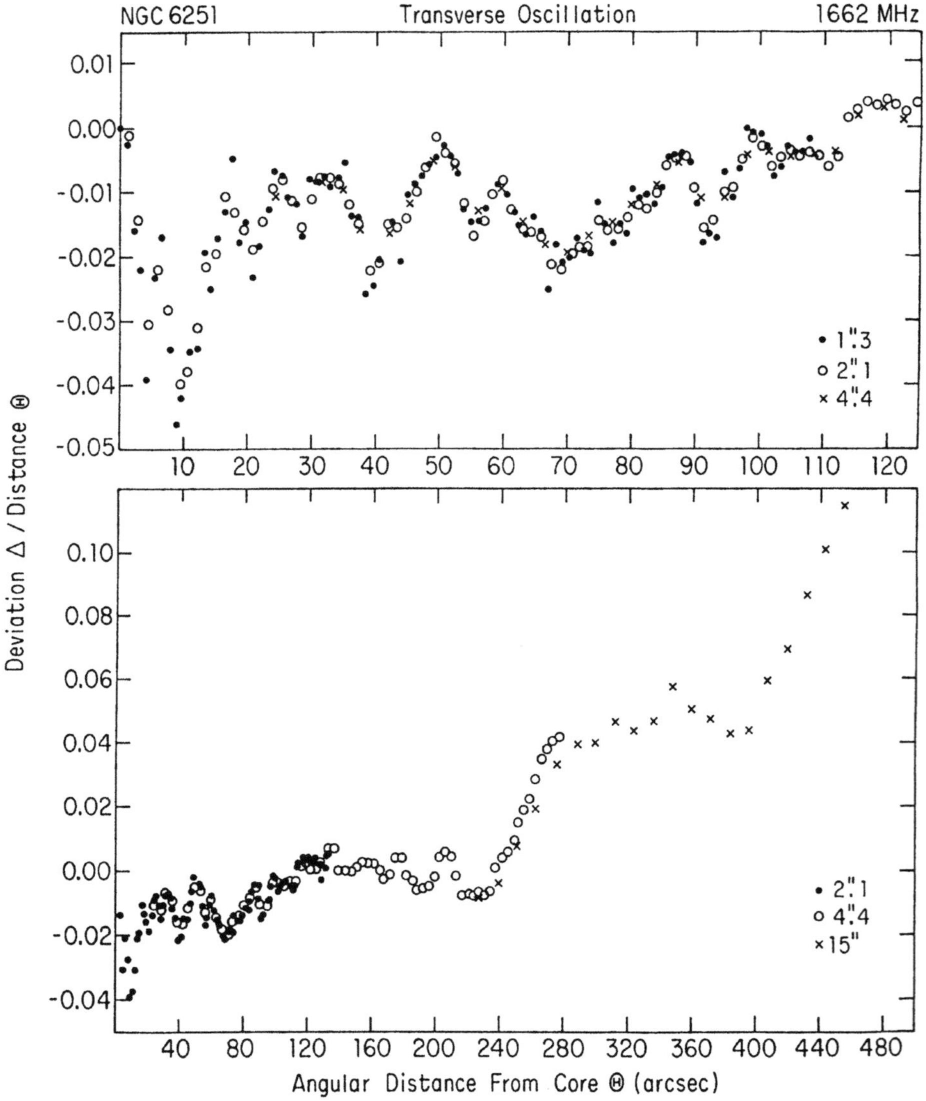

Figure 6: Deviation Δ of the ridge line of the jet from an axis with a position angle of 296.4 degrees, normalised by angular distance Θ from the core. Note that Δ/Θ is a measure of the angular deflection of the ridge line of the jet from the mean position angle, as seen from the core. The noise in these plots decreases from left to right, due to the normalisation by the distance from the core; the scatter between adjacent points indicates the noise in that part of the plot.

perturbation theory is applicable to it. The 143" oscillation could also be ascribed to precession of the central "engine" in NGC6251, as its amplitude grows approximately linearly with distance from the core.

12. Various model-dependent estimates can be made of the velocity of the jet; values in the range 5000 to 15000 km/s appear plausible. The mass flux down the jet would be of order 0.25 to 0.75 solar masses per year according to such estimates. This mass flux seems reasonable in comparison with the gas production rate expected from the stellar population in a giant elliptical galaxy.

These results will be presented in detail in a paper soon to be submitted to the Astrophysical Journal. AHB thanks the National Radio Astronomy Observatory for its hospitality and financial support during a leave of absence from Queen's University in which this work was completed. We thank Jean Eilek, Ralph Fiedler, Phil Hardee, Dick Henriksen, Robert Laing and Fraze Owen for valuable discussions.

REFERENCES

Bridle, A.H., Chan, K.L. and Henriksen, R.N. 1981. J. Roy. Astron. Soc. Canada, 75, 69

Chan, K.L. and Henriksen, R.N. 1980. Ap. J., 241, 534

Ferrari, A., Trussoni, E., Zaninetti, L. 1982. Preprint

Hardee, P.E. 1982. Preprint

Laing, R.A. 1980. M.N.R.A.S., 193, 439

Laing, R.A. 1981 Ap. J., 248, 87

Waggett, P.C., Warner, P.J. and Baldwin, J.E. 1977. M.N.R.A.S, 181, 465

DISCUSSION

P. BIERMANN: With very long Einstein X-ray observations of a giant elliptical galaxy, we find parameters in the density, temperature and total mass of hot gas very similar to those you require for confinement of the NGC6251 jet.

J.-L. NIETO: High resolution optical observations have allowed the detection of a dust lane in NGC6251 quite parallel to the radio jet. It could be responsible for the knot at 18" from the core. Also, assuming (by analogy with NGC5128) this object could be a merger gives a beam velocity lower than your estimates by an order of magnitude (see Nieto, this Workshop).

G. BENFORD: Your inferred "galactic magnetosphere" implies an ordered magnetic field. Could this be a confining B_θ? This would remove the need to explain wiggles with a linear instability theory since magnetic ripples can occur with large amplitudes. Linear theory clearly does not apply since sideways shocks occur when the displacement equals R(jet)/M, or about R(jet)/10 here.

R.A. PERLEY: We cannot say from these radio data whether the external fields are really a large-scale "magnetosphere", or a field structure outside, but still spatially associated with, the radio jet. I don't see how a truly galactic magnetic field could confine the jet unless such a field wraps the jet. In that case the field would be due to currents in the jet, so it would not be termed a "galactic magnetosphere".

L. ZANINETTI: What is your definition of amplification of magnetic field in NGC6251?

R.A. PERLEY: That the field evolves more slowly with increasing jet radius than would be expected if magnetic flux is conserved.

BENT JETS AND TAILED RADIO GALAXIES

Jack O. Burns
Department of Physics and Astronomy
University of New Mexico, Albuquerque, N.M. 87131 USA

ABSTRACT

The observationally defined properties of tailed radio sources are reviewed with particular emphasis on the characteristics of bent jets. A comparison is made between the structures found in straight and bent jets. It is demonstrated that the curvature of jets associated with narrow-angle tailed galaxies can provide a strong constraint on the average velocity of the plasma flow -- a contraint not possible for straight jets. It is also shown that the wide-angle tailed sources must be bent by a mechanism other than dynamic pressure resulting from motion through the ICM.

1. INTRODUCTION

An edge-darkened radio source whose emission peaks near the optical galaxy and trails off to one side has been termed a "tailed" or "head-tail" galaxy. Prototype examples include NGC 1265 in Figure 1 and 3C129 in Figure 2. On the large scale ($\sim$arcmin), head-tail sources do not share the same symmetry as the classical doubles -- the large scale radio plasma flows appear to be unipolar with respect to the galaxy. On the small scale ($\sim$arcseconds), however, there is a C-shaped symmetry created in some instances by dual curved jets which emanate from a compact nuclear core and merge (after significant bending) with the more extended radio tail(s). Tailed radio sources are also at the opposite extreme from classical doubles in terms of their radio power. They are generally weaker Fanaroff and Riley (1974) class I objects.

Tailed sources are nearly always in clusters of galaxies although selection effects may have prevented their detection in more isolated regions of space. In fact, Stocke and Burns (1983) recently found a head-tail galaxy in a very poor environment where only 6 galaxies exist in a 1 Mpc ^{3} region.

A. Ferrari and A. G. Pacholczyk (eds.), Astrophysical Jets, 67–79.
Copyright © 1983 by D. Reidel Publishing Company.

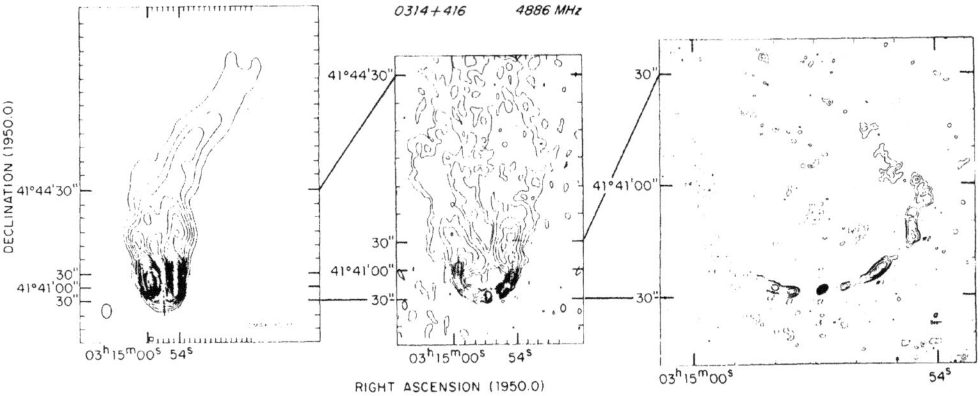

Figure 1. Radio structure of NGC 1265 at 5 GHz adapted from Jones and Owen (1979). The maps, which show the relationship of structure on different scales, are taken from (left to right) Miley (1973), Miley et al. (1975), and Owen et al. (1978).

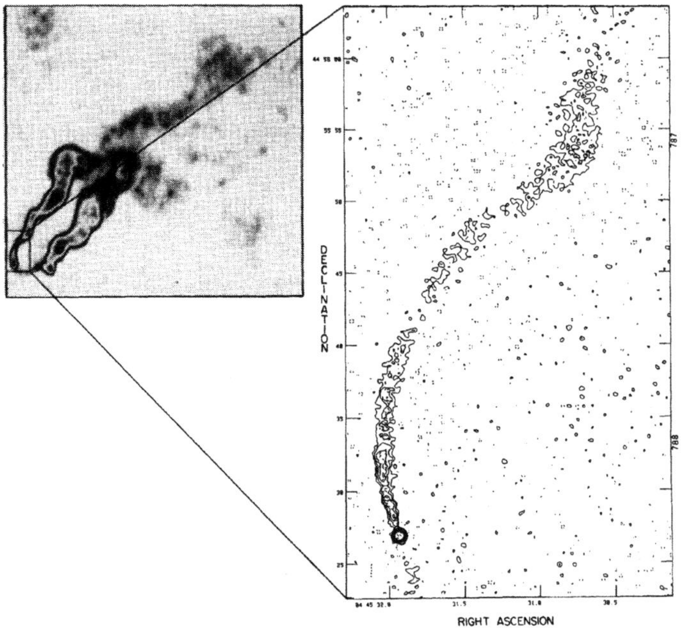

Figure 2. Radio structure of 3C129. Radiograph is from Rudnick and Burns (1981) at a resolution of 6" x 6" and a wavelength of 20-cm. The contour map is a higher resolution view (.5" x .5") of the northern jet at 6-cm (Rudnick and Burns 1983); contours are logarithmic with the lowest contour 74 μJy/(clean beam). Dashed lines are E-vectors.

Among the tailed sources there is a separation in luminosity, morphology, and galaxy type. This important distinction between narrow-angle tailed (NAT) sources such as NGC 1265 and wide-angle tails (WAT) such as 1919+479 (Figure 4) was first pointed out by Owen and Rudnick (1976). NAT sources are identified with average, generally outlying cluster galaxies, and have the lowest luminosity and sharpest bends. WAT sources are identified with dominant, central cluster galaxies, and have bends and luminosities which are intermediate between classical doubles and NATs. As will be shown, this distinction is quite important since it probably delineates two separate environments and mechanisms for bending radio jets and tails.

Jets are a very common feature in tailed radio sources, possibly more prevelent than in any other source type. One of the first discoveries of dual radio jets was in the nearby Perseus cluster galaxy NGC 1265 by Wellington et al. (1973). In spite of their early detection and their abundance there has been comparatively little work on jets, per se, in tailed sources. The focus in jet research has clearly been on the straight jets. This may be because of the misconception that the curvature adds an extra dimension of complexity to the flow problem. It will be argued that this is not only untrue but that head-tail sources can provide an additional constraint on plasma flow parameters that cannot be obtained for the straight jets.

I will begin this review by comparing the observed properties of straight and bent jets. Does the extreme curvature and environment influence the internal source energetics and dynamics of jets? Second, constraints on the velocity of plasma flow in the curved jets will be discussed along with their impact on our general understanding of jets. Third, I will address the problem of bending jets in WAT sources - a problem which will be shown to be quite different from that in NATs. Finally, I will outline some of the current observational projects which are underway with the VLA that may provide important new insights into tailed radio sources.

2. COMPARISON BETWEEN "STRAIGHT" AND BENT JETS

For the properties of straight jets (e.g. 3C31, Centaurus-A, M87), I will use the summaries by Willis (1981) and Bridle (1982). For bent jets in tailed sources, I have compiled data from an inhomogeneous collection of work including that of O'Dea (1983), O'Dea and Owen (1983; NGC 1265), Rudnick and Burns (1982, 1983; 3C129), van Breugel (1979), Burns (1981), Burns et al. (1983; 1919+479), Eilek et al. (1983; 3C465), and Burns and Gregory (1982). The properties of jets in tailed sources listed below should be treated as very preliminary, since most of this work is still in progress and the sample is not statistically complete.

(1) Fraction of sources with Jets - Referring to the Fanaroff and Riley's classification, about 70% of Class I sources have straight

jets. There is some indication that far fewer Class II sources have jets (∿ 25%) although this may partly be due to a dynamic range selection effect. About 90% of the tailed sources have jets when observed at high enough resolution to resolve the head. This may not be surprising since there is a tendency for low-luminosity sources (of which NATs are the lowest) to preferentially contain jets.

(2) Sizes - Bent jets run the full gamut in lenght from < 2 kpc to ∿ 350 kpc (1919+479, Fig. 4), comparable to that of the straight jets. Bending does not seem to inhibit the size to which a jet can grow. To date, no VLBI jets have been detected although a MK III experiment was successfully performed on several tailed sources including NGC1265, 1919+479, and IC310 (Cohen et al. 1983).

(3) Spectra - Spectral indices (defined by $S_\nu \propto \nu^{-\alpha}$) in straight jets are typically 0.6 ± 0.15 with little or no gradient down or across the jet. The same magnitude and lack of α-gradient appears to be true for the few bent jets where two frequency spectra are available. There are no significant jumps in α at knots in bent jets.

(4) Magnetic Field Structure - There is a tendency for projected magnetic fields in luminous jets (e.g. quasars) to lie along the jet axis. For lower luminosity jets (e.g. NGC6251; see Perley and Bridle, this workshop) the field configuration flips from a parallel regime near the core to a perpendicular regime in the outlying parts of the jet. This structure becomes more complex in jets with knots (e.g. M87 in Owen et al. 1980). There is some suggestion from preliminary data for NGC1265 that the B-field remains parallel to the jet axis until the jet expands dramatically to produce the tail. This potential difference from the straight jets may be attributable to the strong dynamic pressure (shearing?) which is exerted on curved NAT jets.

(5) One or Two Sided Jets? - Luminous sources have a strong tendency toward one-sided jets, whereas low luminosity sources such as 3C449 have a preference for dual jets. The tailed sources also seem to follow this trend with a preference for two-sided jets in the majority of tailed sources; the NATs in particular seem nearly always to have dual jets when observed with high sensitivity.

(6) Wiggles - Numerous straight jets (e.g. M87, 3C449) have nonlinear oscillations about the jet axis which have been interpreted as hydrodynamic instabilities (Hardee 1981, Trussoni et al. in these proceedings) or procession of the nuclear engine (Ekers et al. 1978). Similar oscillations can be seen in the structure of bent jets (Figures 1-5) and in the tails. In the case of 3C129, Icke (1981) has very successfully reproduced the wiggles in the tails

using a nuclear precession model coupled with dynamic pressure bending.

(7) Jet Stability - Tailed radio sources illustrate the fact the jets are remarkably stable flows. Jets maintain their collimation and integrity over curvatures of as much as 70°. Although the preponderance of straight jets found in all types of sources suggests that these structures are stable, bent jets unequivically demonstrate the extreme stability of jets even in the presence of strong distortion forces. Clearly, this must be considered in models of hydrodynamic turbulence which grow at jet boundaries.

To summarize, it appears that the characteristics of bent jets are not significantly different from those of the straight jets with the possible exception of the magnetic field structure.

3. CONSTRAINTS ON THE JET VELOCITY OF NAT GALAXIES

NAT galaxies move at large velocities (~ 1000 km s^{-1}) comparable to the velocity dispersions within clusters (e.g. Ulrich 1978). This motion is through a relatively dense intracluster medium (ICM) which is known to pervade clusters; the hot cluster gas is seen directly in the thermal bremsstrahlung x-ray emission which it emits (Forman and Jones 1982). Such supersonic trajectories through the ICM produce the strong curvature in the jets and deposit radio plasma in the galaxy wake marking the trail of the galaxy in the cluster.

Both the continuous beam models (Begelman et al. 1979; Jones and Owen 1979; Baan and McKee 1982) and the plasmoids-in-a-channel model (Christiansen et al. 1981) utilize dynamic pressure to distort the path of the jet. For a nonrelativistic, hydrodynamic jet, the equation of internal and external pressure gradients (applicable in the first order to all the models) produces a simple but quite important constraint on the jet flow velocity in the form

$$\frac{\rho_{ICM} v_g^2}{h} = \frac{\rho_j v_j^2}{R} , \qquad (1)$$

where v_g and v_j are the galaxy and jet speeds, respectively, ρ_{ICM} and ρ_j are the cluster gas and internal jet densities, respectively, and R is the radius of curvature. The parameter h is the scale height through which the dynamic pressure is conducted. In the naked jet model of Begelman et al., the jet is directly in contact with the ICM so that h $\sim$ radius of the jet (r_j). In the Jones and Owen picture, an interstellar medium (ISM) in the galaxy insulates the jet from the ICM so that h $\sim$ radius of galaxy. Equation (1) follows directly from Euler's equation for a flow of constant velocity.

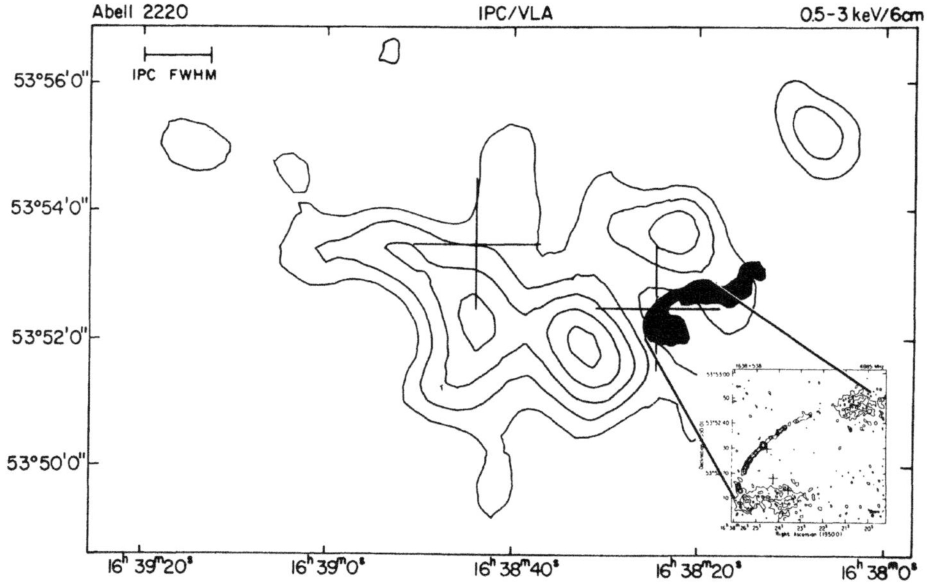

Figure 3 - A comparison of the Einstein Observatory IPC X-ray emission in A2220 with the radio emission of 1638+538, associated with the westernmost of two dominant cluster galaxies (Burns and Balonek 1982).

Note that in equation (1) all the parameters except the jet velocity can be determined observationally. This means, then, that the average jet velocity can be obtained from (1) with a minimal number of assumptions. The radius of curvature and the jet radius can be found from radio maps. The NAT galaxy velocity can be estimated from cluster galaxy redshifts if enough objects are measured to properly compute the cluster mean. The internal density of the jet can be determined from multifrequency depolarization and Faraday rotation observations.

The density of the surrounding ICM can be computed from x-ray observations of the cluster gas. Such a calculation was performed for the Einstein Observatory x-ray emission within Abell 2220 as shown in Figure 3. Burns and Balonek (1982) assumed a constant density and a thermal bremstrahlung emissivity for the 2 x-ray clumps in A2220 to compute a value of 2×10^{-3} cm^{-3} for the region near 1638+538. For smoother gas distributions, isothermal models are found to match the observed distributions of x-ray emission (Jones et al. 1979). More sophisticated polytropic models (e.g., Fabian et al. 1981) have been fit to some cluster x-ray data such as in the Perseus cluster. All the models produce density estimates which agree to within a factor of $\sim$ 3-5. The well-defined cluster environment, unlike that near many straight jets, can be used to accurately determine the external pressure which bends the jets.

At present, the full potential of equation (1) has not been realized because of data lacking at one of the wavelengths noted above. For both NGC1265 and 1638+538, only upper limits of $n_j \lesssim 10^{-3}$ cm^{-3} for the density of the jets could be determined from Faraday depolarization. For a naked jet, then, this limit plus the optical and x-ray data indicate jet speeds in excess of 10^4 km s^{-1}. For a jet embedded in an ISM, the lower limit on the velocity is 1000 km s^{-1}.

It should be emphasized that when better multifrequency radio data are available (see Section 5), the bending criteria in eq. (1) should tightly constrain (comparatively) the average velocity of a curved jet. In spite of the uncertainties in h and the naive assumption of a constant v_j value, the resulting estimate of the curved jet velocity should be far superior to that of the straight jets. For the straights jets, previous estimates of v_j range from 10^3 Km s^{-1} (jet opening angle argument) to 0.99 c (relativistic beaming for one-sided jet). A bending analysis of curved jets should produce estimates of v_j that are good to within a factor of $\sim$ 5-10. In my judgement, bent jets in NAT sources offer the best opportunity to determine the velocity of a jet in a low-luminosity source.

4. BENDING WIDE-ANGLE TAILED GALAXIES

There is a serious problem if one assumes that WAT sources are bent to their observed shape by the same dynamic pressure mechanism which bends NATs. The nature of the problem lies in the velocity and distance that such galaxies can move and still be consistent with other observational constraints.

The most severe test of the radio trail model for WATs is in the case of 1919+479 (Fig. 4). The radio tails are $\sim$ 0.8 Mpc in length, thus making 1919+479 one of the largest known tailed sources. In the standard trail picture (e.g. Jaffe and Perola 1974), the western tail of 1919+479 is viewed as the remnant of radio plasma ejected by the galaxy and decelerated by ram pressure. Thus the tail marks a distance of 0.8 Mpc which the galaxy has traveled in a north-south direction. Furthermore, a fit to the curvature of the jet in the eastern half of the source with either an independent blob model (Valentijn 1982) or a continuous beam model (Burns 1981; eq. (1)) reveals that the galaxy must be moving at a velocity of >400 km s^{-1} - 700 km s^{-1}.

Both the extreme distance traveled and velocity of the galaxy predicted by the trail model are inconsistent with the dynamics of

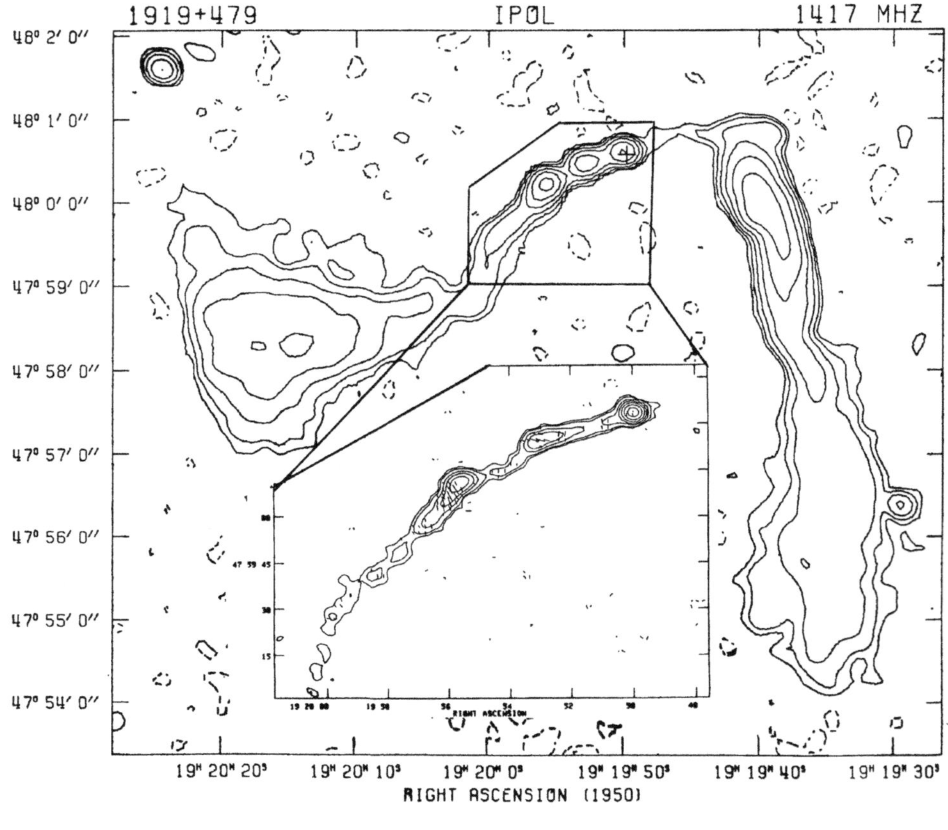

Figure 4 - The radio structure of 1919+479. The low resolution 20-cm map was restored with a clean beam 14".9 x 12".6. The insert is a 6-cm VLA map at a resolution of 4".5 x 3".6. The dashed lines are E-vectors. The cross marks the position of the cD galaxy.

galaxy in the cluster. 1919+479 is associated with a supergiant cD galaxy which appears to lie at the optical center of a rich cluster with z=0.103 (412 Mpc for H_o = 75 km s^{-1} Mpc^{-1}). Identification with the central dominant galaxy in a cluster is a common feature among all WATS. There are three reasons why such a massive galaxy should not move with the characteristics required by the radio trail model:

(1) Optical redshift observations for cD galaxies in Abell clusters demonstrate that these galaxies have velocities that are very small with respect to the cluster means. Quintana and Lawrie (1982) showed that the velocity of dominant galaxies in 28 clusters tended to be within ∿ 100 km s^{-1} of the cluster averages. Since the optical data generally have an accuracy of ∿ 100 km s^{-1}, the dominant galaxies are consistent with little or no motions. In the specific case of the WAT galaxy 3C465 in A2634, Eilek et al. (1983) noted that the systemic velocity of the radio ID, NGC 7720, was zero within the errors.

(2) X-ray emissions from clusters with dominant galaxies are always centered on the massive galaxies (see e.g., Forman and Jones 1982). Since the x-ray emission originates from hot ICM gas and that gas is a good tracer of the cluster gravitational potential well, it would appear safe to say that the dominant galaxies are at or very near the center of the wells. Similarly, the optical light distribution from the cluster galaxies are usually centered on the dominant galaxy (e.g., Geller and Beers 1982). If these massive galaxies were swinging through clusters to distances >0.8 Mpc from the centers, then there should be numerous cases where the x-ray and optical emission are significantly offset from the current location of the D galaxy.

(3) Theoretical models (e.g., Malumuth and Richstone 1981) for clusters with dominant galaxies indicate that the D galaxies should have velocities < 200 km s^{-1} and do not move more than 0.3 of a cluster radius (∿ 50 kpc) from the center. Such a configuration will develop after only a few standard cluster crossing times. Furthermore, cannibalism models for the formation of cD galaxies (e.g., Ostriker and Hausman 1977) require their presence in the dense center of the clusters. These ideas plus the fact that D and cD galaxies can be ten times more massive than the second brightest cluster galaxies argue against large velocities and large traversals as required by the ram pressure trail model.

If dynamic pressure produced by motion in the cluster does not bend the jets/tails in WATs, then what mechanism produces the dual tail structure which looks similar to the NAT sources? The answer to this question is unclear at the moment. Eilek et al. (1983) reviewed all the major possibilities for bending the structure in 3C465 (Figure 5), and found them all wanting. Among the models which they considered were:

(1) Orbital motion of the radio nucleus about the barycenter of the two nuclei which form NGC 7720 (e.g., Wirth et al. 1982). The

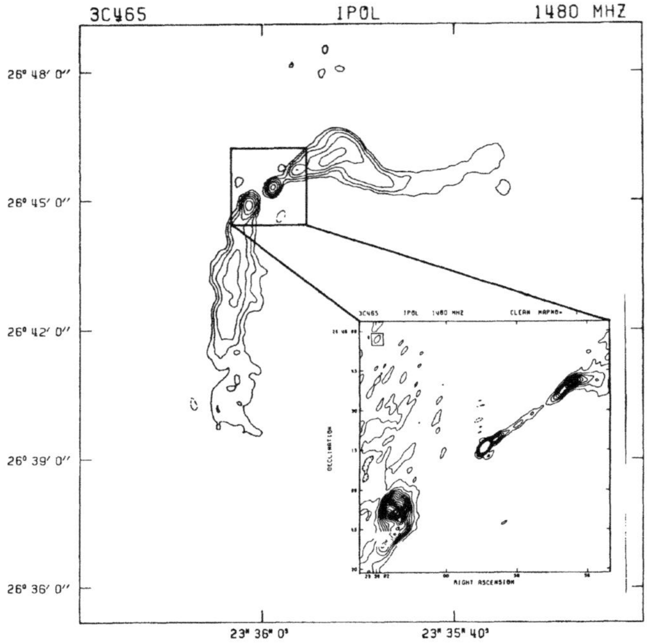

Figure 5 - The 20-cm radio structure of 3C465 from Eilek et al. (1983).

short orbital period ($4x10^7$yrs) requires a large flow speed ($>10^4$ km s^{-1}) in order to bend the tail in ½ an orbital period. From equation (1), this velocity leads to a tail density of $< 10^{-7}$ cm^{-3}, which is clearly inconsistent with the radio polarization data.

(2) Buoyancy effects produced by an asymmetrical distribution of cluster gas. There is indeed an asymmetry in the x-ray emission but that asymmetry is in a direction opposite that expected from buoyancy (i.e., radio tails are observed to bend toward emission and density enhancements in the ICM).

(3) Bending by gravity. For this to work, a mass asymmetry of $10^{13}M_{\odot}$ would be required on the tails' side of the cluster. Such a large asymmetry is not clearly visible in galaxies or gas.

(4) Motion through a Supercluster. A2634 (3C465) and A2666 do appear to form a supercluster system. However, for any reasonable parameters for the supercluster gas environment, the dynamic

pressure produced by the motion of A2634 is far short of that needed to bend the radio tails.

(5) Large Scale Mass Motions. If the ICM of A2634 exhibits large scale ($\gtrsim$200 kpc), high velocity ($>10^3$ km s^{-1}), asymmetric fluid flows then the tails in 3C465 could be conceivably bent to their observed orientation. Eilek et al. consider this to be unlikely, however, since the fluid cells should be visible as clumps in the x-ray maps. No such clumps near the radio galaxy are present.

(6) Electrodynamic Effects. If the plasma beam carries a current and if there is a net B-field in the ICM of the proper orientation, then Lorentz forces would bend the flow. Interestingly, the cluster magnetic fields and jet/tail currents seem to be of the right order needed to bend the 3C465 structure. However orienting the magnetic field in this manner for all WATs may be implausible.

Although we seem to now understand the basic bending mechanism for NAT galaxies, the situation is far from clear for the Wide-Angle Tails. Much more theoretical and observational work is needed in order to solve the problem.

5. CURRENT VLA OBSERVATIONAL PROGRAMS

Several large survey programs of clusters of galaxies are currently underway using the VLA "snapshot" mode at 20-cm. First, over 200 Abell clusters with $z \lesssim 0.1$ and with integrated flux densities (determined by a previous survey with the NRAO 91-m telescope by Owen et al. 1982) $\gtrsim$100 mJy are being mapped in the VLA C and B configurations. This survey is finding many new tailed galaxies which are needed to statistically define the properties of NAT and WAT sources. A detailed comparison of the radio emission in the Abell clusters with the Einstein Observatory x-ray emission is planned; this will permit further study of the origin, confinement, and bending of tailed sources. The radio work is being performed by F. Owen, R. White, C. O'Dea, and J. Burns, and the x-ray project is in collaboration with C. Jones, W. Forman, and D. Harris. Second, a VLA survey of nearby ($z \lesssim 0.5$) poor clusters (richness class $\lesssim$ 0) is also underway at 20-cm. This detection experiment will find cluster radio sources which lie in the weak part of the radio luminosity function. The effects of environment on the formation and evolution of cluster tailed sources should be discernible. This work is by J. Burns, E. Nelson, R. White, and F. Owen.

In addition to the radio surveys, full synthesis observations at 2 and sometimes 3 frequencies are being made of the prototype tailed sources (Figures 1-5). The redundant baselines in the different configurations of the VLA are being used to map distributions of spectral index, Faraday rotation, and depolarization. This should help us to better estimate the curved jets's flow velocities (Sec. 3) and to model the curvature of NAT and WAT sources.

ACKNOWLEDGMENTS

I whish to thank my coworkers T. Balonek, J. Eilek, C. O'Dea, F. Owen, and L. Rudnick for useful conversations in which many of the ideas in this review were conceived. I also wish to thank the University of New Mexico and NASA for travel support.

REFERENCES

Baan, W. A. and McKee, M.R., 1982, preprint
Begelman, M.C., Rees, M.J., and Blandford, R.D., 1979, Nature, 279, 770
Bridle, A.H. 1982, I.A.U. Symposium No. 97 on Extragalactic Radio Sources, edited by D. Heeschen and C. Wade, (D. Reidel), p. 21
Burns, J.O., 1981, M.N.R.A.S., 195, 523
Burns, J.O. and Balonek, T.J., 1982, Ap. J., 263, in press
Burns, J.O. and Gregory, S.A., 1982, A.J., in press
Burns, J.O., Balonek, T.J., and MacCallum, C.J., 1983, in preparation
Christiansen, W.A., Pacholczyk, A.G., and Scott, J.S., 1981, Ap. J., 251, 518
Cohen, N., Geldzahler, B., Gorenstein, M., Marcaide, J., Burns, J., and Bääth, L., 1983, in preparation
Eilek, J.A., Burns, J.O., Owen, F.N., and O'Dea, C.P., 1983, submitted to Ap. J.
Ekers, R.D., Fanti, R., Lari, C., Parma, P., 1978, Nature, 276, 588
Fabian, A.C., Hu, E.M., Cowie, L.L., and Grindlay, J., 1981, Ap. J., 248, 47
Fanaroff, B.L. and Riley, J.M., 1974, M.N.R.A.S., 167, 31
Forman, W. and Jones, C., 1982, Ann. Rev. Astr. Ap., 20, 547
Geller, M.J. and Beers, T.C., 1982, P.A.S.P., 94, 421
Hardee, P.E., 1981, Ap. J. (Lett.), 250, L9
Icke, V., 1981, Ap. J. (Lett.), 246, L65
Jaffe, W.J. and Perola, G.C., 1973, Astr. Ap., 26, 423
Jones, C., Mandel, E., Schwarz, J., Forman, W., Murray, S.S., and Harnden, F.R., 1979, Ap. J. (Lett.), 234, L21
Jones, T.W. and Owen, F.N., 1979, Ap. J., 234, 818
Malumuth, E.M. and Richstone, D.O., 1981, Bull. A.A.S., 13, 506
Miley, G.K., 1973, Astr. Ap., 26, 413
Miley, G.K., Wellington, K.J., and van der Laan, H., 1975, Astr. Ap., 38, 381
O'Dea, C.P., 1983, Ph.D. Thesis, Univ. of Massachusetts (in preparation)
O'Dea, C.P., and Owen, F.N., 1983, in preparation
Ostriker, J.P. and Hausman, M.A., 1977, Ap. J. (Lett.), 217, L125
Owen, F.N. and Rudnick, L., 1976, Ap. J. (Lett.), 205, L1
Owen, F.N., Hardee, P.E., and Bignell, R.C., 1980, Ap. J. (Lett.), 239, L11

Owen, F.N., White, R.A., Hilldrup, K.C., and Hanisch, R.J., 1982, A.J., 87, 1083
Quintana, H. and Lawrie, D.G., 1982, A.J., 87, 1
Rudnick, L., and Burns, J.O., 1981, Ap. J. (Lett.) 246, L69
______________. 1983, in preparation
Stocke, J.T. and Burns, J.O., 1983, in preparation
Ulrich, M.-H., 1978, Ap. J., 221, 422
Valentijn, E., 1982, preprint
van Breugel, W., 1979, Ph.D. Thesis, University of Leiden
Wellington, K.J., Miley, G.K., and van der Laan, H., 1973, Nature, 244, 502
Willis, A.G., 1981, Optical Jets in Galaxies (Proc. 2nd ESO/ESA Workshop, 18-19 February 1981), p. 71
Wirth, A., Smarr, L., and Gallagher, J.S., 1982, A.J., 87, 602

DISCUSSION

P. BIERMANN: Very long Einstein X-ray observations which we made of two DeVaucouleurs-Turner-Gott groups establish the continuity of the intergalactic medium properties from very loose groups to big clusters over five powers of ten in L_x. For the NAT sources this is exactly what is needed.

J. BURNS: The problem for the particular group near B2 0800+24 (Stocke and Burns 1982, in press) is that more mass in gas than in galaxies may be needed in order to provide the ρv^2 required to bend the NAT source. Where does this gas come from? Previously, it was believe that gas in rich clusters is processed through galaxies. In this case, the gas must be primordial.

G. BENDFORD: If $\vec{J} \times \vec{B}$ force bends 1919 + 479, the implied B_θ - self-field confines the jet. A helical field of this strength will also keep the jet together as it bends (added, but flexible, "back bone"), so your picture seems consistent. Is there any way to see the implied hoop-like extragalactic field encircling the galaxy?

J. BURNS: Unfortunately the answer is no. There have been several detection of radio halos in clusters: this is diffuse emission associated with the cluster as a whole, not a single galaxy. The observations are difficult, however, and polarization observations, needed to determine the B-field orientation, are impractical at present.

JET-LIKE STRUCTURES IN LOW FREQUENCY VARIABLE SOURCES

L. Padrielli
Istituto di Radioastronomia C.N.R., Bologna

J.D. Romney
Max- Plank institute für Radioastronomie, Bonn

SUMMARY

We present 18 cm VLBI observations of a sample of compact radio sources, selected on the basis of low frequency variability.

Preliminary results show that a large number of these sources display a jet-like structure on angular scale of milliarcseconds. Furthermore, observations on arcsecond scale show that very often the sources have large scale structures in addition to their compact core.

1. INTRODUCTION

In the second half of 1975, we started a systematic program with the Bologna Radiotelescope to monitor the flux density, at 408 MHz, of about one hundred extragalactic radio sources (Fanti et al., 1979). Several of them showed large flux density variations with time scales from a few months to about one year. Difficulties in the interpretation of these phenomena have been pointed out by several authors (see for instance, Jones and Burbridge 1973). The main problem is that the linear dimensions, deduced from the variability time scale (on the basis of causality arguments), imply severe difficulties with the self inverse Compton process. The deduced brightness temperatures often exceed the 10^{12} K limit for incoherent synchrotron process. Such a situation implies very large X-ray fluxes (which are not observed) and extremely short lifetimes for the relativistic electrons, which are supposed to radiate at radio frequencies.

A variety of alternative models have been proposed, abandoning or modifying the conventional model of a cloud of relativistic plasma expanding adiabatically. Recently, a general consensus emerges in favour of the interpretation of low-frequency variability in terms of bulk relativistic motion, in a direction near the line of sight. This offers the advantage of explaining by a single mechanism both the flux variations and the directly observed "superluminal" structural changes.

A. Ferrari and A. G. Pacholczyk (eds.), Astrophysical Jets, 81–90.
Copyright © 1983 by D. Reidel Publishing Company.

Therefore, in order to extend the amount of information available on these objects and to determine their physical size and structure directly, in 1980 we started VLBI observations of a subsample of 23 sources which have shown flux variations. We repeated the observations in 1981 to investigate possible correlations between flux variations and structural changes. The reduction of these second observations is still in progress.

2. THE OBSERVATIONS

The first observations were made in February 1980 using 7 telescopes at Simeis (USSR), Onsala (Sweden), Effelsberg (W. Germany), Hartebeesthoek (South Africa), Green Bank, Fort Davis, and Owens Valley (USA) with a resolution of a few milliarcseconds (mas). Many people from different Institutes were involved in the first run of observations, namely: J.D. Romney, N. Bartel, K.W. Weiler, A. Ficarra, F. Mantovani, L. Padrielli, M. Massi, L.I. Matveenko, V.I. Kostenko, L. Kogan, L. Baath, G. Nicolson.

The choice of the observing frequency of 1670 MHz represented a compromise between the lower frequencies where the observed variations are difficult to understand, and the higher frequencies where satisfactory resolution is obtainable. The 23 program sources were observed in 10 minute blocks, from 3 to 8 times during the 48-hour observing period. This multi-baseline spot-observation technique produced visibility measurements at, typically, 40 to 50 different locations in the u-v plane.

The analysis was done independently in Bonn, Bologna and M.I.T. and the data reduction is now complete. Hybrid maps were obtained and the major characteristics of the sources were well determined, although the u-v plane was poorly covered due to the spot-observation technique.

3. RESULTS

The morphologies found in this sample confirm that a large number of compact sources show on milliarcsecond scale a rather compact dominating component associated with a diffuse elongated emission region. These low brightness regions are sometimes curved, expanding away from the bright "core" in one or two directions. Numerically the results can be summarized as follows:

- 40% of the sources show strongly asymmetric structures (one-sided jet like) with a linear size ranging from 5 to 50 pc.;

- 9% present an elongated emission region symmetric about the core (two-sided jets) with a size up to 100 pc;

- 35% are still compact or slightly resolved on the scale of few mas with physical dimensions less or equal to 10-20 pc;

- 16% of the cases show more complex structures difficult to classify in such a simple scheme.

Observations at VLA (Perley, 1982) and at MERLIN (Browne et al., 1982) on a scale of few arcseconds show that the majority of the sources (60%) has "large" scale structures showing, in addition to a compact core which coincides with the optical identification, one or more weaker components.

In Fig. 1 we report the distribution of the differences between the position angles of the large and small scale structures.

The comparison between arcsecond and milliarcsecond structures shows a misalignment, that is often very great and contrasts with the alignment, over many orders of magnitude in size, of some radio cores of extended radio galaxies with the outer lobes.

Fig. 2 shows the 18 cm maps of several sources compared with those obtained at 20 cm at VLA and with the MERLIN 408 MHz maps, when available.

4. INDIVIDUAL SOURCES

0224+67 - This source shows a strong core and a very weak diffuse emission (9 mas), elongated northwards. The secondary component, detected in large scale observations, is in the south direction (Perley et al., 1980).

0333+32 - Marscher and Broderick (1982) on the basis of 3 epoch VLBI observations at 2.8 cm discovered that this source is expanding at a rate 0.10-0.14 milliarcsec/yr. In the 18 cm VLBI measure we can see the "jet" very well. The large scale map shows a secondary weak component which is misaligned relative to the core small structure of about 15°. There is no evidence for extended emission connecting the core with the outer component.

0605+08 - Interesting case of misalignment between small and large scale structure (50°).

0723-00 - This source is completely unresolved in the VLA map. The elongated structure corresponds to a jet of about 20 pc.

0736+01 - There is no information on the arcsecond scale structure of this source. This is an example of rather symmetric emission about the core.

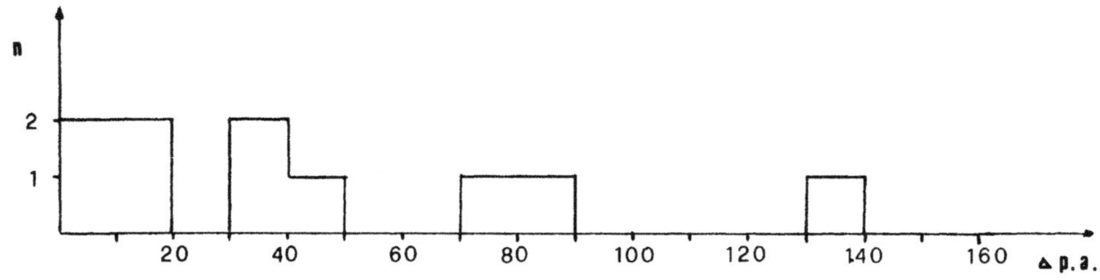

Figure 1: Distribution of the differences between the position angles of large and small scale structures.

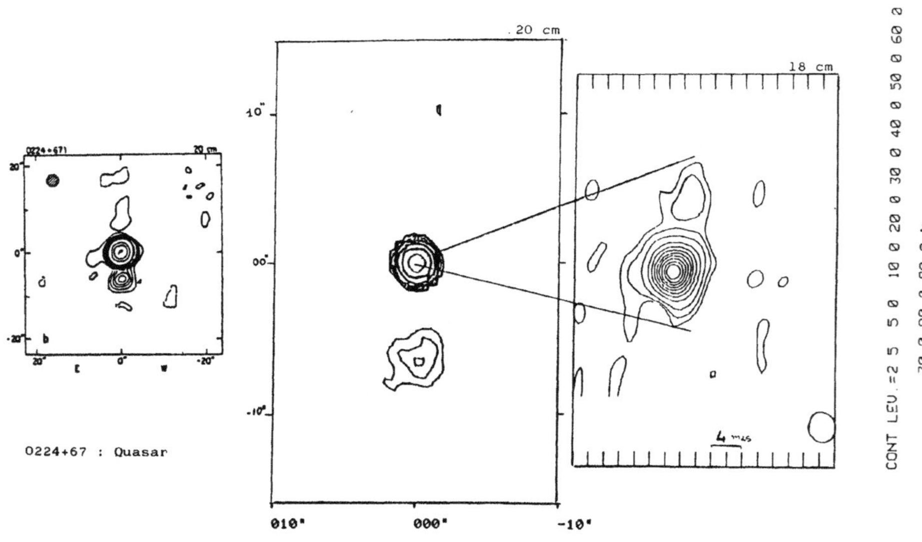

2a.

Figure 2: VLBI 18 cm maps of several sources compared with the VLA 20 cm maps and with the 408 MHz maps obtained with MERLIN.

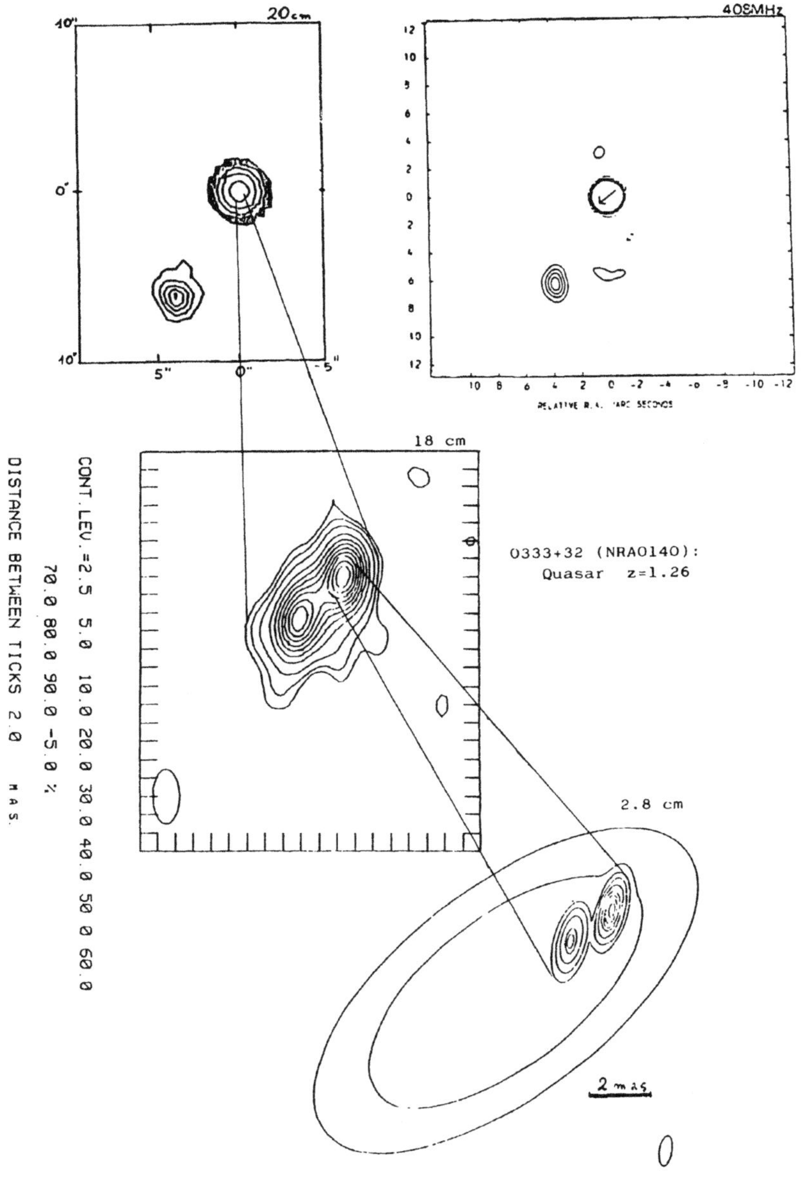
20cm
408MHz
18 cm
0333+32 (NRAO140):
Quasar z=1.26
2.8 cm
2 mas
CONT.LEV.=2.5 5.0 10.0 20.0 30.0 40.0 50 0 60.0
70.0 80.0 90.0 -5.0 %
DISTANCE BETWEEN TICKS 2.0 MAS

2b.

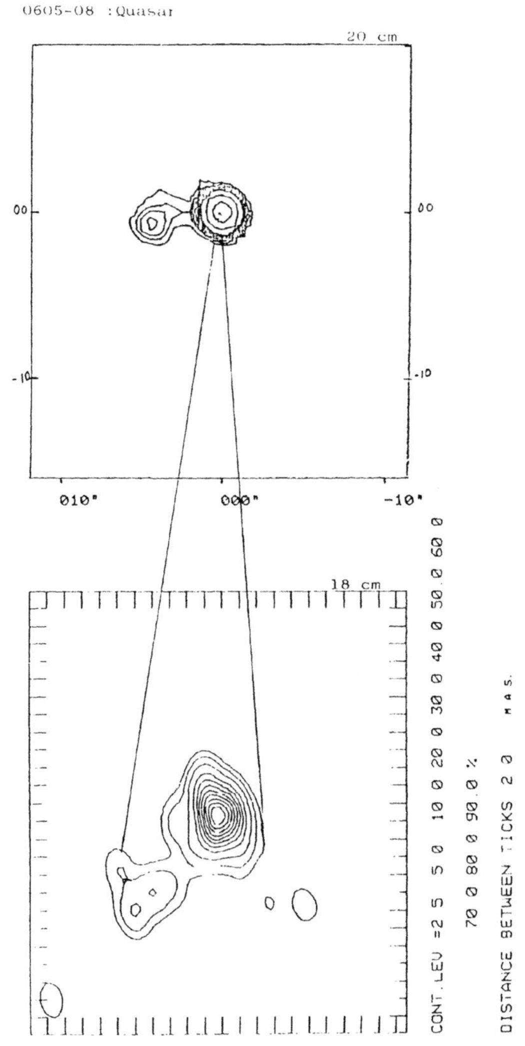
0605-08 :Quasar
20 cm
00
-10
010"
000"
-10"
18 cm
CONT.LEV =2 5 5 0 10 0 20 0 30 0 40 0 50 0 60 0
70 0 80 0 90 0 %
DISTANCE BETWEEN TICKS 2 0 MAS.

2c.

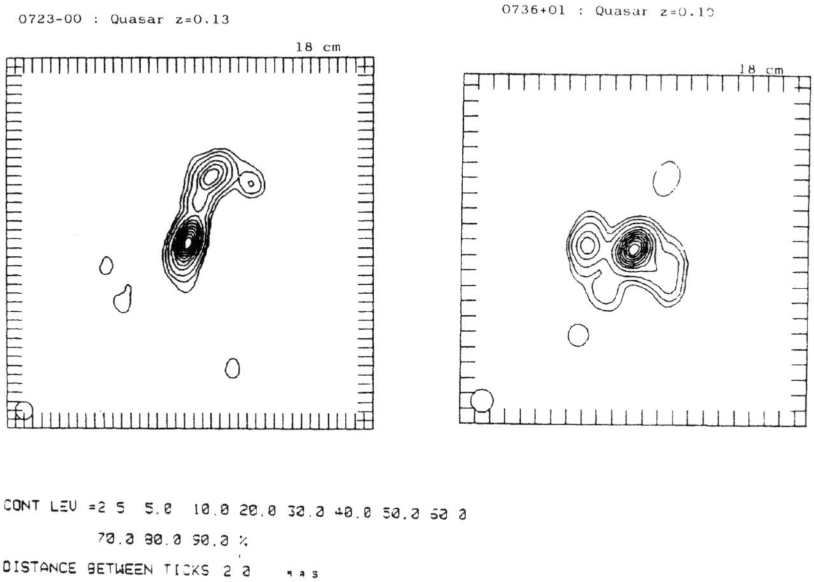
0723-00 : Quasar z=0.13
18 cm
0736+01 : Quasar z=0.19
18 cm
CONT LEV =2 5 5.0 10.0 20.0 30.0 40.0 50.0 60 0
70.0 80.0 90.0 %
DISTANCE BETWEEN TICKS 2 0 mas

2d.

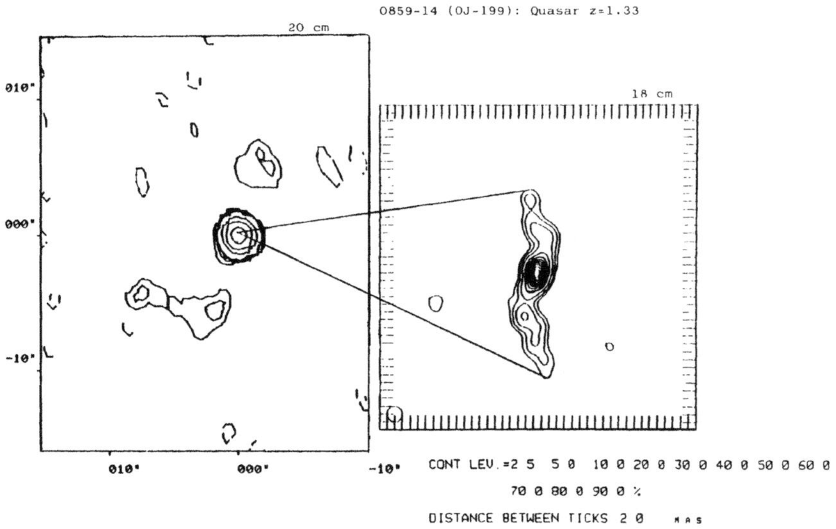
0859-14 (OJ-199): Quasar z=1.33
20 cm
18 cm
010"
000"
-10"
010"
000"
-10"
CONT LEV.=2 5 5 0 10 0 20 0 30 0 40 0 50 0 60 0
70 0 80 0 90 0 %
DISTANCE BETWEEN TICKS 2 0 mas

2e.

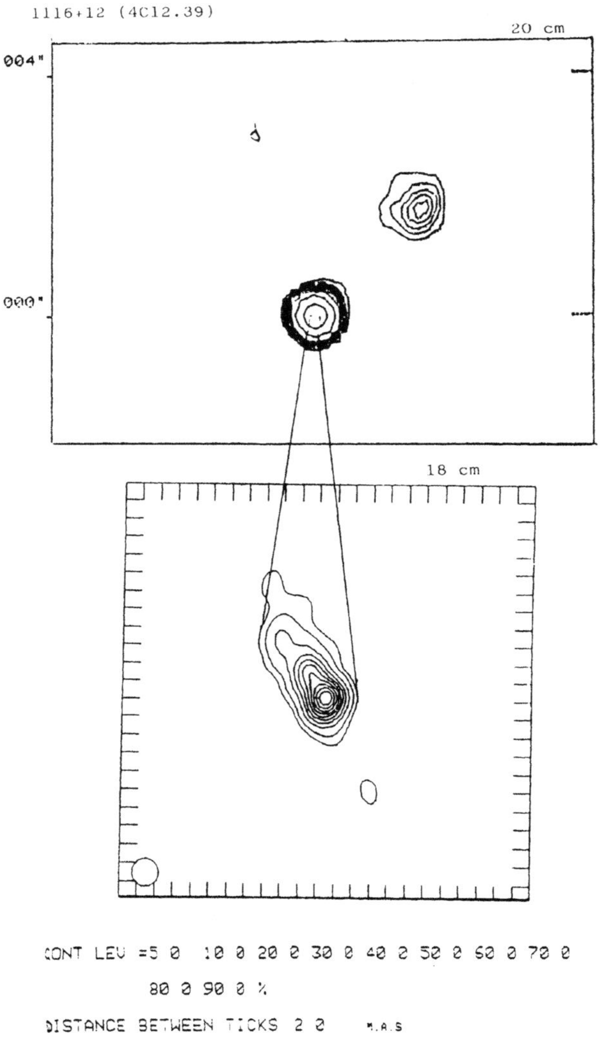
1116+12 (4C12.39)
20 cm
004"
000"
18 cm
CONT LEV =5 0 10 0 20 0 30 0 40 0 50 0 60 0 70 0
80 0 90 0 %
DISTANCE BETWEEN TICKS 2 0 M.A.S

2f.

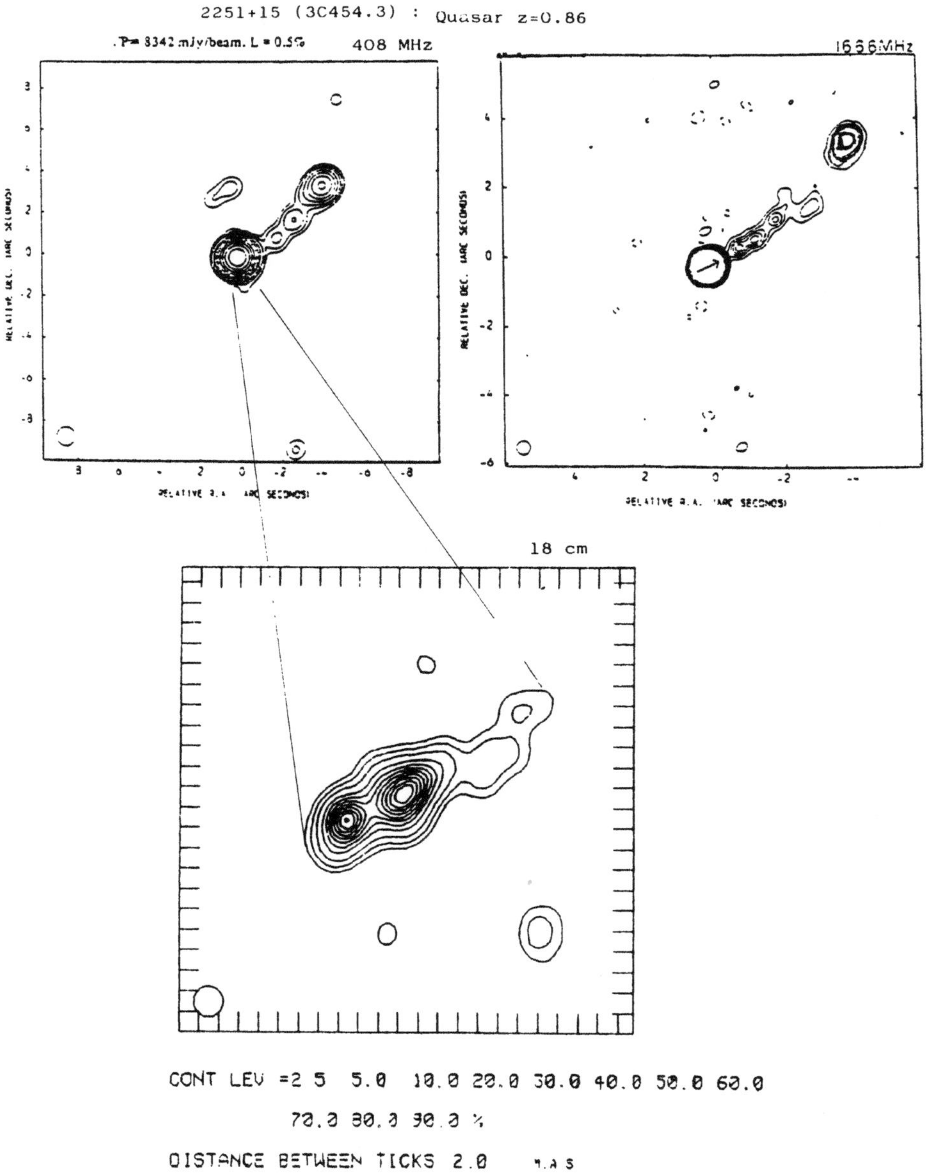
2251+15 (3C454.3) : Quasar z=0.86
P= 8342 mJy/beam. L = 0.5%
408 MHz
1666MHz
RELATIVE DEC. (ARC SECONDS)
RELATIVE R.A. (ARC SECONDS)
18 cm
CONT LEV =2.5 5.0 10.0 20.0 30.0 40.0 50.0 60.0
70.0 80.0 90.0 %
DISTANCE BETWEEN TICKS 2.0 M.A.S

2g.

0859-14 - This source shows a very interesting structure with two symmetric jets extending to about 100 pc. This morphology is quite unusual for a compact source. On a scale thousand times greater, the VLA map shows a symmetric structure with bent features. In this case it is difficult to determine the alignment between the VLA and VLBI components, because both jets curve along their entire length.

1116+12 - In the VLBI map this source shows a jet-like structure of about 10 mas in position angle 30°. The large scale structure is strongly misaligned (about 80°)

2251+15 - This is one of the strongest low frequency variable sources. It is unclear whether this source has to be considered a one-sided or two-sided jet. There is good alignment between the VLBI structure and the very weak slightly curved jet leading to the relatively bright component 4 arcseconds away.

REFERENCES

Browne, I.W.A., Clark, R.R., Moore, P.K., Muxlow, T.W.B., Wilkinson, P.N., Cohen, M.H., Porcas, R.W., 1982, Nature, in press

Cornwell, T., Wilkinson, P., 1981, M.N.R.A.S. 196, 1067

Marscher, A.P., Broderick, J.J., 1982, Extragalactic Radio Sources, IAU Symp. N. 97, p. 359, ed. Heeschen, Wade, Reidel, Dordrecht

Fanti, R., Ficarra, A., Mantovani, F., Padrielli, L., Weiler, K., 1979, Astron. Astrophys. Suppl. 36, 359

Jones, T.W., Burbidge, G.R., 1973, Astrophys. J. 215, 474

Perley, R.A., Fomalont, E.B., Johnston, K.J., 1980, Astron. J. 85, 649

Perley, R.A., 1982, NRAO preprint

CURVATURE IN HIGH REDSHIFT QUASARS

P.D. Barthel
Leiden Observatory
P.O. Box 9513
2300 RA Leiden
The Netherlands

Abstract

A preliminary analysis of the radio morphologies of powerful, steep spectrum quasars suggests that these sources are more distorted at high redshifts.

Quasars provide a unique tool for tracing the history of the universe, since we see high-redshift quasars in a situation when the universe was much younger. The radio morphology of extended radio sources associated with quasars is believed to be influenced strongly by the ambient medium, and it could well be that this influence is epoch-dependent. We have begun a program to study such effects by comparing similar samples of steep-spectrum quasars at high and low redshift.

Before we began this project, the extended radio structure of only about 15 (large) high redshift QSOs, mainly 3C sources, was known. To extend this number we mapped 86 quasi-stellar radio sources with steep or unknown radio spectrum, from the 4C, PKS, MC, B2 and W Catalogues, at a resolution of 0"5, using the VLA at 4885 MHz (Barthel, Miley & Schilizzi, in preparation). The quasars observed had redshifts between 1.5 and 3.2 and monochromatic luminosities at 408 MHz (emitted frequency) between 10^{26} and 10^{29} Watt/Hz, assuming H_o = 75 km/sec and q_o = 0.5.

First inspections of the radio maps indicate that the high-redshift quasars have systematically different morphologies than low-redshift quasars. In particular there appears to be a smaller fraction of symmetric edge-brightened double sources, and a larger

A. Ferrari and A. G. Pacholczyk (eds.), Astrophysical Jets, 91–94.
Copyright © 1983 by D. Reidel Publishing Company.

fraction of curved structures.

For many of the high-z sources, the optical position is not well defined. Therefore we have measured the curvature with reference to the outer extremities of the source E and E' and the highest peak between these extremities P, and define the curvature c as the angle EPE'. We have compared two samples:

I. Low redshift sample
QSRs/RGs, with
- $z < 0.7$
- > 5 beams/source on Westerbork/Cambridge maps
- $\delta_{1950} > 25^{o}$
- $^{10}\log P_{408} > 27.0$ W/Hz
- $\alpha_{5000}^{14,900} \lesssim -0.7$ $(S_\nu \propto \nu^\alpha)$

II. High redshift sample
QSRs, with
- $z > 1.5$
- $\gtrsim 5$ beams/source (VLA)
- $^{10}\log P_{408} > 27.0$ W/Hz
- $\alpha_{1420}^{5000} \lesssim -0.7$

We mention three important properties of the samples:
a) As there are only 11 powerful low-redshift quasars we added 21 powerful radio galaxies to the first sample.
b) For the low-redshift sources with published Westerbork and/or Cambridge maps we excluded low-δ sources, in order to have a fairly circular restoring beam.
c) Our luminosity constraint of $P_{408} > 10^{27}$ Watt/Hz includes (at low redshift) only Fanaroff & Riley Class II sources, i.e. edge-brightened doubles.

For both samples there are typically between 10 and 20 beams across each source. Sample I consisted of 32 sources, which were compared with 38 high-redshift quasars in sample II. Figure 1 shows histograms of the curvature c in both samples. The most distorted low-redshift sources are 3C 275.1 (Stocke et al., 1982) and the radio galaxy 3C 123 (Kronberg & Strom, 1977), both having curvature greater than 20°. The high-redshift sample however contains 9 sources with curvature greater than 20°, and two exceeding 50°.

So far we merely draw attention to this effect as suggestive. Further analysis is necessary before we can state with confidence that it exists. We intend to:
i) make maps having a constant number of beam areas per source;

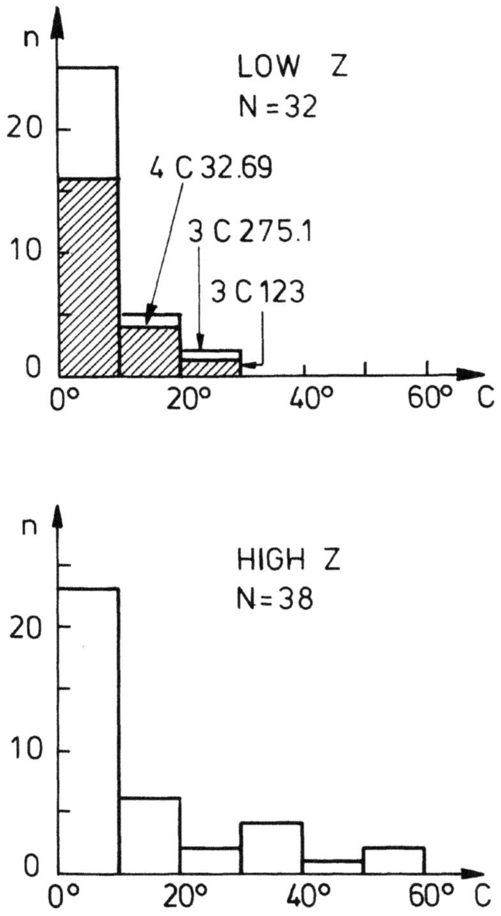

Figure 1
Histograms showing the distribution of curvature c in the low- and the high-redshift sample. Radio galaxies are depicted shaded.

ii) investigate spectral effects by making observations at λ2 and 20 cm.

Concerning the physical reasons for these possible high-redshift distortions we can only speculate. One possibility is that the high-z quasars are in a more turbulent environment due to differences in intracluster or circumgalactic media. Alternatively, since these are quasars at epochs close to the redshift-cutoff the axes of the nuclear machines may be less stable.

The above comparison will be described and discussed elsewhere in more detail (Barthel & Miley, in prep.).

Acknowledgments

This investigation was supported by the Netherlands Foundation for Astronomical Research (ASTRON) with financial aid from the Netherlands Organization for the Advancement of Pure Research (ZWO). The author would like to thank Drs. G. Miley and R. Schilizzi for discussion on the subject. Some travel support from the Leids Kerkhoven-Bosscha Fonds is also acknowledged.

References

Kronberg, P.P., Strom, R.G. 1977, Ap. J. 215, 438
Rudnick, L., Owen, F. 1977, A. J. 82, 1
Stocke, J. et al. 1982, IAU Symp. 97, p. 39, Reidel, Dordrecht.

Rapid Variability of 3C273 at 300 GHz

W.A. Sherwood, E. Kreysa, H.-P. Gemünd, P. Biermann
Max-Planck-Institut für Radioastronomie
Bonn, West-Germany

ABSTRACT

With a broadband bolometer mounted on the 3.6 m ESO telescope we have measured 3C273 at 300 GHz 6 times in 6 days. During this series of measurements we observed a sudden drop in flux density by 11.6 ± 2.9 Jy (spectral index assumed to be zero, and all known errors included). In an interpretation with the model of bulk relativistic motion we can combine the brightness temperature inferred with the observed superluminal motion and derive strict limits on the model parameters.

Rapid variability of compact extragalactic sources has provided the first argument that bulk relativistic motion is present in active galactic nuclei (Rees 1966). The subsequent observation of superluminal motion (for a review see Kellermann and Pauliny-Toth 1981) has given support to this model. The recent observation of compact sources both in the mm range and in soft X-rays (Owen et al. 1981) has strengthened the case for a model of Synchrotron-Self-Compton emission from relativistically moving blobs (Blandford and Königl 1979). Similarly, the first coherent attempt to combine the statistics of flat and steep spectrum radio sources in an interpretation of bulk relativistic motion (Orr and Browne 1982) has underlined the merits of such a model in spite of some difficulties (Browne et al. 1982).

Here we come back to the original idea of using the evidence contained in flux density variations to constrain models of compact nuclei. We have measured the source 3C273 on consecutive days at 1 mm; the data are presented elsewhere (Sherwood et al. 1982). The most conspicuous feature is a sudden decrease by 11.6 ± 2.9 Jy in flux density within a span of one day; here we assume a spectral index of $\alpha = 0$ across the broad band of the bolometer and give the combined error from measurement, extinction, calibration and pointing. For calibration we used the planets and as reference the active nucleus Cen A (see Sherwood et al. 1982). Using a conservative lower limit to the drop in flux density of 6.4 Jy we obtain an inferred brightness temperature (e.g. Blandford

A. Ferrari and A. G. Pacholczyk (eds.), Astrophysical Jets, 95–96.
Copyright © 1983 by D. Reidel Publishing Company.

and Königl 1979) of 1.9 10^{15} K requiring a boosting factor, D_j, of 14.3. Assuming the apparent superluminal motion to be caused by bulk relativistic motion with the same Lorentz factor γ_j and angle to the line of sight θ we can derive a lower limit for γ_j of 11.2 and an upper limit for θ of $3\overset{\circ}{.}8$ (H_o = 50 km/sec/Mpc).

For sources, for which we have not observed superluminal motion but do have evidence of abrupt changes in flux density at high frequencies (Epstein et al. 1982) we can again use the model of bulk relativistic motion to predict a range for γ_j and θ. We discuss their observations of the compact source OJ287: They give a boosting factor of $D_j = 12.5$, a lower limit to $\gamma_j = 6.25$, an upper limit to θ of $\theta_{max} = 4\overset{\circ}{.}6$ and an expected superluminal velocity at $\theta = \theta_{max}$ of $v_{\perp}/c = 12.5$. Again we have used a conservative lower limit to the drop in flux density, an adopted spectral index of $\alpha = 0$ across the band, and a Hubble constant of $H_o = 50$ km sec^{-1} Mpc^{-1}. On the same basis we do not require bulk relativistic motion for Cen A and so no superluminal motion is expected on the basis of the mm-data presented in Epstein et al. (1982).

Obviously, the model of bulk relativistic motion is just the one most often discussed at present and so we expect to be able to constrain the parameters of any other model for sudden changes in flux density and superluminal motion. In order to finally check the model of Synchrotron-Self-Compton emission from relativistically moving blobs, VLBI observations at mm wavelengths together with simultaneous X-ray observations are highly desirable and should be possible in the future.

This work is partially supported by the Deutsche Forschungsgemeinschaft (SFB131).

REFERENCES

Blandford, R.D., Königl, A.: 1979, Astrophys. J. 232, 34
Browne, I.W.A., Clark, R.R., Moore, P.K., Muxlow, T.W.B., Wilkinson, P. N., Cohen, M.H., Porcas, R.W.: 1982, Nature 299, 788
Epstein, E.E., Fogarty, W.G., Mottmann, J., Schneider, E.: 1982, Astron. J. 87, 449
Kellermann, K.I., Pauliny-Toth, I.I.K.: 1981, Ann. Rev. Astron. Astrophys. 19, 373
Orr, M.J.L., Browne, I.W.A.: 1982, Monthly Notices Roy. Astron. Soc. 200, 1067
Owen, F.N., Helfand, D.J., Spangler, S.R.: 1981, Astrophys. J. 250, L55
Rees, M.J.: 1966, Nature 211, 468
Sherwood, W.A., Kreysa, E., Gemünd, H.P., Biermann, P.: 1982, Astron. Astrophys. (in press)

ARE THERE JETS IN SPIRAL GALAXIES?

C. Kotanyi (1), E. Hummel (2), J. van Gorkom (3)

(1) ESO, Garching, FRG
(2) Max Planck Institute for Radioastronomy, Bonn, FRG
(3) NRAO, Socorro, NM, USA

The absence in spiral galaxies of large-scale collimated jets of radio emission is still little understood. Collimated emission in elliptical galaxies in some cases exceeds by far the optical extent of the galaxy (e.g. in NGC6251). Westerbork observations of NGC4258 (van der Kruit, Oort and Mathewson, 1972) and NGC3079 (de Bruyn, 1977) have provided some evidence for linear features of radio emission in spiral galaxies. However in both cases the emission is strictly confined to the optical disk. The features were interpreted as lying in the disks of the galaxies, possibly explaining their different morphology as compared to the jets in ellipticals, which are in an essentially gas-free environment.

Further possible examples of jets in spiral galaxies were found using the VLA. The Seyfert galaxies NGC1068 and NGC4151 show indications for the presence of linear radio features (see Wilson, 1981). Normal spirals have also shown faint linear features. Table 1 lists the luminosity, size and orientation of the anomalous radio emission as well as the optical major axis orientation in these galaxies. Most of the features are nearly along the minor axes of the galaxies. Therefore lying within the gaseous disk of the galaxy is probably not a suitable single explanation for the low luminosity of these radio components as compared to jets in ellipticals.

On other hand several of these galaxies show Seyfert-like properties, e.g. stronger than average radio nuclei (Hummel et al., 1983). It is suggested that the linear features are related to a Parker instability triggered by the high pressure in the galaxy centre, rather than to the collimation mechanisms thought to operate in elliptical galaxies. The buoyant nature of the anomalous features could readily explain why their size is confined to the optical disk even when they are perpendicular to it.

A. Ferrari and A. G. Pacholczyk (eds.), Astrophysical Jets, 97–98.
Copyright © 1983 by D. Reidel Publishing Company.

REFERENCES

De Bruyn, A.G., 1977, PhD Thesis, Leiden University
Hummel, E., van Gorkom, J., Kotanyi, C. 1983, Ap.J. Lett. (in press)
Van der Kruit, P.C., Oort, J., Mathewson, D.S., 1972, A.A. 21, 169
Wilson, A. 1981, in Extragalactic Radio Sources, IAU Symp. 97, 179 (Reidel)

Table 1: Anomalous features in spiral galaxies

NGC	Type	Radio luminosity at 1.4 GHz (10^{21} W/Hz)	Radio size (kpc)	Mean Radio orientation (N thru E)	Optical major axis orientation
2992	Sa	6.3	8.3	115°	6°
3079	Sm	2.5	2.3	65	165
4388	Sb	0.8	1.9	15	92
4438	Spec	1.3	0.5	123	27
5728	Sa	5.5	0.9	132	35
6500	Sa	15.8	1.1	113	50

OPTICAL EMISSION FROM JETS

George Miley
Sterrewacht, Postbus 9513, 2300 RA Leiden, The Netherlands

1. INTRODUCTION

Until a few years ago the only information available about extragalactic radio sources came from measurements of the radio emission itself and of the optical objects from which the sources were presumed to originate. Relatively few source parameters could be determined. This was depressing for theorists but attractive to people like me who find detailed physics difficult! I am afraid that the situation has changed. First, it was demonstrated that some, maybe all, radio jets have M87-type optical/X-ray counterparts (Butcher et al., 1980; Schreier et al., 1979; Feigelson et al., 1981). It followed that the spectra of synchrotron jets can extend over at least seven decades of frequency. These discoveries of optical and X-ray emission provided compelling arguments that localized particle acceleration takes place in jets. It would be impossible to transport the particles down the jet within the extremely short lifetimes (few years) of particles which radiate in the optical and X-rays. Moreover the evidence for the occurrence of in situ particle acceleration has weakened severely if not finally laid to rest those ritualized arguments which we all used to make about the ages of radio sources. If localized acceleration occurs in some places, can it be excluded anywhere in extended radio lobes?

More recently several morphological associations have been found between extended radio emission and optical emission lines. It is on these objects that I focus this talk. The discussion will not be about jets in the strict sense but will involve knots, hot spots and boundaries of radio sources. To justify my title, these various radio components will be regarded as the interface of the jets with their surroundings. I will make no attempt to give an exhaustive review. Instead, I will concentrate on the work with which I am most familiar - that carried out by "our group". The other members of "our group" are Wil van Breugel (Leiden → Kitt Peak → Steward), Tim Heckman-(Leiden → Steward → Maryland) and Harvey Butcher (Kitt Peak). We were

A. Ferrari and A. G. Pacholczyk (eds.), Astrophysical Jets, 99–112.
Copyright © 1983 by D. Reidel Publishing Company.

augmented by Bruce Balick (Seattle) for the work on 3C305 and by Alan Bridle and Ed Fomalont (NRAO) for the radio measurements of Coma A.

2. WHY AND HOW?

From radio measurements alone one can determine only a handful of physical quantities which characterize radio sources (e.g. Miley 1980). Besides information about luminosities and projected linear dimensions, radio intensity distributions yield spatial data about the minimum energies and corresponding pressures and magnetic fields within the synchrotron sources. Depending on what holds the sources together, these minimum energy requirements place upper limits on nT or ρv^2. Here n is the number density and T the temperature of a hot gas that thermally confines the radio source and ρ is the density of a medium assumed to confine the source (moving with velocity v) by ram pressure.

In addition radio polarization distributions and their wavelength dependence tell us the projected magnetic field directions within sources, as well as something about the distribution of $N_e H_{11}$ along the line of sight where n_e is the number density of ionized gas and H_{11} is the line-of-sight component of the magnetic field which permeates it.

If line-emitting thermal gas is associated with the radio source, considerable more information can be derived. The emission lines are a powerful tool for measuring temperature density ionization and, most important, velocity. Comparison of these observables with the parameters derived from the synchrotron emission open several new avenues for approaching radio source physics. First, the spatial distributions of the various line parameters and their relation with the radio morphologies give clues to how the two regimes are linked. Secondly, the pressures and kinetic energy of the line gas can be compared with the pressures needed to confine the radio source and the minimum energy requirements. Thirdly the densities and mass required for the line emission can be related to the properties of the medium needed to depolarise the radio source. Fourthly, as we shall see, the optical line widths give indirect evidence about the mechanism which accelerates the radio-emitting relativistic particles.

Many techniques embracing a multitude of instruments have been used to study the optical line/radio continuum associations. In the optical domain broad and narrow band imaging have been used as well as spectroscopy through apertures and slits. An essential complement to the optical observations has been imaging of the radio intensities and polarizations at several frequencies with various resolutions. These many observational requirements tempt us, addicts of the data acquisition drug to uncontrollable fits of proposal writing.

The diversity of instrumentation and the complexity of the associated reduction systems require team efforts reminiscent of the early days of VLBI. The papers which result are almost invariably too long, since not only do the different observations need separate descriptions but also the "cross-product" information must be discussed.

3. SOME EXAMPLES

Morphological associations between extended radio and optical line emissions are seen in both radio galaxies (e.g. Miley, 1981; van Breugel and Heckman, 1982) and in Seyfert galaxies (e.g. Wilson, 1981). A list of some radio galaxies with known or suspected cases of line/radio associations is given in Table 1. I shall confine myself to two of the most spectacular examples, 3C305 and Coma A/3C277.3.

TABLE 1

Some Radio Galaxies with Extended Radio/Optical Line Associations

Redshift	Object	Size of line Association (kpc)	Reference
.0016	NGC5128/Centaurus A	∿ 40	Graham and Price, 1981; Brodie (these Proceedings)
.0043	NGC1275/Perseus A/ 3C84	∿ 10 ?	Pedlar et al.1982
.026	NGC7385	∿ 11	Simkin and Ekers, 1979; Hardee et al. 1982
.041	3C305	∿ 3	van Breugel et al., 1983
.05	PKS0531-36	∿ 10	Danziger (these Proc.)
.063	4C26.42	∿ 8	van Breugel et al., 1983
.069	4C29.30	∿ 50	van Breugel et al., 1983d
.086	3C277.3/Coma A	∿ 80	Miley et al.,1981 van Breugel et al., 1983a

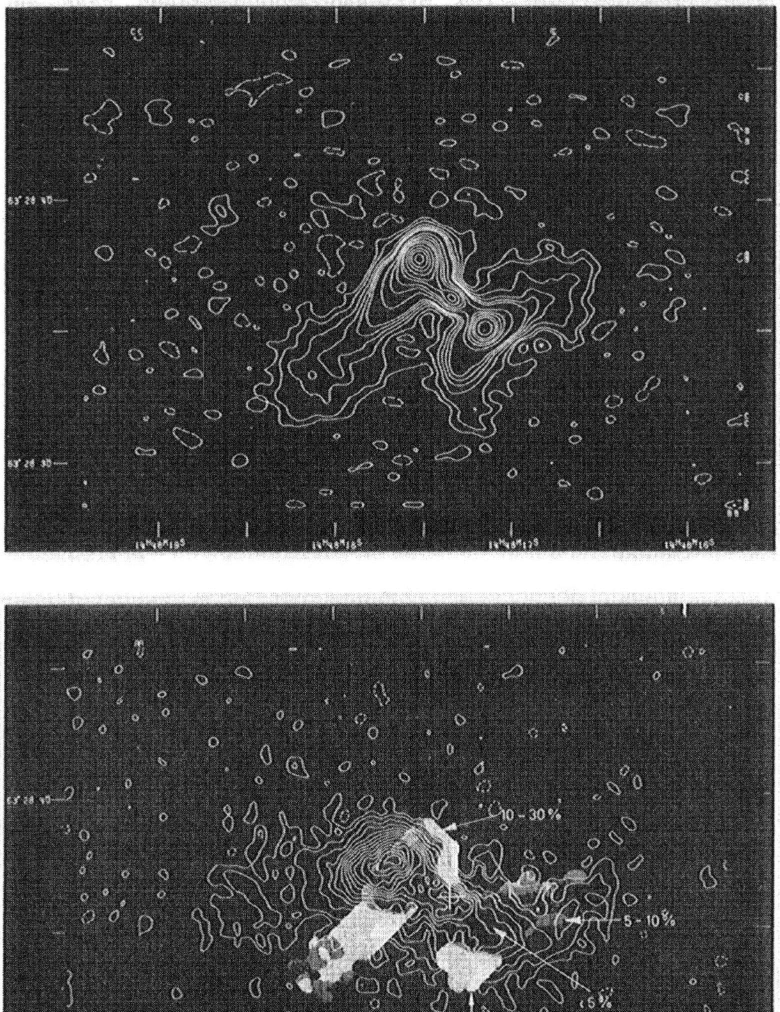

Figure 1: 3C305. Radio continuum and optical emission line structures. Above (Fig. 1a) is a contour representation of λ 6 cm VLA radio intensity map. Below (Fig. 1b) is a contour representation of a Kitt Peak ($H\alpha$ + NII) map superimposed on a gray-scale representation of radio polarization percentage. Note (i) the agreement of the ($H\alpha$ + NII) and radio intensity morphologies and (ii) the anticorrelation of the emission line intensities with the percentage polarization (from Heckman et al., 1982).

First 3C305 (heckman et al. 1982). This is a relatively small radio source. Note the similarity between the structure of the radio emission (Fig. 1a contours) and that of the Hα emission (Fig. 1b contours). You see also that in overlapping radio/line regions of the source the percentage radio polarization (Fig. 1b grey levels) is very small. In these overlapping regions the widths of the strongest observed narrow line, [OIII] λ5007, are broader (∿ 650 km s^{-1} FWHM) than in the outer (≲ 130 km s^{-1}).

The second and most revealing example that I will show is Coma A (3C277.3) (Miley et al. 1981, van Breugel et al. 1982). Do not be deceived by name. The associated galaxy is not a member of the Coma cluster. It has a redshift of 0.09. This radio source has an extent of ∿ 60 kpc. Its radio structure (contours in Figs. 2a, b, c) is fairly typical for a radio galaxy with such an intermediate radio luminosity. However, the map of Hα emission (gray levels in Figs 2a, c, d) is exciting because of its remarkable similarity to the radio morphology. The southern radio knots and the outer northern hot spot are clearly reflected in Hα. In addition, line emitting gas appears to extend from the hot spot along the eastern boundary of the northern lobe. Note that the Hα emission is again anticorrelated with the radio polarization (Fig. 2c).

In the region of the southern knot the data suggest that the radio jet is bent towards the southwest and decollimated by the line-emitting gas. The bright radio knots may well be formed in this interaction. It is interesting (Fig. 2b) that beyond the bend a jet like structure is visible in the (gray scale) radio depolarization. This immediately suggests that thermal gas from the knot may become entrained in the jet.

4. SO WHAT?

Many of the observations that have been made of the objects listed in Table 1 are as yet only partially analysed. However several trends have emerged. Here I summarize some of these, together with conclusions which we have tentatively drawn from the data.

A. Morphologies

The line/radio associations are usually seen (a) just outside bright radio regions and (b) close to bends in the radio sources.

This leads naturally to a picture in which the line-emitting gas in pushed outwards by the radio jets and in which this gas occasionally has sufficient inertia to deflect the jets. In addition, the morphologies provide some evidence (e.g. high ionization of the southern knot in Coma A - see Fig. 3) that the jet is a source of heating for the gas.

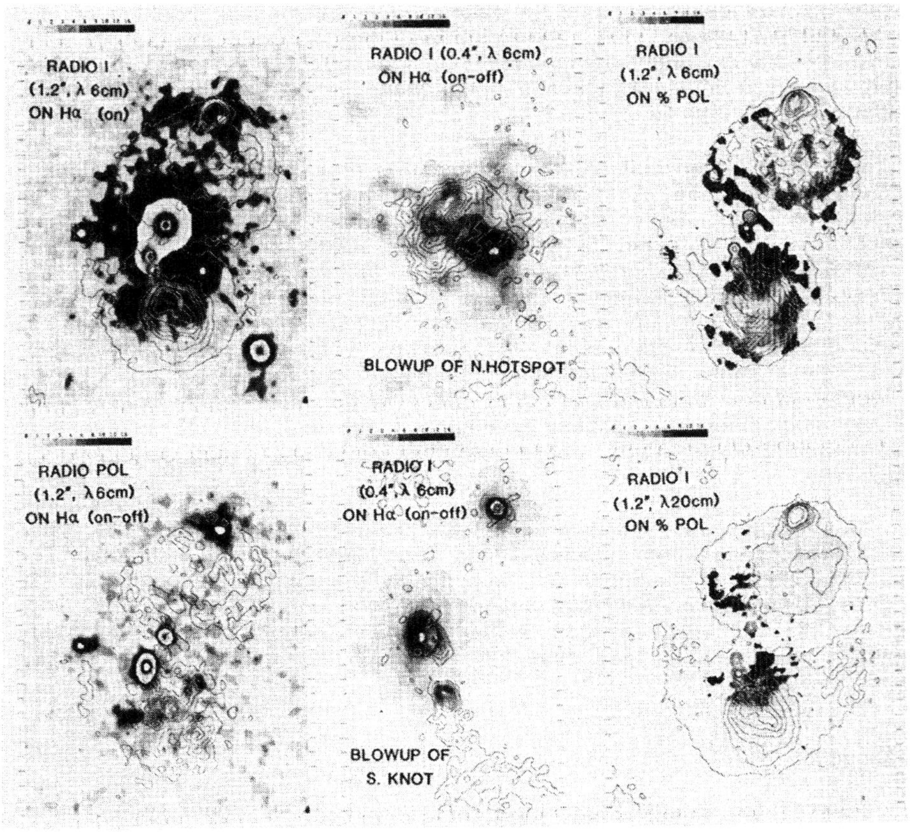

Figure 2: 3C277.3/Coma A. Radio continuum and optical emission line structures. Above left (Fig. 2a) is a 6 cm VLA radio intensity contour map superimposed on a gray-scale representation of the [Hα + NII + continuum] distribution. Below left (Fig. 2b) is the 6 cm polarized intensity contour map superimposed on the [Hα + NII] distribution. The middle two figures (Fig. 2c and d) show blowups of these radio continuum/optical line superpositions for the northern hot spot and southern knot regions. Right are the radio intensity contours superimposed on gray-scale representations of the fractional radio polarization at λ 6 cm (above/Fig. 2e) and λ 20 cm (below/Fig. 2f) (from van Breugel et al. 1981a).

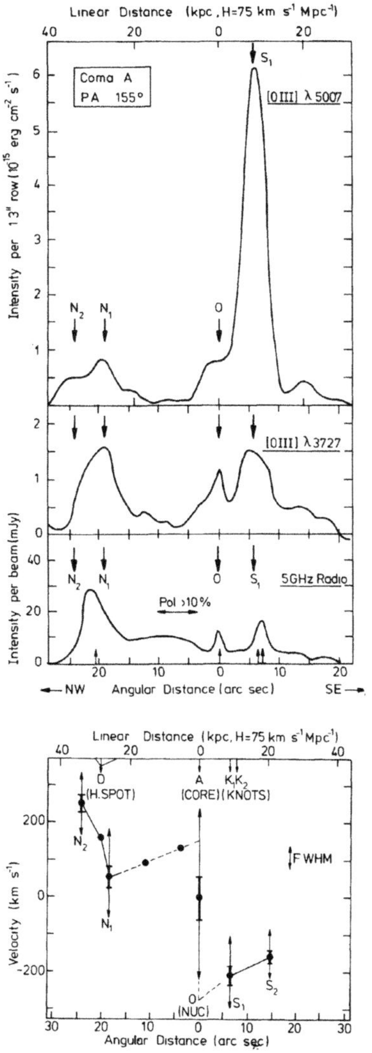

Figure 3: 3C277.3/Comma A. Spectroscopic data obtained along a slit aligned in PA 155° through the nucleus and the northern hot spot. Fig. 3(a) (above) shows intensity profiles of [OIII] λ5007, [OII] λ3727 and the λ 6 cm radio continuum. Fig. 3(b) (below) shows the velocity of the [OIII] emitting gas with respect to the nucleus (from Miley et al. 1981).

B. Depolarization

Where the radio and line emissions overlap, the radio polarization is almost invariably low (< 5% at $\lambda = 6$ cm). This suggest that the radio depolarization is at least partially caused by differential Faraday rotation through the line-emitting gas or an invisible tenuous gas associated with it. Indeed if one takes a density of $n_e \simeq 10^2\,cm^{-3}$ as derived from the [SII] lines, the filling factor of $\sim 10^{-5}$ deduced from Hβ luminosities, the minimum-energy value for the magnetic field in the synchrotron source of $H_{ME} \sim 10^{-5}$G and a path length of L $\sim$ 1 kpc, the plane of polarization would be Faraday-rotated through several radians at $\lambda = 6$ cm. Multifrequency polarization maps have been made with the VLA for most of the objects listed in Table 1. A comparison of these depolarization data with the optical results will in the near future provide more sophisticated information on the geometrical relationship between the relativistic and thermal components in the sources.

C. Luminosities - Energy Transport

The sources listed in Table 1 radiate much more energy in optical emission lines than in the radio by factors of up to 20. If the line emitting gas is powered by the jet, the kinetic energy in the jet must exceed the total radiated energy. With an efficiency of < 1% for conversion from kinetic to radiated energy this condition enables a lower limit of a few times 10^4 km s^{-1} to be placed on the speed of energy transport.

D. Line Widths-Particle Acceleration

A statistical correlation has been found between the whidth of the [OIII]λ5007 line and the integrated radio luminosity of the kiloparsec-scale emission from active galaxies (Wilson and Willis, 1980, Heckman et al., 1981). In addition, in a few cases within an individual source (e.g. 3C305) the broader lines are seen to coincide with the brightest radio regions. We have seen that there must be localized acceleration of relativistic particles in radio sources and this line-width/radio correlation argues that turbulent-induced particle acceleration processes are to be preferred.

E. Pressures

In most cases the synchrotron pressure (derived from the usual minimum energy conditions) roughly equals the thermal pressure exerted by the line-emitting gas. This may reflect a real pressure balance and would then be circumstantial evidence that the minimum energy condition is realistic. However this could well be merely an artifact of the combination of sensitivity and angular resolution attainable with the present generation of radio and optical instruments.

F. Velocities - Origin of the Gas

In all cases the observed velocities in the extranuclear emission line gas are within a few hundred km s^{-1} of the nucleus. Interpretation of the measured velocities is critically dependant on the origin of the gas.

One possibility is that the gas is moving outwards with the radio source. Then the gas velocities would directly trace the kinematics of radio source expansion. Note that this is not necessarily the velocity at which energy is supplied from the nucleus to the lobes (see 4.C.). On this basis the kinematic age of the Coma A radio source would be $\sim 10^8$y. The situation would in some ways resemble that which exists in some supernova remnants such as Tycho (Strom and Duin, 1973) on the edge of which gaseous filements are seen to move outwards with the same speed as the radio source (Strom et al. 1982). The gas would then be interstellar matter swept up and entrained by the radio source as it expands (De Young 1981) or matter ejected from the nucleus. In this case the line-width at the northern boundary, ΔV, might be interpreted as being the velocity of lateral expansion of the radio lobe, while the velocity difference of the northern boundary from the nucleus, δV, would be viewed as the longitudinal radial velocity with which the edge of the radio boundary is moving outwards from the nucleus. Note that $\Delta V \sim \delta V \sim 300$ km s^{-1}, consistent with the round appearance of the radio lobe. Another possible clue to the origin of the gas is provided by the strong Hα knot which lies to the southeast (Fig. 2) beyond the bend in the radio source, on the extrapolated "fundamental" radio axis of PA $\sim 155°$. This is circumstantial evidence in favour of a model in which the gas has been ejected from the nucleus with the radio jet but does not share its deflection. Models of radio sources in which relativistic and thermal components become separated have been considered previously (Cowie and McKee, 1975; Gisler and Miley, 1979), and the separation was attributed to buoyancy effects.

In an alternative picture the morphological line/radio association may be produced through compression of a hot ambient medium by a shock associated with the moving radio source. Instabilities in this tenuous medium would cool to form the line-emitting gas. No strong dynamic coupling would then be expected between the radio source and the gas. The velocities would then be much closer to those of the medium (interstellar or circumgalactic) through which the radio source is moving. Under reasonable assumptions (medium à la M87, ram pressure confinement of the knot) we can show that the cooling time could be comparable with the "crossing time", the time that the radio source would take to move the distance of the optical to radio peak ($V \sim 1000$ km s^{-1}). Although such a model is tenable, it would be difficult to explain simply the divergence of the velocity behaviour for the gas, shown in Fig. 3, from the rotation curve expected for an elliptical galaxy. Note for example the rapid increase in velocity just outside the northern boundary of the radio source.

5. THE PARSEC SCALE - SUPERLUMINAL EXPANSION

I will here inject a speculative note. Just as steep-spectrum radio cores on the kiloparsec scale are sometimes associated with narrow-line emission regions, I think it likely that broad-line regions may be intimately associated with the parsec-scale flat-spectrum radio cores. There are several pieces of circumstantial evidence for this. First, the flat-spectrum components have the same spatial scale as the broad line emission regions. Secondly, the distinction between the broad and narrow line regions of active nuclei is less clear cut than had been supposed previously. In a few cases (e.g. 3C305) the line/radio coincidence is closer and the [OIII] lines are broader for the regions immediately surrounding the nuclei. Thirdly, if turbulent-induced particle acceleration plays a role in producing the kiloparsec-scale radio emission, isn't it likely to be more important in the presumably turbulent broad-line regions? Fourthly, such a picture might provide a plausible "screen" alternative to ballistic explanations of superluminal expansion.

Look again at the blowup of the southern knot in Coma A. The picture provides evidence that (i) collision with the line-emitting gas excites the observed radio knots and (ii) the line mission undergoes morphological distortion due to the impact with the radio jet. Excitation of such a curved screen, formed say by a cloud in the broad line region compressed by the walls of the jet could result in an expanding (bent) superluminal source (Blandford et al. 1977, Morrison and Sartori, 1968, Lynden-Bell, 1977). Thermal absorption by such a cloud would also provide a natural explanation for observations of variability at low frequencies.

A schematic model for superluminal sources incorporating these ideas in the light-echo pitcture of Lynden-Bell (1977) is shown in Fig. 4. Broad line clouds are swept up by the jet as it moves outwards. Eventually they bend the jet as we have discussed for the knot in Coma A (Fig. 2). The ploughed-up gas forms a skin around the bend of the jet and this skin serves as the screen. A signal from the nucleus N excites a point P on the skin which emits radio emission. P moves out along the skin with a velocity c. In the one-sided case shown here the observed apparent velocity of NP is $V_a \sim c \sin\theta\ (1-\cos\theta)^{-1} \sim 6c$ for $\theta \sim 20°$, where θ is the angle between the initial jet directions and the line of sight. The apparent velocity would be 11c for $\theta \sim 10°$. Several variants of this simple model are possible. Both components of the expanding double might be produced by exciting the same gas cloud displaced from the nucleus. Also the screen could well be moving outwards from the nucleus with the jet.

Models where the superluminal motion is caused by screen effects in the broad-line region could also account for several other properties of compact radio sources. Thermal absorption by the screen would occur below a frequency $\nu_T(\text{MHz}) \sim \frac{1}{2} N_e(\text{cm}^{-3})\ \phi\ t^{\frac{1}{2}}$ (pc) where N_e is the electron density, ϕ is the filling factor and t is the screen

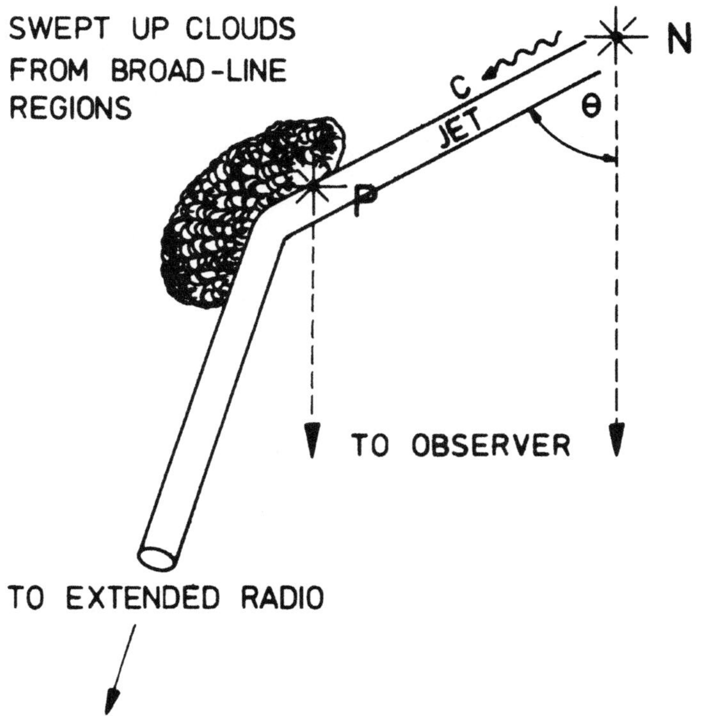

Figure 4: A schematic diagram showing how the situation shown in Fig. 2d when extrapolated to the parsec scale might result in superluminal expansion.

thickness. For values of N_e and ϕ close to those of the broad-line clouds screen, thicknesses of a few hundreds of a parsec could account for the observed variability at low frequencies. Also some of the spectroscopic differences between compact and extended radio quasars such as the presence of FeII emission (Miley and Miller, 1979) might well be attributed to modification of the conditions in the broad-line region by the outward-moving jet.

I contend that such models are at least as plausible as the "Doppler enhancement" ballistic models for superluminal sources. They have the added attraction of being the generally accepted explanation of another observed superluminal phenomenon - the expansion of novae (e.g. Lynden-Bell, 1977).

6. FINALLY

I have tried to convince you that extranuclear emission lines are unique tools for studying extended radio sources. We are at present conducting a survey of a flux-limited sample of extended radio sources for further examples of radio/line associations on scales of a few kiloparcsecs or greater. The hope is that as well as producing new laboratories for studying radio source physics the statistical results of this survey will provide information about the relation of the thermal plasma to the general phenomenon of nuclear activity. With ground based telescopes the line emission from the knots and hot spots of Coma A could be seen beyond a redshift of unity. The advent of the Space Telescope will enable morphological associations on the kiloparsec scale to be studied even for the most distant quasars (Miley, 1981). If such associations are found, the resultant parameters that could be determined would be of significant importance for observational cosmology.

ACKNOWLEDGEMENTS

As I stated at the outset the work descibed here is the collective effort of a large group. I am particularly grateful to Wil van Breugel, Tim Heckman and Colin Norman for recent discussions and to Harry van der Laan for commenting on the manuscript. The research was partially financed by NATO Research Grant 1828.

REFERENCES

Blandford, R.D., McKee, C.F., and Rees, M.J., 1977, Nature, 267, 211
Butcher, H., van Breugel, W.J.M., Miley, G.K., 1980 Ap. J., 235, 749
Cowie, L.L., McKee, C.F., 1975, Astron. Astrophys., 43, 337
De Young, D.S., 1981, Nature, 293, 43
Feigelson, A.E.D., Schreier, E.J., Delvaille, J.P., Giacconi, R., Grindlay, J.E., and Lightman, A.P., 1981, Ap. J., 251, 31

Gisler, G.R., Miley, G.K., 1979 Astron. Astrophys., 76, 109
Graham, J.A., Price, R.M., 1981, Ap. J., 247, 813
Hardee, P.E., Eilek, J.A., Owen, F.N., 1980, Ap. J., 242, 502
Heckman, T.M., Miley, G.K., van Breugel, W.J.M., Butcher, H.R., 1981, Ap. J., 247, 403
Heckman, T.M., Miley, G.K., Balick, B., van Breugel, W.J.M., and Butcher, H.R., 1982, Ap. J., in press
Lynden-Bell, D., 1977, Nature, 270, 396
Miley, G.K., 1980 Ann. Rev. Astron. Astrophys., 18, 165
Miley, G.K., Heckman, T.M., Butcher, H.R., van Breugel, W.J.M., 1981 Ap. J. (Lett), 247, L5
Miley, G.K., Miller, J.M., 1979, Ap. J. (Lett), 228, L55
Miley, G.K., 1981, Proc. ESO/ESA Workshop: Optical Jets in Galaxies, p. 9
Morrison, P., Sartori, L., 1968, Ap. J., 152, L139
Schreier, E., Feigelson, E., Delvaille, J., Giacconi, R., Grindlay, J., Schwartz, D., and Fabian, A., 1979 Ap. J. Lett., 234, L39
Strom, R.G. and Duin, R.M., 1973, Astron. Astrophys., 25, 351
Strom, R.G., Goss, W.M. and Shaver, P.A., 1982, Mon. Not. Roy. Astron. Soc., 200, 473
van Breugel, W.J.M., and Heckman, T.M., 1982, Proc. IAU Gen.Assembly, Patras
van Breugel, W.J.M., Butcher, H.R., Miley, G.K., Heckman, T.M., Bridle, A., and Fomalont, E., 1983, in preparation
van Breugel, W.J.M., Heckman,T.M., Butcher, H.R., Miley, G.K., 1983b,c,d, in preparation
Wilson, A.S., and Willis, A.G., 1980 Ap. J., 240, 429
Wilson, A.S., 1981, Proc. ESO/ESA Workshop: Optical Jets in Galaxies, p. 125

DISCUSSION

P. BIERMANN: The light-echo model to explain superluminal motion predicts most sources to be close to either one or two times the speed of light, contrary to observations. How do you reconcile this?

G. MILEY: I do not believe this to be serious problem so far. The statistics of superluminal expansion are very uncertain. We heard today that about 50% of the strongest few compact sources are observed to be superluminal. The simple light-echo model would predict that about 50% of compact sources would have apparent velocities in excess of 2.4 c. Much higher velocities would be observed for θs smaller than 30° and these higher velocities are of course easier to observe.
Even so I caution against placing too much weight on the statistics or on the simple light-echo model I have sketched. Several variants are possible in which due to their geometry, sources of preferred orientation (and hence velocity) would be selectively attenuated by the (absorbing) screen. Also the "screens" may well be moving outwards with the jet.
My main purpose was to draw attention to the fact that our data suggest

that an exciting screen exists. In view of the increased difficulty encountered by ballistic models I suggest that screen models should be considered more seriously.

M. BEGELMAN: In order to deflect the jets your gas clouds would have to intercept much of the jets cross-sectional area. Would you expect the resulting interaction to excite significant line emission from the clouds, in which case the appearance of a new superluminal "blob" might be associated with changes in the broad emission line spectrum of the quasar?

G. MILEY: Possibly. However, it is not clear what fraction of the total broad line emission would be affected. Also, there are a large range of temperatures and densities for possible gas screens whose heating would not result in broad lines.

ASTROPHYSICAL JETS:
OPTICAL MORPHOLOGIES OF RADIO JETS
AND THEIR PARENT GALAXIES

Jean-Luc Nieto
Observatoires du Pic-du-Midi et de Toulouse
Laboratoire associé au CNRS # 285
65200 Bagnères de Bigorre, FRANCE

ABSTRACT

We discuss the relevance of optical observations for a better understanding of the nature of cosmic jets and of the phenomena related to them. We insist in particular on the morphology of the optical jets known as yet, especially that of M87, from high-resolution observations. We also discuss the new insights that the detection of interstellar dust can bring on the jet phenomenon as related to the star and dust system of the parent galaxy.

INTRODUCTION

It is difficult to supersede the philosophy that a few major papers or review articles have recently brought on the matter of optical emission from radio jets, essentially in the ESO/ESA workshop of Feb. 1981 in Garching: "Optical Jets in Galaxies" that I will as much as possible refer to, and in Miley (1980). Many discussion in the 1981 workshop were oriented towards the Space Telescope (ST) and its 0.1 arcsec resolution necessary to solve some crucial questions related to a clear understanding of jets in general. Since, in hardly more than one and a half year, attempts to fill the gap between the best resolution that regular ground-based observations can reach and the resolution of the ST have started to be successful, bringing new elements on the possibility of ground-based observations. The aim of this paper is to present the relevance of optical observations for the comprehension of cosmic jets and all phenomena related to them. This will be discussed in Sec. A. I shall concentrate the rest of my talk on morphologies and broad-band optical results, where the different aspects mentioned in Sec. A will appear. Sec. B will concern the structure and the morphology of optical jets with a special section on the M87 jet. Sec. C will concern the morphology of the parent galaxy and its interstellar medium. All along, when possible, I shall indicate the resolution of the different observations to investigate the possibilities of high-resolution ground-based observations before the ST. I shall not

A. Ferrari and A. G. Pacholczyk (eds.), Astrophysical Jets, 113–130.
Copyright © 1983 by D. Reidel Publishing Company.

mention, except in a very few instances, results concerning emission lines in jets, since they are discussed in detail by Miley in this workshop. In the same fashion, I shall generally avoid infrared and polarisation considerations.

A. RELEVANCE OF OPTICAL OBSERVATIONS OF EXTRAGALACTIC COSMIC JETS

Assuming that the jet is made of a continuous fluid beam expelled from the nucleus (Rees, 1971, Blandford and Rees, 1974), its overall structure may be controlled by several processes (see Trussoni, this workshop):

i) Precession of the power source located in the nucleus of the galaxy (Miley, 1976, Rees, 1978a, van Breugel and Miley, 1977...)
ii) Kelvin-Helmholtz instabilities (helical mode) as the jet goes through the interstellar medium (Hardee, 1979, 1981, 1982a, b, Ray, 1982, Ferrari et al., 1978, 1981a, 1981b)
iii) Gravitational perturbations due to the presence of a nearby companion galaxy (Blandford and Icke, 1978, Wirth et al. 1982).

The inner structure of the source, namely the presence of gaps and knots, might be due to several causes, like shock waves created as the beam goes through the medium (Blandford and Königl 1979, Norman et al. 1982) variability of the source power in the nucleus (Rees, 1978b), Kelvin-Helmholtz instabilities (pinching mode), internal structure of the magnetic field...

Along with the important parameters determining the nature of the beam, like inner pressure, magnetic field, velocity, etc., the angle of sight is crucial for the nature of the ejection: are jets intrinsically one-sided (they would be issued from some "flip-flop" mechanisms (Willis et al. 1978)), or have they a two-sided nature hidden by beaming effects (see Rees 1978b, Blandford and Königl 1979, van Groningen et al. 1980)?

If our present knowledge on the physics of jets is essentially due to the constant improvement of radio detectors, it turns out that optical data bring some fundamental information to a better understanding of the problem, beside the obvious point that they represent a wavelength range in which information is always useful (to derive spectral indices, for instance). We shall review them briefly:

1. If we assume that the emission mechanism is synchrotron, as polarisation observations suggest, probably the most important characteristics of optical observations is their ability to act as an instant snapshot on the structure of the jet and then to determine with a high accuracy the location of the electron acceleration within the jet. In particular the detection of proper motions would be a step towards direct evidence of the velocity of the beam, since optical spectra fail in showing any feature, in the

most characteristics jets at least. SS433 is, to my knowledge the only jet where the velocity of the fluid has been determined (Mammano et al., 1980, Margon et al., 1980, Schilizzi this meeting).

2. The optical data probably bring the best compromise between time variation scale and resolution, since the resolution or the availability of shorter wavelength instruments are quite limited.

3. The interstellar medium is considered as playing a significant role in determing the morphologies and (probably) the observed properties of jets. The properties of ionised gas may bring some new constraints. Presence of dust brought some new insights.

4. Along with the dust which is, in its stationary state, a tracer of the stellar potential, data on the photometry and the dynamics concerning the stellar (and the gas) system reveal interesting features to jets, such as the role of the angle between the jet axis and the rotation axis, and the role of gravitational interaction between galaxies.

Other aspects of the relevance of optical observations that will not be discussed here are the following:

5. The photometry and the dynamics of the central regions of the parent galaxy are also informative of the degree of strangeness in the source (Young et al., 1978, 1979, Sargent et al., 1978, de Vaucouleurs and Nieto 1979a), in contrast with calm nuclei of galaxies whose best known representants are the closest ones, in the Local Group namely M31, M32 and M33 (Light et al., 1974, Lallemand et al., 1960, Walker, 1962, Nieto and Aurière, 1982). The study of these nuclei themselves, because of their nearness, gives us some ideas of what a nucleus might be expected to be (see Nieto et al., 1983).

6. Polarisation observations in particular bring the zero-point of the Faraday rotation (see Schmidt et al., 1978).

B. THE STRUCTURE AND MORPHOLOGY OF OPTICAL JETS

If the problem of accurate location of the electron acceleration in the jet has been quite a theoretical and academic one as long as the resolution of optical observations was seriously limited, this idea becomes much more realistic with the strong improvement in the resolution of regular ground-based observations, as well as the future launching of the Space Telescope. For this particular problem also the higher capabilities in the UV of optical telescopes bring (and will bring) a major contribution, since jets exhibit very strong blue and UV continua.

1. The morphology of the M87 jet.

The best candidate for such a purpose is, because of its nearness, the M87 jet. A first attempt to completely analyse in two dimensions the detailed structure was made by de Vaucouleurs and myself (1979b), after a previous one-dimensional study by de Vaucouleurs et al. (1968). The resolution of the material was 1.5 arcsec. Several steps were needed to complete an accurate analysis: a) first, the intensity ratio knot/continuous component adopted at different points of the jet (I call "knot" a compact region presenting a well-defined maximum of emission); b) second, once we made the assumption of a complete knot structure (which was checked afterwards), the disentanglement of several parameters, like the number of knots, their optical brightness their size etc...

Taking into account the Point-Spread-Function (PSF) of the observations, we were able to count 11 knots for which we gave magnitude, size, and relative position, with an accuracy believed to be 0.2 arcsec.

We found that the knots were resolved so that the jet has a diameter proportional to the distance to the central source, within the accuracy of the data, defining a cone having an opening angle of 3.°6 sin(ψ), where ψ is the inclination of the jet to the line of sight. These results are nevertheless relatively model-dependent, more particularly number-of-knot dependent. See for comparison Arp and Lorre (1976).

The publication of the VLA map (Owen et al., 1980) with a resolution of 0.6 and 1.0 arcsec at almost the same time as the availability of the Canada-France-Hawaii (CFH) telescope (and its UV corrector) were good reasons to improve on this previous study (Nieto et al., 1981, Nieto and Lelièvre, 1981, 1982). The resolution reached was about 0.5 arcsec, from the FWHM of the stellar profiles of our plates. Fig. 1 illustrates the results, as sketched on Fig. 2. If a simple visual inspection confirmed the existence of the three new knots, as found by taking into account the PSF in our 1979 study, the splits in two or more of knots that were believed to be single, calls into question the validity of the previous model. We also confirm the absence of a continuous component between the knots as stressed in our 1979 paper. Another major change brought by the high-resolution in the interpretation of the structure of the jet comes from the "wiggles" that were not detected before, when the jet was assumed to be straight. In addition, a periodicity of 2.9 arcsec appeared between ensembles of knots that we decided to call "complexes". Image processing of these plates are underway (Lorre and Nieto, 1983). Quite preliminary images were obtained with the maximum entropy procedure and the pixel compensation method on a composite of a few plates (Fig. 3). The similarity of our material with the VLA 2 cm map of Owen et al. (1980), the MERLIN maps (Charlesworth and Spencer, 1982) and the recent VLA 0.2 arcsec resolution map (Fomalont, this workshop) is striking. Note also

Figure 1: The M87 jet. Plate CFH 1031. U filter. 30 minute exposures with the CFH telescope (see Nieto and Lelièvre 1982).

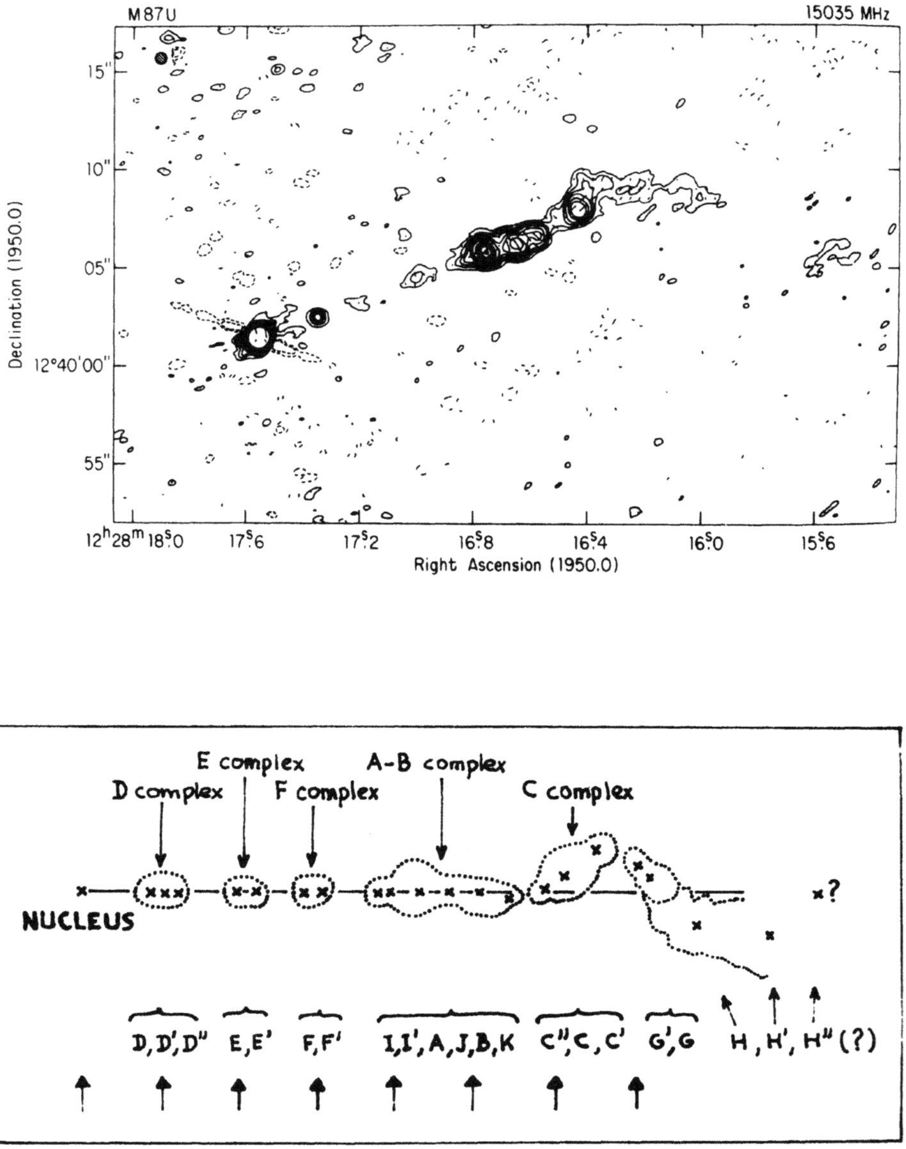

Figure 2: The M87 jet. Sketch of the jet as seen from the high-resolution material by Nieto and Lelièvre (1982) compared with the VLA map at 2 cm (Owen et al. 1980). The arrows show the 2.9 arcsec period.

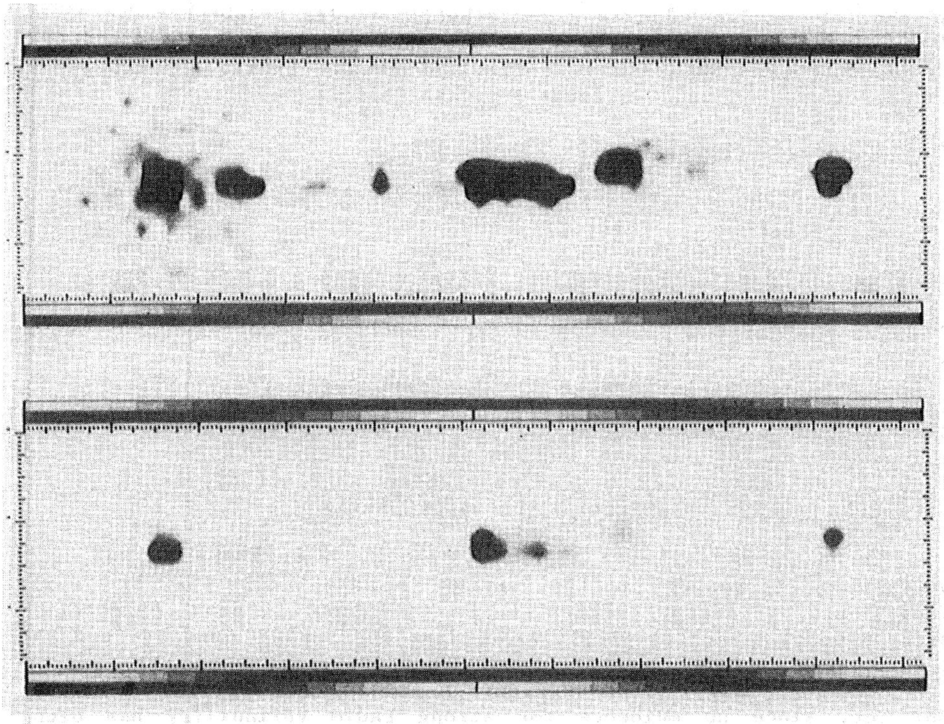

Figure 3: The M87 jet. Preliminary images restored with the pixel compensation method by Lorre and Nieto (1983), from a composite of several blue and UV CFH plates. Reference star is on the right.

that the X-ray image of the M87 jet (Schreier, 1981) presents a morphology quite consistent with both the radio and the optical data (see Feigelson, this workshop).

A surprising feature here should be mentioned: although there are some arguments for and against the optical nucleus, if not being a part of the jet, at least having a contribution to its brightness due to a "jet-like" feature in it (Young et al. 1978, de Vaucouleurs and Nieto 1979a, Tarenghi 1981), the fact that the nucleus fits quite well with the 2.9 arcsec periodicity frame supports such an assumption.

2. Other optical properties of the M87 jet.

Most of the optical properties of the M87 jet are summarized in Tarenghi (1981). So, I shall not repeat them here. I want to insist on the problem the variability of the knots, since it is of the highest importance at optical wavelengths. Seeing effects are a major source of error, as stressed by Sulentic et al. (1979), who gave some evidence for change in brightness in knot A, by comparison with Baade's plates (1956) taken in similar seeing conditions. For this reason, a simple visual inspection to compare our CFH plates to the plates used in the 1979 study can be misleading. Note that, in principle, a change in brightness in a complex region made of several entangled knots does not mean necessarily brightness variations only, but might mean also (or instead) the motion of a knot relative to another. If so, detection of brightness variability might not be quite conclusive.

If we assume that the nucleus of M87 is a part of the jet, the variability suggested in the nucleus between 1964 and 1977 (de Vaucouleurs and Nieto, 1979a), can be considered in good qualitative agreement with Sulentic et al.'s results, but such a result should be considered with care. If this is confirmed, it would be the first real evidence for a difference between radio and optical observations in the M87 jet, since no evidence for strong radio variability has been observed in the nucleus during that period (Andrew et al. 1978, Cotton et al., 1981).

3. Theoretical and numerical investigations on the shape of the M87 jet.

We have mentioned in Sec. A the different processes controlling the overall and the inner structures of the jets. Among those, the effects of perturbations leading to Kelvin-Helmholtz instabilities have been studied in detail. In particular, Hardee (1982a, b) has performed an analysis for high Mach number jet flows and investigated the influence of jet expansion on the growth rate of the different wave modes. The pinching mode and the helicoidal mode were studied in detail. In the case of M87, he finds a pinching wavelength of 3.2 arcsec, in excellent agreement with our 2.9 arcsec measurement. Norman et al., 1982 (see also Norman, this workshop) have analysed the Blandford and Rees (1974) model and conducted numerical experiments to study the structure and the stability of the jet component for various Mach numbers and temperature ratios with respect to a constant intergalactic medium. They find that perturbations produce internal shock waves in the beam with a rather regular spacing of 1-2 jet diameters quite independent of the beam Mach number, such as a Kelvin-Helmholtz beam. This prediction is in quite good agreement with our estimates of the dimension of the knots (1979 paper) and of the periodicity.

4. The other optical jets.

4.1) Another jet with non thermal emission: 3C 273.

The optical jet in 3C 273 was discovered 45 years after that of M87 from inspection of the Sky Survey plates (Hazard et al. 1963). Like that of M87, to which it is always compared (although their sizes differ by a factor of 30), it presents a feature-less spectrum (Greenstein and Schmidt, 1964, Schmidt et al. 1978). Note that it is the only optical jet attached to a QSO, and that, amongst the optical jets, it is the furthest one. Because of its faintness (its integrated magnitude is 20.0 according to Kron et al., 1973, see also Tyson et al., 1982) and the proximity of the bright QSO image, the jet has been little investigated at optical wavelengths. Kron et al. (1973) made integrated measurements in a slot 5.6 arcsec long, but no morphological considerations have been discussed in detail. The optical jet appears to have an approximately constant surface brightness, except at the end nearest the QSO, where there is a slight peak of emission, probably a "knot". The optical structure then does not seem to agree with the radio data (see Conway et al., 1981, Davis, this workshop, Sol, this workshop), but the surface brightness is quite faint and the optical jet resolution limited. Note that the gap of 12 arcsec observed between the QSO and the beginning of the jet is quite enormous at such a distance. Also, the jet seems to be hardly polarised in the optical, unlike M87 (see Schmidt et al., 1978 and Arp, 1981 for discussions).

4.2) Jets with thermal emission and some others.

Although there is some thermal emission intimately linked to jets, but located in general in their periphery (Ford and Butcher, 1979, see also Miley, this workshop), the principal property of the M87 and 3C 273 jets is to have a pure synchrotron emission. Different cases of optical jets have been reported in the literature, whose nature is still quite uncertain, and which might not be due to the same phenomena as those we are dealing with here. These jets present some significant amount of thermal emission like those of DA240 or NGC5128, or have uncommon characteristics like that of NGC1097 (see Wolstencroft 1981). As mentioned by Miley (1980), Ghigo (1978) has presented a list of objects candidates for having a jet-like structure, from direct examination of the Palomar Sky survey plates. Wehinger and Wyckoff (1978) also have detected a visible extension in PKS 0837-012. I just wonder whether some asymmetries seen in the resolved nebulosities around QSOs at small redshift could not be due to a jet-like phenomenon. I note in passing the suggestion of a jet structure in NGC

3310, that VLA observations fail to show (Bertola, 1982, and this workshop) and another one in a galaxy near the center of the Coma Cluster (Thompson, 1982). Note that, also gravitational interactions between galaxies may produce some jet-like features, which would consist in stars or/and possibly thermal gas.

Both NGC5128 and DA240 should for now be taken into account since they present a direct relationship with the large scale radio morphology, although they seem to contain a large amount of thermal emission with no relativistic motions.

The history of the jet in the peculiar dust-lane galaxy NGC5128 started quite recently (Blanco et al. 1975, see Danziger, 1981, and Cannon, 1981). The optical jet is made of gaseous filaments and of resolved blue emission regions, which morphologically do not correspond to our main idea on jets. Anyway, this jet was the first one to be discovered in X-rays (Schreier et al., 1979, Feigelson et al., 1982, and this workshop). Radio emission was difficult to detect because of the very southern location of the galaxy (Schreier et al., 1981). Nevertheless, we note that the optical jet features do not correspond to the radio or X-ray knots, until recent work brought new insights (see Brodie, this workshop). The radio galaxy DA240 presents an optical jet almost visible on the Palomar Sky Survey. The features are quite similar to those of NGC5128, although and not for the same reason, no radio component has been detected. See review in Strom and Willis (1981).

To this category of objects, we should also add the galaxy NGC 7385 where just a blue knot was found by Simkin and Ekers (1979) to coincide perfectly with a peak of radio emission inside the jet-like structure of the radio source. They show evidence for some similarity with the knots in the M87 jet. This knot however has an emission-line spectrum.

I do not know what to say about the galaxy 3C120, which exhibits several emission-line objects such as those described just above (Baldwin et al., 1980). Some of them might just represent the NW extension noticed by Arp (1981) and Wlérick et al. (1981), who report it as a continuum jet (see also Wlérick (1981). Another jet-like feature has been detected in the BL Lac object PKS 0521-06 (Danziger et al., 1979) and confirmed by Wlérick (1981). The nature of this jet is still unknown: Danziger et al. suggest the possibility of a chance superposition with a background galaxy (see Danziger, this workshop).

Because jets are in general quite faint against a strong background, the knowledge of their morphology requires a certain minimum of image treatment, like the removal of the galaxy image. Otherwise, little can be said about them. It is clear also that the morphology is more apparent in the UV or in the B band, since optical jets are blue, and the galactic background (rather faint at these wavelengths) much more easily removable.

4.3) A systematic search for optical jets.

The technique of image treatment has been used in a systematic search for optical counterparts of radio jets more likely to resemble those of M87 and of 3C273 (Butcher et al., 1980, Miley et al., 1981, see summary in Butcher, 1981). The resolution of the observations (made in the B band) was about 1.5 arcsec. 33 nearby objects were considered (M87 and 3C273 were not included). Three jets were detected, two have proven to be too difficult, two were not seen with certainty. The main results are that:

i) All jets seem to have a continuous nonthermal spectrum with a radio-optical index of 0.7 ± 0.1.

ii) Most of the jets, except that of M87 and one knot in Coma A (see below), are too faint to be studied with regular ground-based observations:
In 3C66B (Northover, 1973), even if the jet is quite apparent since it extends to 16 arcsec, it is very faint (B=21.5). See also a confirmation in Arp (1981). The case of 3C31 is worse since it is fainter (B=22.0), smaller (5 arcsec), and not immediately apparent because of the presence of a ring of dust in the galaxy image which makes the galaxy removal difficult. The case of Coma A (=3C277.3) deserves some special attention: Miley et al. (1981) and Bridle et al. (1981) have respectively obtained optical and radio (VLA) data for this object. The optical data clearly exhibit a jet for which continuum-, line-, and polarisation results are presented, showing quite a complex alignment of continuum sources and of line features with the nucleus. The positions of the blue continuum sources agree quite well with those of the radio knots. They are strongly polarised. One of them, called K1, seems to be slightly resolved, "at 1 arcsec level", and relatively bright (B=20.6). It is estimated to be a good candidate for ST observations. It also strongly emits in optical lines, with a higher degree of ionisation than anywhere else in the source. A second one is much weaker and seems to be just a continuum feature. The emission-line sources are very similar to those found in NGC5128, 3C120, NGC7385, or in the vicinity of the M87 jet. I refer to Miley (this workshop) for such an object (as well as for some new ones), which seems intermediate between the pure synchrotron jets (with possibly thermal matter beside), and the pure thermal emission of the extranuclear line sources just mentioned above.

4.4) Some conclusions.

i) Table 1 summarizes some comments on the different optical jets and their nature.

ii) The existence of 5 optical jets of probably the same nature (pure(?) synchrotron emission) is well-established. Some other suggestions of possible jets might correspond to different physical phenomena.

TABLE 1: OPTICAL JETS

Pure continuum Jets	Intermediate Jets	Pure emission-line Jets	Others
M87 3C 273 3C 66B 3C 31	3C 277.3 (1)	NGC 5128 (2) DA 240 NGC 7385 3C 120 (?)	NGC 1097 NGC 3310 PKS 0837-012 PKS 0521-36
- Featureless continuum - Possible emission line regions in the vicinity	- Superposition of continuum emission and line emission	- Strong emission line spectrum. No evidence for continuum (?)	- Lack of information
- One-sided	- Synchrotron emission one-sided		

(1) See Miley, this workshop

(2) See Brodie, this workshop, for new results

iii) In all these jets, the optical structure and the radio structure appear quite similar. This is quite well-established for M87 at 0.2 arcsec resolution, but needs to be confirmed with other jets. It is reasonable to assume a knot structure for all of them. In particular, all have the same optical to radio index, with a small dispersion.

iv) All these optical jets, at least the synchrotron part of them, are one-sided. No counter-jet was detected in 3C273 (Tyson et al., 1982, Arp, 1981, Conway et al., 1981). In M87, a counter-jet has been detected (Arp, 1967), but it is not of the same nature as the jet itself. Nieto and Lelièvre (1982) estimated the non-detection in agreement with a counter-jet having the same radio to optical index as that of the jet.

4.5) Some questions (among others).

i) As the optical synchrotron characteristic time is much shorter than the radio one, what does the surprising similarity of the radio and optical maps (at least in the case of M87) mean?

ii) Optical jets are detected in the brightest parts of strongly asymmetrical sources. Is there a plausible explanation in terms of "beaming" effects?

iii) What is the link between the two kinds of optical jets, featureless jets and pure emission-line jets?

C. INTERSTELLAR MEDIUM AND MORPHOLOGY OF THE PARENT GALAXY

A second high interest of optical observations for the physics of jets is related to the morphology of the parent galaxy and especially to its interstellar medium.

1. The role of the interstellar medium.

This role has been recently stressed as of major influence on the morphology and other properties of jets. See Sec. A for references. More precisely, the detection of ionised gas (see Miley, this workshop) and of interstellar dust might be very informative about the presence and the influence of interstellar medium.

Bertola and Galletta (1978), Kotanyi and Ekers (1979) have suggested a correlation between the presence of the dust in the elliptical galaxy and the radio emission. Butcher et al. (1980) also report on the detection of dust in some of the galaxies that they have investigated. Although dust in elliptical galaxies is probably not such a rare phenomenon, since its detection is biased towards objects very inclined on the line of sight, we also have detected a thin dust lane in NGC6251 (Nieto et al., 1983), which is very likely to be responsible for the bend of the jet (see Perley, this workshop). As the structure of the dust lane is very reminiscent of that of NGC4374 (Capaccioli and Nieto, 1983) and that of NGC5128, which is suggested to be a merger although recent observations (Marcellin et al., 1982) do not support this hypothesis, it is tempting to assume (see Rees, 1978a) that the merger hypothesis could count for the fuelling of the active nucleus emitting the jet and that the dust lane is a remnant of the merging. Gravitational interactions are suggested to explain complex shapes like that of 3C31 (Blandford and Icke 1978) or that of 3C449 (Bystedt and Högbom 1979). See also Wirth et al. (1982). In such a frame of work, it is not surprising to find dust and interstellar medium in 3C31. The merging hypothesis in the case of NGC6251 is reinforced by the complex shape of the very outer lobes. Since the jet presents a remarkable linearity on a scale $1:10^5$, we may think that the dust lane has reached a stable configuration in the equatorial plane defined by the stellar potential that it should have reached in two or three rotation periods (see Rees 1978a). Then the merging must have happened a few times 10^8 years ago, which, along with the $10^6 x (v/c)^{-1}$ years since the birth of the flow, means a slow non relativistic motion for the jet or a relativistic one only if the jet is very close to the line of sight.

2. "Angles and Things".

Kotanyi and Ekers (1979) have reported that in seven clear cases amongst the eight radio galaxies showing a dust lane, the dust lane is nearly perpendicular to the axis of the radio source (see also Penston and Fosbury, 1978). This point strongly suggests, according to the authors, that the radio axis is related to the rotation axis of the gas and the dust, an hypothesis supported by Simkin's (1977, 1979) results on the dynamics of five radio galaxies. Later, Kotanyi (1981) has found a distribution of the angles separating these two axes more spread out and has suggested a correlation between the power of the source (at 1400 MHz) and the alignment of both axes. Such a suggestion implies that this alignment favors the development of giant radio sources. NGC 6251, which is a powerful radio source at 1400 MHz shows a dust lane quite parallel to the radio axis, and except if the dust is out of equilibrium or is located in a plane parallel to its rotation axis, does not favor such suggestions.

Kotanyi and Ekers' results suggest, with an assumption of a polar ejection from the nucleus, a nucleus oriented perpendicularly to the rotation axis of the dust and the gas. Using also Butcher et al.'s result (1980) on 3C31 where the jet axis presents an angle of 35° with the dust lane, we are tempted to assume no preferred orientation of the nucleus relative to the rotation axis of the gas and the dust (for discussion, see Nieto et al., 1983). Truly, a large sample of objects is necessary here to check in greater details these considerations. Instead of looking very hard for dust lanes in radio galaxies, it may be more successful to map with the VLA the elliptical galaxies which exhibit a dust lane and search for a jet-like structure. Such a project is underway with the dust-lane elliptical galaxies of the southern hemisphere found by Hawarden et al. (1981). Here may lie a possible answer. In fact, we may find many more dust lanes than those known today with some good criteria and a reasonably fair resolution. For instance, the southern galaxy NGC1199 exhibits a dust lane (Nieto, 1983) not detected by Hawarden et al.. Mianes (1982) has given a long list of ellipticals presenting a large V-R excess, and a program is now starting to cover the whole northern sky with a high resolution in search for dust lanes.

ACKNOWLEDGMENTS

This paper was prepared while I was a guest at the Osservatorio Astronomico in Padova with a grant from both CNRS and CNR. I express my thanks to Prof. L. Rosino for providing me working facilities and M. Calvani for several discussions during the preparation of this paper.

REFERENCES

Andrew, B.H., Mac Leod, J.H., Harvey, G.A., Medd, W.J., 1978, Astron. J., 83, 863
Arp, H.C., 1967, Astrophys. Letters, 1, 1
Arp, H.C., 1975, Pub. Astron. Soc. Pac., 87, 545
Arp, H.C., 1981, ESO/ESA workshop on Optical Jets in Galaxies, 53
Arp, H.C., Lorre, J.J., 1976, Astrophys. J., 210, 58
Baade, W., 1956, Astrophys. J., 123, 550
Baldwin, J.A., Carswell, R.F., Wampler, E.J., Smith, H.E., Burbidge, E.M., Boksenberg, A., 1980, Astrophys. J., 236, 388
Bertola, F., 1982, private communication
Bertola, F., Galletta, G., 1978, Astrophys. J., 226, L115
Blanco, V.M., Graham, J.A., Lasker, B.M., Osmer, P.S., 1975, Astrophys. J., 198, L63
Blandford, R.D., Icke, V., 1978, Month.Not.Roy.Astr.Soc., 185, 527
Blandford, R.D., Königl, A., 1979, Astrophys. Lett., 20, 15
Blandford, R.D., Rees, M.J., 1974, Month. Not. Roy. Astr. Soc., 169, 395
Bridle, A.H., Fomalont, E.B., Palimaka, J.J., Willis, A.G., 1981, Astrophys. J., 248, 499
Browne, I.W.A., Clark, R.R., Moore, P.K., Muxlow, T.W.B., Wilkinson, P.A., Cohen, M.H., Porcas, R.W., 1982, Nature, to be published
Butcher, H.R., 1981, ESO/ESA workshop Optical Jets in Galaxies, 23
Butcher, H.R., van Breugel, W., Miley, G.K., 1980, Astrophys. J., 235, 749
Bystedt, J., Högbom, J.A., 1979, unpublished, quoted by Miley (1980)
Cannon, R.D., 1981, ESO/ESA workshop on Optical Jets in Galaxies, 45
Capaccioli, M., Nieto, J.-L., 1983, in preparation
Charlesworth, M., Spencer, R.E., 1982, Month. Not. Roy. Astr. Soc. 200, 933
Conway, R.G., Davis, R.J., Foley, A.R., Ray, T.P., 1981, Nature, 294, 540
Cotton, W.D., Shapiro, I.I., Wittels, J.J., 1981, Astrophys. J., 244, L57

Danziger, I.J., Fosbury, R.A.E., Goss, W.M., Ekers, R.D., 1979, Month. Not. Roy. Astron. Soc., 188, 415
Danziger, I.J., 1981, ESO/ESA workshop on Optical Jets in Galaxies, 143
de Vaucouleurs, G., Angione, R., Fraser, C.W., Astrophys. Letters, 2, 141
de Vaucouleurs, G., Nieto, J.-L., 1979a, Astrophys. J., 230, 697
de Vaucouleurs, G., Nieto, J.-L., 1979b, Astrophys. J., 231, 364
Feigelson, E.D., Schreier, E.J., Delvaille, J.P., Giacconi, R., Grindlay, J.E., Lightman, A.P., 1982, Astrophys. J., 251, 31
Ferrari, A., Trussoni, E., Zaninetti, L., 1978, Astron. Astrophys., 64, 43
Ferrari, A., Trussoni, E., Zaninetti, L., 1981a, ESO/ESA workshop on Optical Jets in Galaxies, 119
Ferrari, A., Trussoni, E., Zaninetti, L., 1981b, Month. Not. Roy. Astr. Soc., 196, 1051
Ford, H.C., Butcher, H., 1979, Astrophys. J. Suppl. 41, 147
Ghigo, F.D., 1978, Astron. J., 83, 1363
Greenstein, J.L., Schmidt, M., 1964, Astrophys. J., 140, 1
Hardee, P.E., 1979, Astrophys. J., 234, 47
Hardee, P.E., 1981, Astrophys. J., 250, L9
Hardee, P.E., 1982a, Astrophys., 257, 509
Hardee, P.E., 1982b, preprint
Hawarden, T.G., Elson, R.A.W., Longmore, A.J., Tritton, S.B., Corwin, H.G., 1981, Month. Not. Roy. Astr. Soc., 196, 747
Hazard, C., Mackay, C.B., Shimmins, A.J., 1963, Nature, 197, 1037
Kotanyi, C.G., Ekers, R.D., 1979, Astron. Astrophys., 73, L1
Kotanyi, C.G., 1981, Thesis, University of Groningen
Kron, G.E., Ables, H.D., Hewitt, A.V., 1973, Pub. Astron. Soc. Pac., 84, 303
Lallemand, A., Duchesne, M., Walker, M.F., 1960, Publ. Astr. Soc. Pac., 72, 76
Light, E.S., Danielson, R.E., Schwarzschild, M., 1974, Astrophys., 194, 257
Lorre, J.J., Nieto, J.-L., 1983, in preparation
Mammano, A., Ciatti, F., Vittone, A., 1980, Astron. Astrophys., 85, 14
Marcellin, M., Boulesteix, J., Courtes, G., Milliard, B., 1982, Nature, 297, 38
Margon, B., Grandi, S.A., Downes, R.A., 1980, Astrophys. J., 241, 306
Mianes, P., 1982, private communication
Miley, G.K., 1976, Proc. NATO Summer School: Physics of non thermal radio sources, 1, Dordrecht: Reidel
Miley, G.K., 1980, Ann. Rev. Astron. Astrophys., 18, 165
Miley, G.K., Heckman, T.M., Butcher, H.R., van Breugel, W.J.M., 1981, Astrophys. J., 247, L5
Nieto, J.-L., 1983, in preparation
Nieto, J.-L., Aurière, M., 1983, Astron. Astrophys., 108, 334
Nieto, J.-L., Coupinot G., Lelièvre, G., Madsen, C., 1983, Mon.Not.Roy. Astron.Soc.
Nieto, J.-L., Laques, P., Vidal, J.-L., Augé, A., Despiau R., 1981 ESO/ESA workshop on Optical Jets in Galaxies, 43

Nieto, J.-L., Lelièvre, 1981, Astronomical Photography 1981, Meeting of the IAU working group on Photographic Problems, J.-L. Heudier and M.E.Sim Ed., 189
Nieto, J.-L., Lelièvre, 1982, Astron. Astrophys., 109, 95
Norman, M.L., Smarr, L., Winkler, K.H.A., Smith M.D., 1982, Astron. Astrophys., 113, 285
Northover, K.J.E., 1973, Month. Not. Roy. Astr. Soc., 165, 369
Owen, F.N., Hardee, P.E., Bignell, R.E., 1980, Astrophys. J., 239, L11
Penston, M.W., Fosbury, R.A., 1978, Month. Not. Roy. Astr. Soc., 183, 479
Ray, T.P., 1982, Month. Not. Roy. Astr. Soc., 196, 195
Rees, M.J., 1971, Nature, 229, 312
Rees, M.J., 1978a, Nature, 275, 516
Rees, M.J., 1978b, Month. Not. Roy. Astron. Soc., 184, 61P
Sargent, W.L.W., Young, P.J., Boksenberg, A., Shortridge, K., Lynds, C.R., Hartwick, F.D.A., 1978, Astrophys. J., 221, 731
Schmidt, G.D., Peterson, B.M., Beaver, E.A., 1978, Astrophys. J., 220, L31
Schreier, E.J., 1981, ESO/ESA workshop on Optical Jets in Galaxies, 107
Schreier, E.J., Feigelson, E., Delvaille, J., Giacconi, R., Grindlay, J., Schwartz, D.A., Fabian, A.C., 1979, Astrophys. J., 234, L39
Schreier, E.J., Burns, J.O., Feigelson, E.D., 1981, Astrophys. J., 281, 529
Simkin, S., 1977, Astrophys. J., 217, 45
Simkin, S., 1979, Astrophys. J., 234, 56
Simkin, S., Ekers, R.D., 1979, Astron. J., 84, 56
Strom, R.G., Willis, A.G., 1981, ESO/ESA workshop on Optical Jets in Galaxies, 83
Sulentic, J.W., Arp, H.C., Lorre, J.J., 1979, Astrophys. J., 233, 44
Tarenghi, M., 1981, ESO/ESA workshop on Optical Jets in Galaxies, 145
Thompson, L.A., 1982, private communication
Tyson, J.A., Baum, W.A., Kreidl, 1982, Astrophys. J., 257, L1
van Albada, T.S., Kotanyi, C.G., Schwarzschild, M., 1982, Month. Not. Roy. Astr. Soc., 198, 303
van Breugel, W., 1981, ESO/ESA workshop on Optical Jets in Galaxies, 63
van Breugel, W., Miley, G.K., 1977, Nature, 265, 315
van Groningen, E., Miley, G.K., Norman, C.A., 1980, Astron. Astrophys., 90 L7
Walker, M.F., 1962, Astrophys. J., 136, 695
Wehinger, P.A., Wyckoff, S., 1978, Month. Not. Roy. Astr. Soc., 184, 335
Wilkinson, P.A., Cohen, M.H., Porcas, R.W., 1982, Nature, to be published
Willis, A.G., Wilson, A.S., Strom, R.G., Astron. Astrophys., 1978, 66, L1
Wirth, A., Smarr, L., Gallagher, J.S., 1982, Astron. J., 87, 602
Wlérick, G., 1981, ESO/ESA workshop on Optical Jets in Galaxies, 29
Wlérick, G., Bouchet, P., Cayatte, V., Michet, D., 1981, Astron. Astrophys., 102, L11
Wolstencroft, R.D., 1981, ESO/ESA workshop on Optical Jets in Galaxies, 49

Young, P.J., Westphal, J.A., Kristian, J., Wilson, C.P., Landauer, F.P., 1978, Astrophys. J., 221, 721

Young, P.J., Sargent, W.L.W., Kristian, J., Westphal, J.A., 1979, Astrophys. J., 234, 76

DISCUSSION

G. BENFORD: The superposition of optical and radio knots implies tight confinement (if they were unconfined shocks, why would these be so many?). Does this suggest the periodicity is connected with a magnetic structure?

J.L. NIETO: We have measured this periodicity without any theoretical consideration in mind. According to recent theoretical investigations (see the references in the paper), a confined jet (having either a constant radius or a small opening angle) and submitted to instabilities of the Kelvin-Helmholtz type, presents a somewhat periodic structure (pinching modes). In these investigations, the jet is confined by the thermal pressure.

J.O. BURNS: Could you briefly describe your image-processing technique used to produce the extraordinary optical images of M87?

J.L. NIETO: Two image-processing techniques were used to restore the images:

(i) the Maximum Entropy Restoration which is a very "conservative" technique, i.e. in a noisy region the non-physical restorations are left relatively smooth;

(ii) the pixel Compensation Method which:
   - compares the blurred image with a blurred image of the latest estimate of the true image,
   - seeks the pixel which differs the most,
   - replaces that pixel in the new estimate image, so that, when the new estimate is blurred, that pixel agrees very well with the original image, etc... until no substantial change occurs.

This last method, according to J.J. Lorre, is the most powerful technique, provided it is applied to images not heavily oversampled, since it completely analyses blurred images plus noise. It is the method used in Fig. 3.

PKS 0521-36, A BL LAC OBJECT WITH AN OPTICAL AND RADIO JET

J. Danziger[1], R.D. Ekers[2], R.A.E. Fosbury[3], W.M. Goss[4], P.A. Shaver[1]

[1]European Southern Observatory, Garching, W. Germany
[2]N.R.A.O., Socorro, New Mexico, U.S.A.
[3]Royal Greenwich Observatory, Herstmonceux, England
[4]Kapteyn Astronomical Institute, Groningen, The Netherlands

The radio galaxy PKS 0521-36 contains a variety of physical features that makes it an object worth studying in some detail. At the present time it is not clear how all of these various features relate to one another and whether they can be unified. It is worthwhile listing them all in the hope that this might provide insight into the evolution of such an object.

A description of this object has been given by Danziger et al. (1979), but more recent observations have added substantially to that picture. The parent galaxy appears to be a giant elliptical galaxy in which there is no evidence for recent star formation. The morphology, broad-band colours, and absorption line indices all attest to this. Although in the neighbourhood there are fainter galaxies whose redshifts have not been measured, it does not appear to be a member of a rich cluster of galaxies.

We know that a blue variable BL Lac object exists at the centre of the galaxy, and its luminosity variations (~1 mag) affect the integrated colours of the galaxy. The optical spectral index of this BL Lac object at maximum light has been reported to be of order -1, while at minimum light over the broader wavelength range UV to IR the spectral index is -1.5 (Danziger et al. 1983a).

There is low density ionized gas giving rise to narrow emission lines concentrated towards the centre of the galaxy. In addition, there is ionized gas of low excitation distributed towards the outskirts of the main body of the galaxy (Fosbury 1982), especially towards the south-east quadrant at a distance $\gtrsim$16 kpc. The detailed distribution of this gas is still being untangled, but it is already clear that it is not symmetrically disposed about the centre of the galaxy. Systematic motion of this gas has not been established, but there is a spread of several hundred km/sec in filamentary structures having sizes as large as 10 kpc.

A. Ferrari and A. G. Pacholczyk (eds.), Astrophysical Jets, 131–133.
Copyright © 1983 by D. Reidel Publishing Company.

A broad Balmer line region exists near the centre of the galaxy and has been discussed by Ulrich (1982) and Danziger et al. (1983a,b). This broad-line region was apparent in the spectra published by Danziger et al. (1979), but became more evident when the BL Lac object faded, making the emission-line equivalent widths larger. Whether there has been a change with time in the profile or absolute intensity of the broad Balmer lines remains to be established. Lyman α has been measured on IUE spectra (Danziger et al., 1983a) taken when the BL Lac object was probably at a reduced luminosity. It has a width (FWOI ~8000 km/sec) compatible with the measurements of the Balmer lines. The interstellar reddening is small, and thus the amount of dust inside or in front of both narrow and broad line regions is low.

The optical jet (Danziger et al. 1979) extending ~8 arcsec to the north-west is probably emitting non-thermal continuum radiation. Emission lines are not evident from slit spectroscopy and narrow-band direct imaging at emission-line wavelengths. There is a pronounced variation of the surface brightness along the jet, in a manner reminiscent of the jet in M 87.

A recent VLA continuum radio map at 21cm shows that the radio structure is complicated and consists of several components. The milliarcsecond flat spectrum source at the centre has been known for some time. In addition, there appears to be an elongated asymmetric structure aligned SE-NW with an average spectral index $\alpha \sim -0.5$. What is particularly interesting about this radio structure is that it is so asymmetrical with respect to the flux carried in each component. The sharp edge of the resolved and predominant SE lobe coincides with a curved filament of the outlying ionized gas discussed above, which suggests an association between the two. A similar association has been proposed for Coma A by van Breugel (1981) and Miley (this conference). To the north-west, the radio contours, which carry much less flux than those of the south-eastern component, appear to follow very closely the position of the optical jet. This again suggests a possible physical association. If both these possible associations were in fact correct, this would be a truly unique object because the optically emitting gas on one side is presumably thermalised in some way, while on the other side it is non-thermal emission that we see.

All of this radio structure is imbedded in a roughly circular halo component ~30 arcseconds in diameter, carrying most of the radio flux. Indeed, the presence at the edge of the halo source of the steep contours associated with the south-eastern radio lobe might suggest movement of the source through the intergalactic medium and ram pressure confinement of particles on the leading (SE) edge. Current ideas on the sources of fuel for the nuclear activity and the extended radio activity might imply that the visible gas comprises the remnants of material in the form of galaxies or gas clouds that has fallen into and coalesced with the elliptical galaxy. While there are no observations that effectively refute this, there is little in the <u>observed</u> dynamical picture that unequivocally supports this picture.

In some ways the variety of distinct physical phenomena present in or associated with this galaxy reminds one of the situation in NGC 1275. Whether this remains true in detail or not, it is clear that better defined physical parameters for the jet are desirable, and these are fortunately attainable at optical and radio wavelengths.

REFERENCES

Danziger, I.J., Fosbury, R.A.E., Goss, W.M., Ekers, R.D., 1979. Mon.Not.Roy.Astr.Soc., 188, 415.

Danziger, I.J., Bergeron, J., Fosbury, R.A.E., Maraschi, L., Tanzi, E.G., Treves, A., 1983a. Mon.Not.Roy.Astr.Soc., in press.

Danziger, I.J., et al., 1983b. In preparation.

Fosbury, R.A.E., 1982. Extragalactic Radio Sources, IAU Symposium 97, 65.

Ulrich, M.-H., 1981. Astron.Astrophys. 103, L1.

van Breugel, W., 1981. Optical Jets in Galaxies, Proceedings of the Second ESO/ESA Workshop, p. 63.

CCD - OBSERVATIONS OF OPTICAL JETS AND EXTENSIONS IN GALAXIES

Hélène Sol
ESO - Munich, FRG, and Observatoire de Paris, France

ABSTRACT

Preliminary results from a survey of peculiar galaxies with jet-like features are reported.

INTRODUCTION

Only a very small number of galaxies with optical jets have been studied in some detail until now. This number fluctuates between 2 and 20, according to the definition used for a jet: A "jet" appears in the literature to vary from a "confirmed" synchrotron jet to any simple narrow extension! Nevertheless, a greater number of galaxies are indicated as having an optical jet in different catalogues. As an example, 13 of the 16000 galaxies of the ESO-Uppsala catalogue have been suspected by visual inspection of Schmidt plates to have a jet (Lauberts, 1982). Other catalogues of peculiar galaxies, such as the Arp (1966) and Arp-Madore (1975, 1977) catalogues, provide also lists of so-called jet-galaxies.

A survey of these jet-galaxies has been undertaken in order to allow a statistical approach to the optical jet phenomenon. It is expected to lead to a more precise definition of the optical jets and to permit a first classification of them, according to morphological and photometric parameters of the jets themselves and of their parent-galaxies, since the wide dynamical range of the CCD used for the observations allows to study simultaneously the low luminosity extensions as well as the bright nuclear regions. It is attempted to distinguish or to find a continuity between the different possible jet-cases as synchrotron jets (M87), 3C120-type jets, radio-quiet NGC1097-type jets (Wolstencroft, 1981), extensions due to collisions or gravitational interactions of galaxies like IC1182 (Bothun et al. 1981).

A. Ferrari and A. G. Pacholczyk (eds.), Astrophysical Jets, 135–141.
Copyright © 1983 by D. Reidel Publishing Company.

The data have been obtained at the 1.5-m Danish telescope of the ESO-La Silla observatory in January and July 1982. The Cassegrain focus of the telescope was equipped with an RCA thinned back-illuminated CCD (type SID 53612) camera. 60 peculiar galaxies have been observed in different colours, using the standard B and V Johnson's filters and the g, r, i, and z filters described by Thuan and Gunn (1976), and Wade et al. (1979). This set of broad-band filters covers the wavelength range of the CCD response, from 4000 Å to 1μ.

FIRST RESULTS

Five jet-galaxies already described in the literature have been first observed to be used as references for the galaxies of the survey: M87, 3C273, 3C120, IC1182 and PKS0521-36.

The data on 3C273 show convincing evidence for coincidence of the optical emission with the radio structure observed at 408 MHz by Conway et al., 1981 (see also Davis, this workshop). The optical jet seems to have 3 regions corresponding to the radio wiggles. Contrary to the UBV photometry published in 1972 by Kron et al., the luminosity of the jet appears to increase with the distance to the nucleus and ends very sharply at its outer extremity in the V and r colours. In the near infrared range (8500-9500 Å) only this outer extremity is detected. CCD-images obtained independently by Tarenghi (1982) show also clearly these features. This radio-optical similarity suggests that the optical emission of the jet might be synchrotron in origin, instead of inverse Compton scattering of infrared radiation from the nucleus as it was proposed in 1975 by Guthrie and Napier to explain the displacement between the radio and optical centroids. These results will be presented in more detail by Grosbøl et al., 1983.

The observations of 3C120 seem to confirm the 4 arcsec optical elongation on the west-side of the nucleus, mainly due to radiation of the continuum, which has been pointed out by Wlérick et al. (1981). It could be associated with the radio component seen at the same position on the VLA maps at 6 and 20 cm (Balick et al., 1982).

New information on the morphology of the jet of PKS0521-36 has been obtained and confirms the jet-nature of the feature described by Danziger et al., 1979, and by Wlérick, 1981. Figure 1 shows the i-band isophotes of this galaxy. Subtraction of the galactic background clearly emphasizes the jet-feature. An NE-SW asymmetry seems to appear, which might be due to the presence of other peculiar features as a dust lane or another extension, or to some gravitational effects since 5 other galaxies are clearly visible in the 2x4 arcmin field around PKS0521-36 (see Table 1). The jet can be described in terms of 3 regions (Figure 2): region 1, "knot" close to the galactic nucleus, dominant on the bluer exposures; region 2, elongated; region 3, "knot" more apparent on the redder exposures. It is 10 arcsec long or 10 kpc for z = 0.055, which is a median value between the 3 kpc jet of M87 and

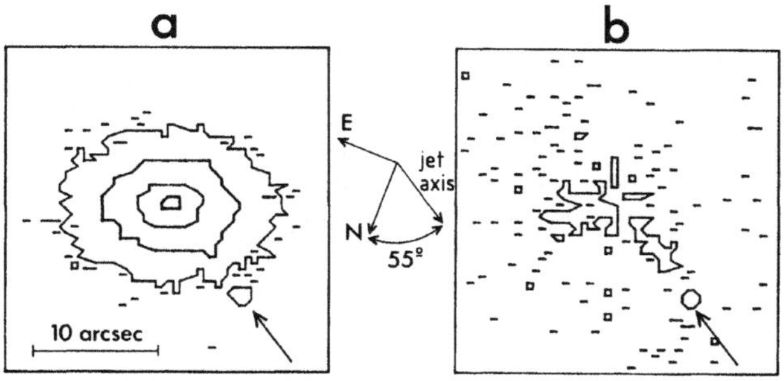

Figure 1 - PKS0521-36 (i-filter)

a. Isophotes at the levels 0.4, 1.2, 5.0, 40.0% of the maximal intensity. Note the deformation of the isophotes in the direction of the jet and the knot at the extremity of the jet-feature
b. Contour obtained after subtraction of the galaxy from its reversed image, symmetrical with respect to its minor axis. The jet-feature is well visible. Difficulties arise in the central regions from unavoidable uncertainty in the position of the center. The NE-structure comes from an NE-SW asymmetry of the galaxy, which requires further investigation.

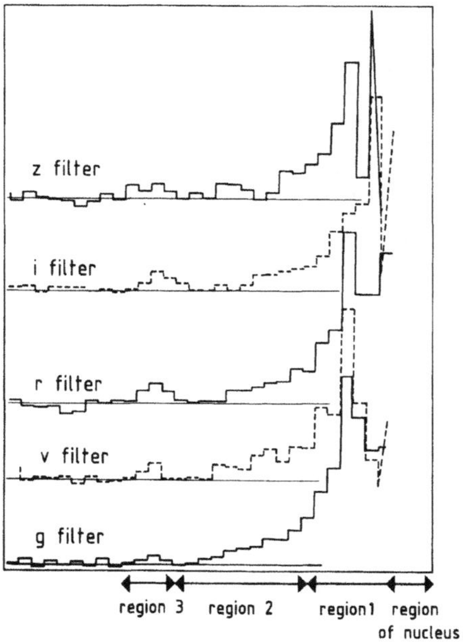

Figure 2 - Profiles of the PKS0521-36 jet in z, i, r, v, g filters, after subtraction of the symmetric galactic contribution (from 15 mn - CCD exposures; January 21st, 1982; 1 pixel = 0.471 arcsec).

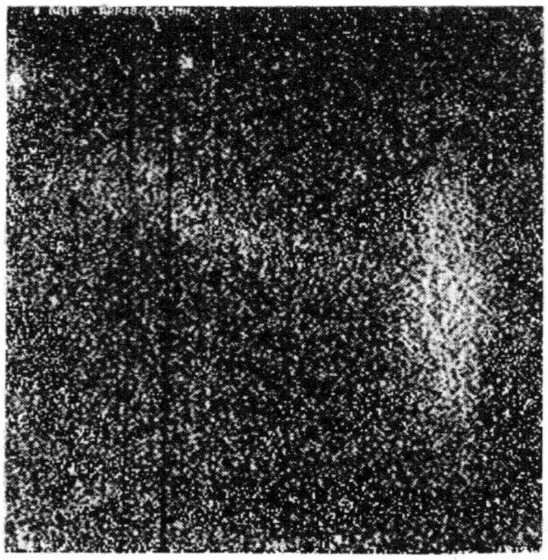

1a

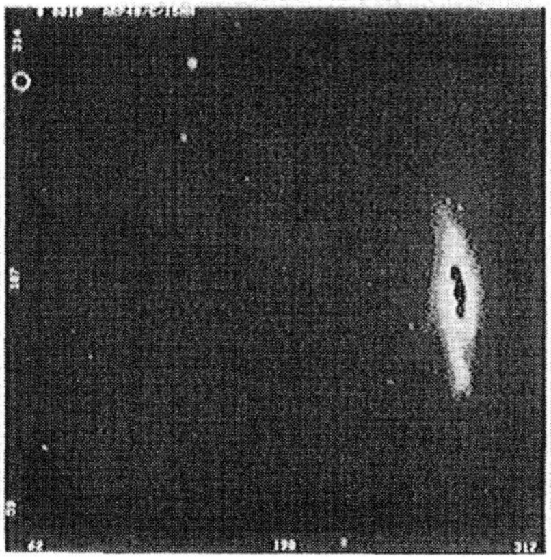

1b

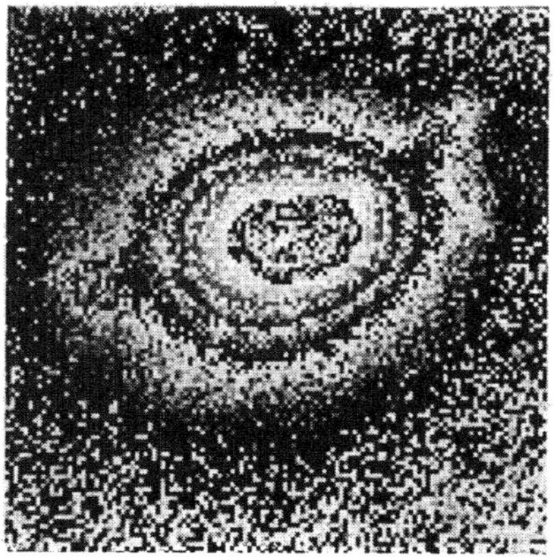

2a

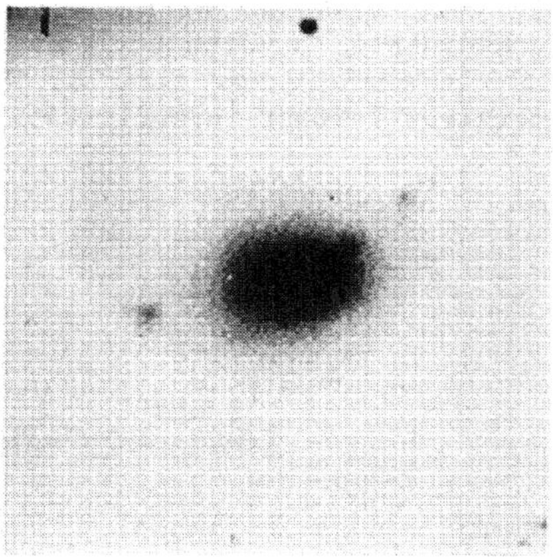

2b

PHOTOS:

2 galaxies of the survey, printed with 2 different contrasts: a) show the bright central regions, b) emphasize the faint jet-like structures.

1 - a,b "galaxy with jet" No. 48 of the Arp-Madore catalogue; described in the ESO/Uppsala catalogue as "S..., streamer or companion PA110". 347G22, α, δ 1950 23:30:49, -38:52.4. (15mn-CCD-exposure, g filter)

2 - a,b "galaxy with jet" No. 7 of the Arp-Madore catalogue; described in the ESO/Uppsala catalogues as "disturbed? or eruptive?. 108IG17, α, δ 1950 22:07:00, -67:06.9. (15mn-CCD-exposure; r filter).

TABLE 1: Position and rough V-magnitudes of the 6 brightest galaxies in the field of PKS0521-36

X (in arcsec) Increasing westwards	Y Increasing northwards	V
0 (PKS0521-36)	0	14.9
-40	-95	17.9
10.5	-23.5	18.4
-46.5	52.5	18.4
82.5	23	19.0
-7	-87	19.1

the 35 kpc jet of 3C273 ($H = 75$ km s^{-1} Mpc^{-1}). The V-luminosity of the entire jet appears slightly greater than 1% of the total V-luminosity of the galaxy. ($m_V \simeq 19.8$ for the entire jet; $m_V \simeq 23.2$ for the outer knot). These properties of the jet in PKS0521-36 bear similarities with the M87 jet. Since there is some evidence of coincidence between the optical and radio structures (see Danziger, this workshop), this jet might well be another case of optical synchrotron jet. No counterjet has yet been detected.

The reduction of the data on the galaxies of the survey is in progress. They can be classified into different morphological groups: (1) 10% of the jet-like features appear with the better resolution and sensitivity of the CCD to be possibly a superposition of classical astronomical objects, as faint stars or edge-on galaxies. (2) However, the majority of the objects remain very peculiar. There are roughly 40% of confirmed straight extensions (see photos), 20% of curved ones and 20% of other peculiar extensions. 20% of the objects are obviously interactive cases and their jet-like structures are probably due to tidal effects (Toomre and Toomre, 1972). 20% show peculiar multiple nuclear condensations which may also suggest mergers (see photos). Thus, there is a great percentage of detection of interactive galaxies (hereafter IG) in a sample of objects selected on the sole basis of the presence of a jet-like structure. The same tendency appears clearly in the ESO/Uppsala catalogue, where there are 10 galaxies classified as IG among 13 "jets", 53 IG among 68 "streams", 3 IG among 6 "elongations", 60 IG among 66 "tails", 23 IG among 25 "plumes" and 9 IG among 11 "filaments". A few classical jet-galaxies have also been interpreted in terms of IG: IC1182 (Bothun et al., 1981), Cen A (Tubbs, 1980); NGC1097 has an elliptical interactive companion (Lorre, 1978). On the other hand, it has been proposed that the nucleus of 3C273 and of other galaxies with bent or precessing (radio) jets contains supermassive binaries (Begelman et al., 1980; Whitmire and Matese, 1981; Roos, 1981). M87 itself has perhaps undergone a strong tidal interaction with its companion NGC4486B (Rood, 1965; Faber, 1973). Although the nature of the material and of the emission mechanisms of jets like the ones seen in IC1182 and in M87 is completely different, these jets could have a common property in that they have been initiated and fuelled by collisions of galaxies. In such an eventuality, intermediate cases would be expected.

CONCLUSION

Rather little is known yet concerning the optical emissions of the astrophysical jets. However, this emission contains information not available in radio wavelengths (see Nieto, this workshop, for the case of synchrotron jets), and the energy radiated in the optical range might be as important as the part radiated in radio (or even more) for the total energetic budget of the jets. Thus, knowledge of their optical properties is expected to lead to constraints on the energetic balance and physical conditions in the jets. This survey intendes to increase the number of detected and studied optical jets. The "new" cases found require radio observations and spectroscopic investigations. They could be interesting candidates to observe later on with the Space Telescope.

ACKNOWLEDGEMENTS

I would like to thank Massimo Tarenghi for initiating this work and for his comments on 3C273. I am grateful to Preben Gosbøl and Andrzej Kruszewski for their help in learning how to use the ESO image processing facilities. Thanks should also go to G. Wlérick who provided me with UBV magnitudes of stars in the field of PKS0521-36.

REFERENCES

Arp, H., Atlas of peculiar galaxies, 1966
Arp, H., Madore, B.F., Observatory, 95, 212, 1975, and Q.J. R. astr. Soc., 18, 234, 1977
Balick, B., Heckman, T.M., and Crane, P.C., Ap. J., 254, 483, 1982
Begelman, M.C., Blandford, R.D., and Rees, M.J., Nature, 287, 307, 1980
Bothun, G.D., Stauffer, J.R., and Schommer, R.A., Ap. J., 247, 42, 1981
Conway, R.G., Davis, R.J., and Foley, A.R., Nature, 294, 540, 1981
Danziger, I.J., Fosbury, R.A.E., Gross, W.M., and Ekers, R.D., MNRAS, 188, 415, 1979
Faber, S.M., Ap. J., 179, 423, 1973
Grosbøl P., Sol H., and Tarenghi, M., 1983, in preparation
Guthrie, B.N.G., and Napier W.McD., MNRAS, 172, 85, 1975
Kron, G.E., Ables H.D., and Hewitt, A.V., PASP, 84, 303, 1972
Lauberts A., The ESO/Uppsala Survey of the ESO (B) Atlas, 1982
Lorre, J.J., Ap. J., 222, L103, 1978
Rood, H.J., A.J., 70, 689 (A), 1965
Roos, N., Astron. Astrophys., 104, 218, 1981

Tarenghi, M., 1982, private communication
Thuan, T.X., and Gunn, J.E., PASP, 88, 543, 1976
Toomre, A., Toomre, J., Ap. J., 178, 623, 1972
Tubbs, A.D., Ap. J., 241, 969, 1980
Wade, R.A., Hoessel, J.G., Elias, J.H., and Huchra, J.P., PASP, 91, 35, 1979
Whitmire, D.P., and Matese, J.J., Nature, 293, 722, 1981
Wlérick, G., ESA SP-162, Optical jets in galaxies, Munich, 1981
Wlérick, G., Bouchet, P., Cayatte, V., and Michet, D., Astron. Astrophys., 102, L17, 1981
Wolstencroft, R.D., ESA-SP-162, Optical jets in galaxies, Munich, 1981

THE JET IN NGC 3310

F. Bertola
Istituto di Astronomia, Università di Padova, Italy

N.A. Sharp
Mount Stromlo and Siding Spring Observatories, Australia

Abstract - A jet like feature in the peculiar spiral galaxy NGC 3310 is pointed out and a possible violent ejection mechanism is invoked to account for it.

On the NW side of the disturbed Sc galaxy NGC 3310 (Fig. 1) a feature consisting of several barely resolved knots runs perpendicularly across a spiral arm at P.A. 300°. This structure is linear, with small "wiggles", around an axis passing directly through the nucleus. A continuous luminous component links the knots. The feature starts at 65" from the nucleus and reaches to a distance of 150" (3.2 to 7.3 kpc assuming a distance of 10 Mpc). The knots have an emission line spectrum (Bertola, Koo and Kron, in preparation) and no strong radio emission (van der Kruit and de Bruyn, 1976) so that they do not shine by synchrotron radiation. On the assumption that the above structure was caused by the passage of a jet, which induced star formation and thus created the observed bright emission line regions, we can deduce from its linearity the age, duration and velocity of the jet. Since the beam is no longer active, the deduced age is a lower limit. We find a velocity of 400 km/s, a duration of $\sim 10^7$ years, and an age of about $2 \cdot 10^7$ years.
The lack of a counter jet requires a strongly one-sided ejection, which is an extreme case of the "flip-flop" mechanism (Wiita and Siah, 1981). The subsequent recoil of the jet-producing active region strongly perturbs the dynamics of the galaxy. From the observed 1.5" displacement of the dynamical centre away from the stellar centre in a direction opposite to the jet (van der Kruit, 1976), we find by conserving momentum that the jet carried away about 1% of the mass of the active region, which is quite sufficient to disrupt the region and stop the activity. This explosive event explains quite simply all the unusual features of NGC 3310.

We are grateful to H. Arp for supplying 200" plates. NAS acknowledges support from the Istituto di Fisica and the Osservatorio Astronomico for visits to Padova.

A. Ferrari and A. G. Pacholczyk (eds.), Astrophysical Jets, 143–144.
Copyright © 1983 by D. Reidel Publishing Company.

References

van der Kruit, P.C. 1976, Astron. Astrophys. 49, 161
van der Kruit, P.C. and de Bruyn, A.G. 1976, Astron. Astrophys. 48, 373
Wiita, P.J. and Siah, M.J. 1981, Ap.J. 243, 710

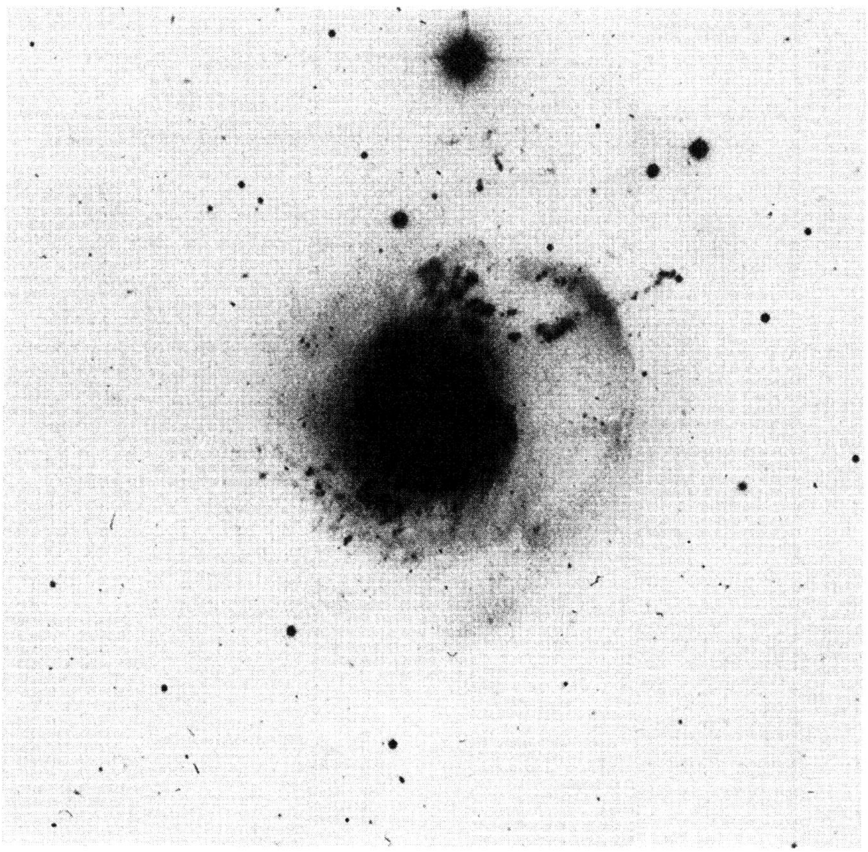

Figure 1

THE DISCOVERY OF OPTICAL EMISSION KNOTS IN THE INNER JET OF CENTAURUS A

Jean Brodie
Institute of Astronomy, Cambridge, England

Arieh Königl
W.K. Kellogg Radiation Laboratory, California Institute of Technology, Pasadena, USA

Stuart Bowyer
Space Sciences Laboratories, Berkeley, USA

Jets have been the subject of increasing optical study in recent years. Because of better instrumental sensitivity, relatively high resolution spectra can now be obtained which are enormously helpful probes of the underlying physical processes, especially when combined with information from other wavelengths. With this in mind we attempted to discover optical counterparts to the X-ray (Feigelson et al. 1981) and radio (Schreier et al. 1981, Burns et. al. 1982) knots recently discovered in the inner jet region of Centaurus A. The coincident X-ray and radio knots lie in the same direction as previously studied optical features in the outer jet (Blanco et. al. 1975; Dufour and van den Bergh 1978, Osmer 1978; Graham and Price, 1981) but the latter are between 4 and 40 times further from the nucleus. Our study concentrated on the inner region, 2 kpc from the nucleus, where dust obscuration and the high level of background brightness discourage optical identification.

The observations were made in the Spring of 1982 using the 4 meter telescope at CTIO and the SIT Vidican detector operating in long-slit mode. By aligning the 2 arcsec-wide slit along the X-ray contour we obtained 9 Å resolution spectra along the jet from close to the nucleus out to a distance of $\sim$ 4 arcmin.

Three regions of intense optical emission were discovered in the inner 70 arcsec of the jet (1' $\sim$ 25 pc). These are shown schematically in Figure 1. No optical features were observed along the rest of the sampled area of the jet. Two of the regions, labelled 1 and 3, coincide with distinct X-ray/radio knots and omit strong optical continuum radiation with superimposed emission lines. Region 2, situated between 1 and 3 and coincident with a low surface-brightness inter-knot zone in

A. Ferrari and A. G. Pacholczyk (eds.), Astrophysical Jets, 145–148.
Copyright © 1983 by D. Reidel Publishing Company.

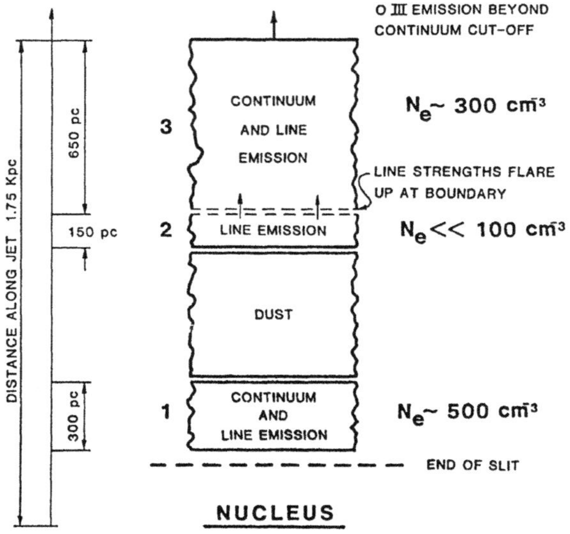

Figure 1: Schematic view of the emission regions with respect to the position of the galaxy nucleus.

the X-ray and radio maps, has little continuum flux but still displays intense line-emission. This discovery marks the first instance that optical line-emission in jets has been detected in continuum emission knots whose measured flux extends all the way from the radio to the X-rays.

Using the strengths of the [SII] λ6717/λ6731 ratio we inferred densities of ~ 500 cm^{-3}, < 10 cm^{-3} and ~ 300 cm^{-3} in regions 1, 2 and 3 respectively. On the basis of the strengths of low-excitation lines we found it unlikely that the line-emission was produced in normal HII regions.

Although our data were consistent with the effects of shock heating in the various line-strengths classification schemes (e.g. Baldwin et al., 1981; Sabbadin et al., 1977) we were unable to exclude the possibility of photoionization by a power-law continuum with a low ionization parameter. Future observations, especially of the temperature-indicative [O III] λ4363 line may help resolve this uncertainty.

Whereas the interknot spectral shape of region 2 is typical of elliptical galaxies dominated by late-type stars, the optical continuum in region 3 is quite different having a relatively flat spectrum which steepens markedly at $\sim 6 \times 10^{14}$ Hz. A similar break is observed in the nonthermal emission knot A in the M87 jet at $\sim 3 \times 10^{14}$ Hz (Schreier et al. 1982) and in both cases may be attributable to synchrotron radiation losses.

A pronounced flare-up of line intensities at the upstream edge of region 3 is interpreted as indicating the presence of shock waves. This conclusion is supported by the general consistency of our data with the theoretical shock models of Shull and McKee (1979) for shock velocities between 80 and 90 km s^{-1}.

We suggest a self-consistent model which satisfactorily explains many aspects of the data including earlier observations of the optical features in the outer jet. This model, first proposed to explain the M87 knots (Blandford and Königl, 1979), involves the interaction of the streaming jet with obstacles, or clumps, in the flow. This encounter is expected to result in the formation of bow-shocks at the upstream edge of the obstacle and the propagation of another shock into the condensation, possibly followed by fragmentation of the clump. If the fragments are sufficiently small they may be accelerated to the velocity of the jet, which might explain the high radial velocities ($\sim$ 400 km s^{-1}) observed in the knot-like regions of the outer jet. Our estimates of the size and density of the emitting condensations are consistent with their being ordinary interstellar clouds.

This general model has been successfully applied to the interpretation of Herbig - Haro objects (Schwartz 1978, 1981) which have many spectral characteristics in common with our Cen A knots. Perhaps this suggests that jets on vastly different length scales may be manifestations of the same general phenomenon.

REFERENCES

Baldwin, J., Phillips, M. and Terlevich, R. 1981, Pub. ASP, 93, 5
Blanco, V.M., Graham, J.A., Lasker, B.M. and Osmer, P.S. 1975, Ap. J. Lett., 198, L63
Blandford, R.D. and Königl, A. 1979, Ap. Lett., 20, 15
Burns, J.O., Feigelson, E.D. and Schreier, E.J. 1982, in preparation
Dufour, R.J. and van den Berg, S. 1978, Ap. J. Lett., 226, L73
Feigelson, E.D., Schreier, E.J., Delvaille, J.P., Giacconi, R., Grindlay, J.E. and Lightman, A.P. 1981, Ap. J., 251, 31
Graham, J.A. and Price, R.M. 1981, Ap. J., 247, 813
Osmer, P.S. 1978, Ap. J. Lett., 226, L79
Sabbadin, F., Minello, S. and Bianchini, P. 1977, Astron. Ap., 60,147
Schreier, E.J., Burns, J.O. and Feigelson, E.D. 1981, Ap. J., 251, 523
Schreier, E.J., Gorenstein, P. and Feigelson, E.D. 1982, preprint
Schwartz, R.D. 1978, Ap. J., 223, 884
Schwartz, R.D. 1981, Ap. J., 243, 197
Shull, J.M. and McKee, C.F. 1979, Ap. J., 227, 131

DISCUSSION

G. BENDFORD: If the M87 knots are regularly spaced as reported, doesn't that argue against a comparison with your cloud-knot model?

J. BRODIE: The accelerated-clump model is attractive for Cen A because it is consistent with many of the features of our data and the X-ray and radio observations. Many of the spectroscopic properties of our knots resemble those observed for Herbig-Haro objects, which are well described by this model, and it provides a natural link between the optical features in the inner and outer jet. In addition we observe no optical knots downstream of our region 3. So the optical periodicity reported in the M87 jet does not seem to be present in Cen A. It is true however that equally spaced knots would not readily be accomodated in this model.

EXTRANUCLEAR ACTIVITY IN MKN 335

K.J. Fricke, W. Kollatschny and
H. Schleicher
Universitäts-Sternwarte Göttingen, FRG

ABSTRACT. On prime focus plates of the radio-quiet Seyfert 1 galaxy Mkn 335 we find a jet-like feature to the north west. The line spectrum of this feature is redshifted with respect to the nucleus and shows line ratios different from those in the nuclear spectrum.

1. INTRODUCTION

Mkn 335 is a well-known FeII Sy 1 galaxy (Phillips, 1977). Shuder (1981) has reported on variability between 1974 and 1976 of the optical continuum particularly in the 3000 Å bump and of the emission lines. Infrared photometry for Mkn 335 has been done by Rieke (1978). An X-ray flux in the 2-10 keV band of 1.6 10^{-32} Wm^{-2} Hz^{-1} has been measured with Uhuru (Tananbaum 1978). The object could not be detected in the radio. De Bruyn and Wilson (1976) give an upper limit of 8 mJy at 21 cm. Adams (1977) has not found any morphological structure in Mkn 335 despite its low redshift cz = 7500 km/s. Some structure, however, indicating extranuclear activity in Mkn 335 may be seen on video camera prints published by Heckman and Balick (1981, 1982).

In this contribution we report on further evidence for activity in Mkn 335 as derived from direct plates and spectroscopy.

2. MORPHOLOGICAL OBSERVATIONS

We have taken direct plates of Mkn 335 in the prime focus of the ESO 3.6 m telescope using IIIaF and IIIaJ emulsions. Exposure times were 40 to 90 min (cf. Table 1).

A. Ferrari and A. G. Pacholczyk (eds.), Astrophysical Jets, 149–155.
Copyright © 1983 by D. Reidel Publishing Company

Table 1: Prime focus observations

plate N°	date	time of exposure	plate emulsion	filter
2131	79/10/20	40 min	IIIa-F B	RG 630
2118	79/10/18	60 min	IIIa-F B	RG 630
2130	79/10/20	40 min	IIIa-J B	GG 385
2117	79/10/18	60 min	IIIa-J B	GG 385
2123	79/10/19	90 min	IIIa-J B	GG 385

The scale on the plate is 18$''$.9/mm. The seeing was ∿ 2" during the observations. The plates were reduced with the PDS 1010A microdensitometer at ESO Garching.

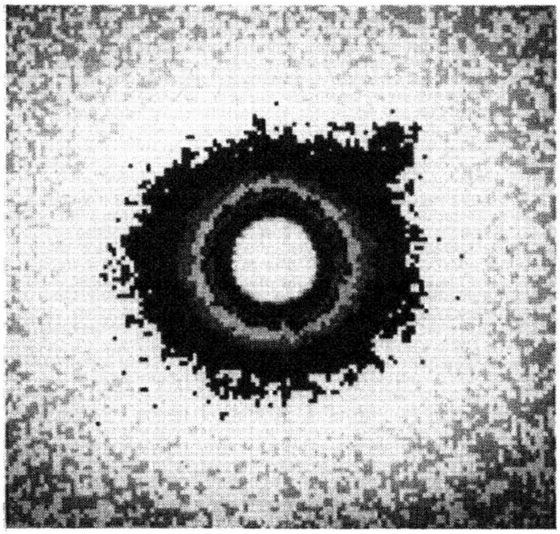

Fig. 1. PDS-scan of Mkn 335 (blue plate, 60 min exposure time, north to the top, east to the left)

Fig. 1 shows an image of Mkn 335 obtained after a 60 min exposure of the blue plate. On this image a jet-like feature to the north-west (P.A. 306°) is visible (hereafter referred to as the "NW-feature").

The NW-feature is present on all of our plates. It can also be identified on Fig. 2b of the paper by Heckman and Balick (1981), who, however, did not pay attention to it. An intensity tracing through this feature along the line joining it with the center of Mkn 335 is represented in Fig. 2. The NW-feature extends radially between 4$''$.5 and 9$''$.5 from the center corresponding to distances 2.25-4.75 kpc

($H_o = \frac{75\ km/s}{Mpc}$). Its width is of the same order as the seeing ($\sim$ 2").

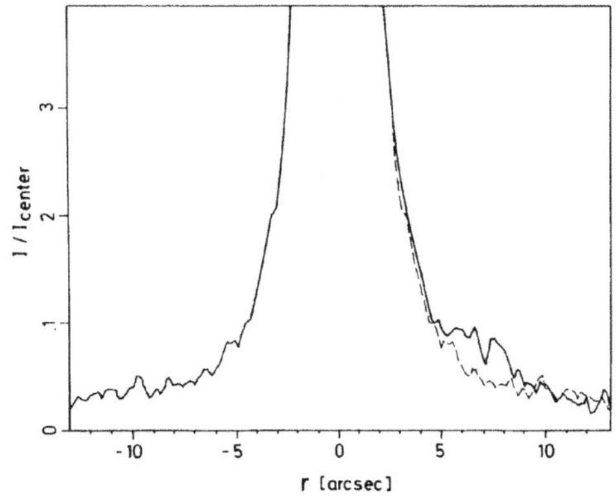

Fig. 2.
Intensity scan across Mkn 335 through the northwest feature. On the right the brightness contours in opposite directions are superposed. Zero intensity corresponds to clear film

Another jet-like structure to the south appearing on contrast-enhanced prints near the background noise level has been described by Heckman and Balick (1981). We refer to this feature hereafter as to the "S-jet". The location and extent of the NW-feature and the"S-jet"with respect to the nucleus are sketched in Fig. 3.

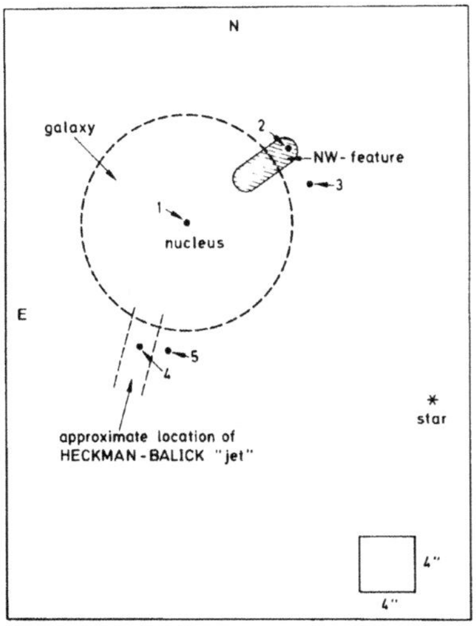

Fig. 3.
Schematical diagram of Mkn 335 showing the location of the NW-feature and of the "S-jet" relative to the nucleus. The numbers label the positions from which spectra were obtained. The square at the lower right represents the projected size of the aperture

3. SPECTROSCOPIC OBSERVATIONS

IDS spectra of Mkn 335 have been obtained on July 18 and 19, 1982 using the Boller and Chivens spectrograph attached to the ESO 3.6 m telescope. We chose an aperture of 4"x4" and a dispersion of 172 Å/mm. The spectra covered the wavelength range ∿ 4000 Å - 7100 Å. The seeing during the spectroscopic exposures was estimated to about 2". The spectra were reduced in the standard way at ESO/Garching and Göttingen. The positions from which spectra were obtained are indicated in Fig. 3. Exposure times were 8-16 min.

We obtained the following spectroscopic results:

a) Nucleus

The spectrum from the nucleus (Fig. 4) has a flat continuum distribution F_λ vs. λ and shows the typical Sy 1 line spectrum. The mean spectral slope is $\alpha \sim -1.7$ ($F_\nu \propto \nu^\alpha$).

b) Off-center emission

The spectrum at position 2 within the NW-feature is displayed in Fig. 5. The continuum rises towards the red ($\alpha \sim -3.5$). The emission line spectrum shows strong permitted and forbidden lines. Hα and Hβ have FWHM ∿ 1600 km/s while the [OIII]4959, 5007 lines have FWHM ∿ 950 km/s.

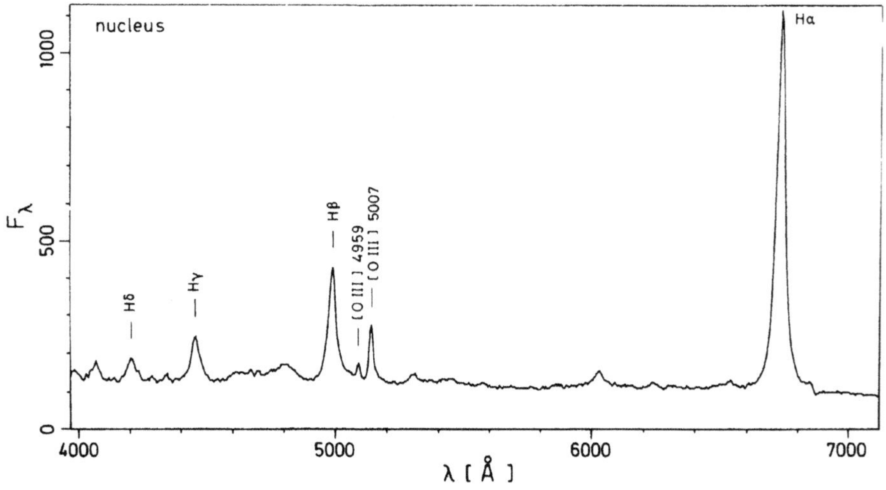

Fig. 4. The spectrum of the nucleus of Mkn 335 (position 1 of Fig. 3)

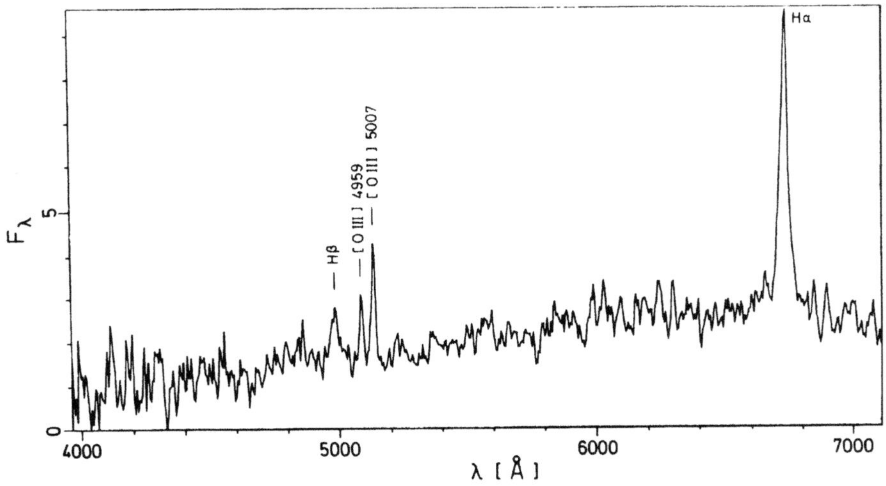

Fig. 5. The spectrum of the NW-feature
(position 2 of Fig. 3)

In Fig. 6a, b the spectra from the NW-feature in the vicinity of the Hα, [OIII], and Hβ lines are superposed on the spectrum of the nucleus. The spectra are normalized to the peak intensity of Hα. The scaling factor for the intensity of the NW-feature is $\sim 10^2$. The line centers in the offcenter spectrum are systematically redshifted by $\sim$ 280 km/s relative to the nuclear spectrum. The emission line profiles suggest the presence of several components with different velocities relative to the nucleus. Such components could be due to discrete, spatially unresolved condensations.This spectrum also shows marginal evidence for the absorption lines Ca H + K and Mg I 5175 Å.

In the offcenter spectrum the Hα/Hβ and [OIII]/Hβ ratios are enhanced relative to the nucleus.

We have taken a spectrum from a position adjacent to the NW-feature (position 3) and at the same distance (9") from the nucleus. This spectrum does not show any lines and its continuum level is by an order of magnitude less than in position 2. We therefore conclude that contamination of the light from the NW-feature by scattering or differential refraction of nuclear light is unimportant.

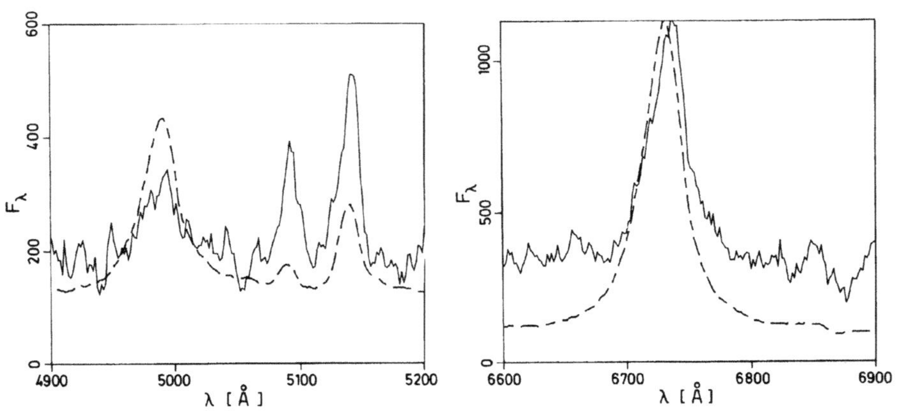

Fig. 6. The spectrum of the NW-feature scaled to the nuclear spectrum (dashed line): a) in the vicinity of Hβ and [OIII]4959,5007 and b) in the vicinity of Hα.

9" south of the nucleus in position 5 we also find a featureless continuum at the same level as in position 3 (10^{-3} of the nuclear continuum). On the other hand, in the spectrum from the adjacent position 4 which is coincident with the S-jet described by Heckman and Balick (1981) the presence of weak emission lines and an increase of the continuum flux toward the UV are indicated.

4. DISCUSSION

The line spectrum of the NW-feature showing emission line widths comparable to those in the nuclear spectrum may arise from a secondary weak Seyfert nucleus located in the outskirts of Mkn 335 as a result of merging or ejection. The bulk velocity of the feature relative to the nucleus of Mkn 335 of ∿ 300 km/s would be compatible with both options. Merging seems unlikely since (i) Mkn 335 is not located in a group or cluster and (ii) there is little sign of galactic material in the vicinity of the NW-feature. The steep Balmer decrement could be an indication for dust associated with such material not belonging to Mkn 335.

The possible velocity spread within the emission lines (cf. Sect. 3b) favours the physical connection of the underlying object with Mkn 335 and its origin by ejection from the nucleus. The steep Balmer decrement could then be due to obscuration by dust within Mkn 335. Alternatively, the broad and narrow emission lines could originate from a gaseous jet and its confining dense surroundings. The conditions might be not unlike those in NGC 1275 and DA 240. In these galaxies different velocity systems and jet-like structures are observed and their large scale morphology bears some resemblance to that of Mkn 335.

ACKNOWLEDGEMENTS

This work was supported in part by the Deutsche Forschungsgemeinschaft under grants Fr 325/12 and Fr 325/15-2.

REFERENCES

Adams, T.F., 1977, Astrophys. J. Suppl. Ser. 33, 19
de Bruyn, A.G., Wilson, A.S., 1976, Astron. Astrophys. 53, 93
Heckman, T.M., Balick, B., 1981, Astrophys. J. 247, 32; 1982, Errata, Astrophys. J. 256, 798
Phillips, M.M., 1978, Astrophys. J. Suppl. Ser. 38, 187
Rieke, G.H., 1978, Astrophys. J. 226, 550
Shuder, J.M., 1981, Astron. J. 86, 1595
Tananbaum, H., Peters, G., Forman, W., Giacconi, R., Jones, C., Avni, Y., 1978, Astrophys. J. 223, 74

JETS IN SS433

R.T. Schilizzi
Netherlands Foundation for Radio Astronomy, Dwingeloo, NL

J.D. Romney
Max-Planck-Institut für Radioastronomie, Bonn, FRG

R.E. Spencer
Nuffield Radio Astronomy Laboratories, Jodrell Bank, UK

I. Fejes
Satellite Geodetic Observatory, Penc, Hungary

SS433 has been observed extensively in virtually all wavelength regimes. In the X-ray (Seward et al. 1979) and the radio (Hjellming and Johnston, 1981a, b, 1982; Spencer 1979, Schilizzi et al. 1981, 1982; Niell et al. 1981, 1982) there is clear evidence of jets emanating from the central star system in the direction of the well-known extensions in the surrounding shell of emission, W50. Our VLBI observations of SS433 in 1980 and 1981 were made at wavelengths ranging from 6 to 21 cm (beamsizes of 10 to 100 milli-arcsec or $\sim 10^{15}$ to 10^{16} cm at the distance of W50) and had sufficient resolution to distinguish individual blobs of emission in the jets. An example is shown in Fig. 1.

Conclusions that can be drawn from this work are:

(1) The shapes of the radio structures at the different epochs of observation are in good agreement with the trajectories predicted from the kinematic model (Abell and Margon, 1979), provided one assumes the centres of symmetry of the radio structure and kinematic model coincide. The precise registration of the VLBI radio maps on the sky is unknown because astrometric information is not available from hybrid mapping.

(2) Assuming the centres of symmetry do agree, there does not appear to be any delay between the radio and the optical, but the blobs in the jet do not lie completely on the predicted trajectory, implying some jitter may occur in the pointing of the beam.

(3) The decay times of blobs in our data range between ~ 5 and ~ 50 days.

A. Ferrari and A. G. Pacholczyk (eds.), Astrophysical Jets, 157–159.
Copyright © 1983 by D. Reidel Publishing Company.

(4) There is probably a smooth component of radio emission in the jet which underlies the blobs and which remains even during periods of quiescence when no blobs are produced.

(5) Assuming equipartition fields, and measuring component sizes and flux densities from the VLBI data, synchrotron radiation cannot be the dominant loss mechanism on these and larger angular scales.

(6) There may be a correlation between the extrapolated birth time of a pair of blobs and a flare in the radio flux density, with the blob production occurring one to two days before the flare.

A paper giving full details of this work has been submitted for publication by Romney et al.

REFERENCES

Abell, G.O. and Margon, B. (1979) Nature, 279, 701
Hjellming, R.M. and Johnston, K.J. (1981a) Nature, 290, 100
(1981b) Ap.J., 246, L141
(1982) Proc. IAU Symp. 97, p. 197
Niell, A.E., Lockhart, T.G., Preston, R.A. and Backer, D.C. (1982) Proc. IAU Symp. 97, p. 207
Schilizzi, R.T., Miley, G.K., Romney, J.D. and Spencer R.E. (1981) Nature, 290, 318
Schilizzi, R.T., Fejes, I., Romney, J.D., Miley, G.K., Spencer, R.E. and Johnston, K.J. (1982) Proc. IAU Symp. 97, p. 205
Seward, F.G., Grindlay, J.E., Seaquist, E.R., and Gilmore, W.S. (1980) Nature, 287, 806
Spencer, R.E. (1979) Nature, 282, 483

Figure 1 - VLBI map of SS433 made on 15 February 1981 at 18 cm with three European telescopes. The axes are graduated at intervals of 50 milli-arcsec; the ellipse in the lower left hand corner represents the interferometer beam. The predicted trajectory of the optically emitting material (Abell and Margon 1979) is shown as a full line through the structure; the dots are spaced at intervals of two days.

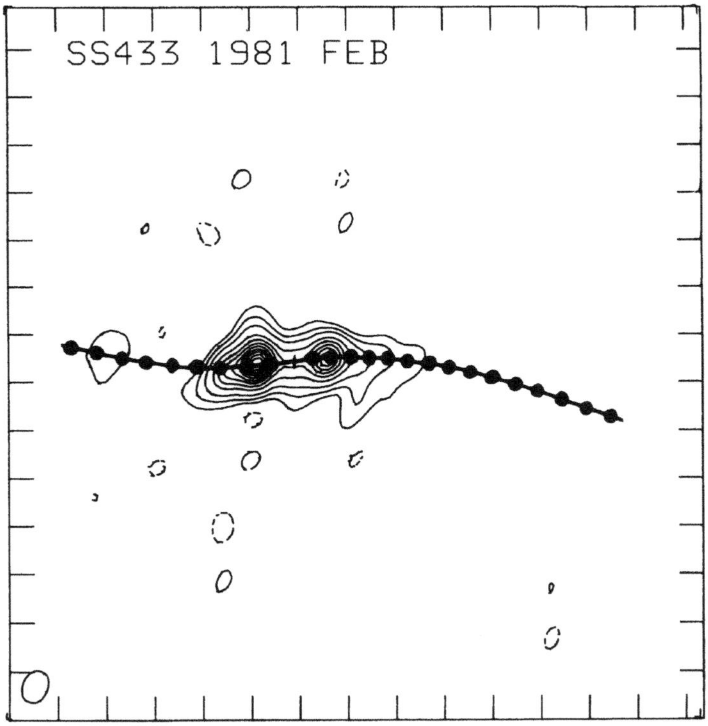

RAPID RADIO-VARIABILITY AT 408 MHz IN SS433

S.R. Bonsignori-Facondi
Istituto di Astronomia - Università - Bologna
Istituto di Radioastronomia - C.N.R. - Bologna, Italy

SUMMARY

Preliminary results of two years (24 April 1980 -30 June 1982) almost daily 408 MHz monitoring of peculiar source SS433 are presented.

1. INTRODUCTION

In this communication we present preliminary results of a 2 years monitoring program, still in progress, of SS433 with the 408 MHz Radiotelescope in Bologna. During the working period of the instrument, data have been taken, daily if possible, with the fan beam E-W arm of the telescope. The main features of the observing system are the following: the 408 MHz Northern Cross East-West arm radiotelescope in Bologna is operating as transit instrument with the three beams available; each of them has an HPBW of about 2 deg in declination and 3.5 arcmin in right ascension and the delay time between beams is 21/cosd sec. Data taken with this procedure provide three independent observations for each meridian transit of the source. Because of the high resolution of the instrument in the E-W direction, the emission of the compact source in SS433 is well separated from that of the surrounding SNR W50. The flux scale has been calibrated using seven different calibrators (3C386, 4C4.68, 3C403, 3C410, 3C424, P2128+04).

2. THE OBSERVATIONS (The data analysis)

In Fig. 1 we present our flux data during the about two years of the observing periods. An immediate remark is that the source at 408 MHz presents strong variations in flux density within a factor of about 2 between maxima and minima. Several one-day flares are evident in Fig. 1. We limit ourselves to present here the two most striking and best documented outbursts which occured in 1981, April 27 (Fig. 2a) and 1981 July 8 (Fig. 2b) respectively. (See also Bonsignori & Braccesi 1981 for similar outbursts on the same time scale). The observed variability on

A. Ferrari and A. G. Pacholczyk (eds.), Astrophysical Jets, 161–164.
Copyright © 1983 by D. Reidel Publishing Company.

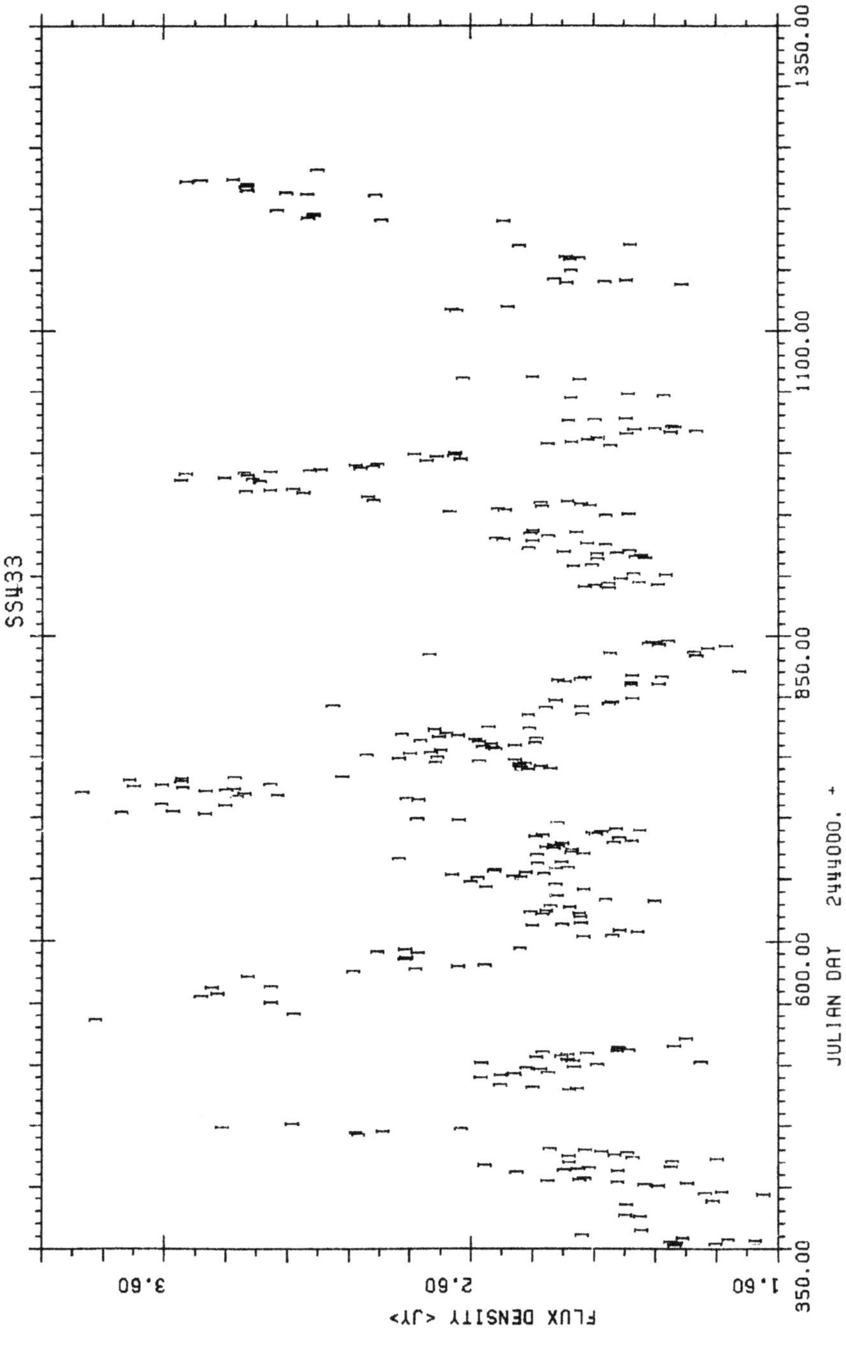

Fig. 1 - Observed flux density of SS443 during the observing period at "Northern Cross" observatory. The error for each single measurement is smaller than the size of symbols used in this plot.

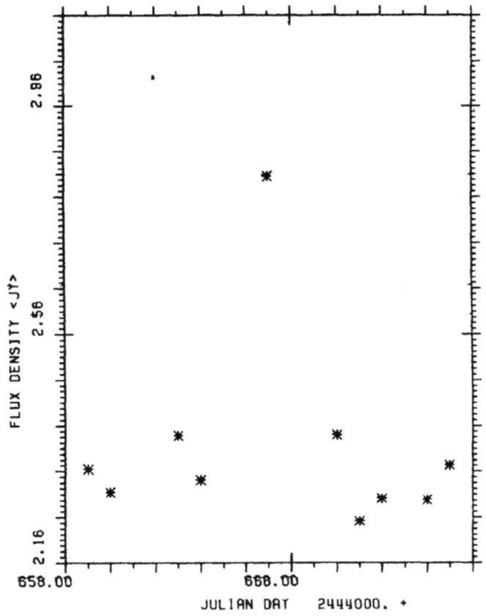

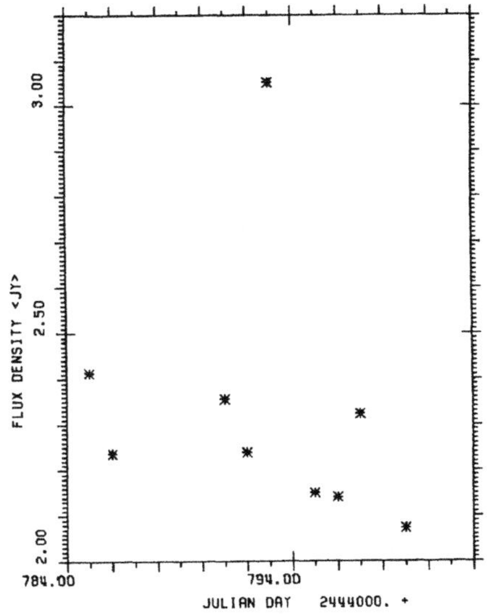

Fig. 2 - Two outbursts of one day: (a) on 1981, April 27 (2444721 J.D.) (b) on 1981, July 8 (2444795 J.D.).

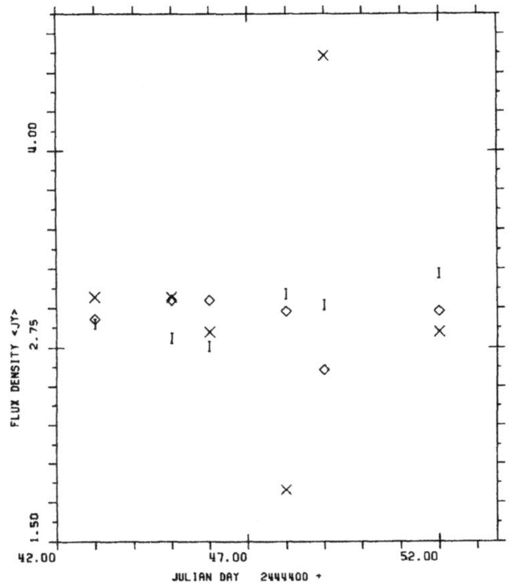

Fig. 3 - Example of a possible variation on 1 minute time-scale on 1980, July 29 (2444449 J.D.). The different symbols represent the three fan beams of the radio telescope.

time scale of about or less than one day implies a structure with a linear size smaller than 2.6 exp15 cm. In addition to these variations on time scale of about one day our data shows also some indications of variability on a much shorter time scale. An example is shown in Fig. 3, where the fluxes measured in the three beams are shown separately with different symbols; here we note a significant beam to beam variability on a time scale of about one minute corresponding to a structure of size less than 6.0 exp11 cm.

Similar flux variations have been observed at least 5 times during our two years monitoring program of SS433. A more detailed analysis of the statistical significance of this short time scale variability is in progress.

3. CONCLUSIONS

1) At 408 MHz SS433 present strong flux density variations of more than a factor two.

2) It is certain that the source flares on time scale of one day as shown by the several strong outbursts observed.

3) There is some evidence for flux variations on time scale of about one minute.

REFERENCE

S.R., Bonsignori Facondi and A. Braccesi. 1981 "Vistas in Astrónomy", 25, 91

X-RAYS FROM JETS AND LOBES

Eric D. Feigelson
Department of Astronomy, The Pennsylvania State University, University Park, Pennsylvania 16802, and
Center for Space Research, Massachusetts Inst. of Technology
Cambridge, Massachusetts 02139

Abstract. The current status of X-ray detections (and non-detections) of radio jets and lobes is presented. High resolution images obtained with the *Einstein* Observatory reveal X-ray emission associated with jet and lobe components of Virgo A, Centaurus A, 3C273, and possibly other radio structures. The Vir A and Cen A jet emission is probably synchrotron in origin, requiring *in situ* particle acceleration to extremely high energies. Inverse Compton emission is the likely cause of asymmetrical X-ray features in the halo of Vir A, supporting the assumption of equipartition.

I. INTRODUCTION

Historically radio galaxies have not played a very prominent role in the development of X-ray astronomy. A number of radio galaxies appeared in early X-ray surveys of the sky, but were largely subsumed under other classes of objects. Centaurus A, 3C120, 3C273, 3C390.3 and several N galaxies were found by the Uhuru, Ariel V and HEAO-1 satellites and were considered to be examples of Seyfert-like active galaxies. Virgo A, Cygnus A, 3C129, 3C264, 3C338, 3C318.1, and others lay in X-ray error boxes, but their X-ray emission was found to be associated with a hot intracluster medium rather than with their radio emitting structures. Early arguments linking the X-ray and radio emission through inverse Compton or synchrotron processes in these lobes (see for instance Hayakawa et al. 1965; Shklovskii 1967; Felten 1968; Lampton et al. 1972) were not upheld as more data concerning their X-ray structure, variability and spectra accumulated. One of the last possibilities for inverse Compton emission, the report of X-ray emission from the outer lobes of Cen A (Cooke et al. 1978), has recently been shown to be incorrect due to confusion by an unrelated X-ray source (Marshall and Clark 1981). Thus, prior to the launch of the *Einstein* X-ray Observatory there was not a single substantiated case (with the exception of galactic supernova remnants) of X-rays produced in a region of extended radio emission.

The *Einstein* satellite, the first equipped with X-ray focussing

A. Ferrari and A. G. Pacholczyk (eds.), Astrophysical Jets, 165–172.
Copyright © 1983 by D. Reidel Publishing Company.

optics, provided a dramatic improvement in resolution (FWHM-3" for the HRI-High Resolution Imager, and FWHM=1' for the IPC-Imaging Proportional Counter) and sensitivity (about 0.01 μJy at 2 x 10^{17} Hz) compared to previous missions. The high resolution is particularly important when weak X-ray emission from jets and lobes must be distinguished from strong emission in the active nucleus or intracluster gas. The following sections summarize the findings and interpretation of several detections of jet and lobe X-ray emission made in the *Einstein*. It must be remembered, however, that there have been considerably more failures than detections. Various radio galaxies with jets (e.g. NGC315, Feigelson and Schreier 1980; Her A, E. Feigelson, G. Clark and J. Dreher, in preparation), hot spots (e.g. 3C295, Henry et al. 1979; Cyg A, Feigelson and Schreier 1980; 3C111, Feigelson and Berg 1983), diffuse lobes (e.g. 3C318.1, Kriss et al. 1980; 3C264, Elvis et al. 1981; DA240, Feigelson and Berg 1983), and tails (e.g. 3C465, Burns 1983; 1638+538, Burns and Balonek, 1982; and several others discussed by Harris, 1982) have been examined with sufficient X-ray resolution to resolve the radio structure without success. But the issue is not closed as *Einstein* results for many radio galaxies have not yet been reported. Nearly 100 3C radio galaxies, or 2/3 of all galaxies in the 3C catalog, appear in *Einstein* IPC or HRI images. About a dozen of these galaxies have HRI exposures $\gtrsim 1 \times 10^4$ sec in duration and offer the best opportunities for X-ray jets and lobes to be found. Many other radio sources not explicitly catalogued as radio galaxies, such as radio-loud quasars and head-tail galaxies in rich clusters, have also been observed.

II. THE VIRGO A JET

Virgo A = M87 in the Virgo cluster was the first extragalactic object to be detected in X-rays. More than 98% of its soft X-ray emission comes from $\sim 10^{12}$ $M_\odot$ of hot gas extending several hundred kiloparsecs from the galaxy (Fabricant et al. 1980). The gas is very dense near the nucleus, resulting in rapid cooling and inflow of ~ 3 $M_\odot$/yr (Canizares et al. 1982). We are interested here in a comparatively weak component of X-ray emission that is not generated by the diffuse thermal plasma. This work is presented in detail by Schreier et al. (1982). Color photos comparing the radio and X-ray images of Vir A and Cen A appear in Feigelson and Schreier (1982).

The HRI image of the center of Vir A (Figure 1) shows two peaks of X-ray emission of similar intensity with the same separation and orientation as the nucleus and brightest knot of the radio/optical jet. Both the nucleus and jet are slightly resolved; the former consists of a gaseous cusp and/or a compact source, while the latter structure is consistent with emission from the outer knots of the jet in the same ratio as seen at lower frequencies. The X-ray emission from Knots A and B is 0.2 μJy at 4 x 10^{17} Hz with a factor of two uncertainty due primarily to the unknown X-ray spectrum. The spectral index between the optical and X-ray bands is then $\alpha_{ox} = 1.4 \pm 0.1$ and the radio/X-ray index is $\alpha_{rx} = 0.77 \pm 0.04$. The synchrotron spectrum is known to be flat between the radio and

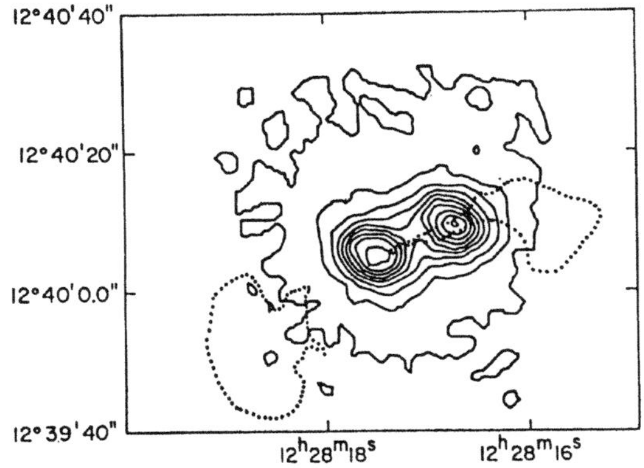

Figure 1. X-ray contour map of inner region of Vir A from the *Einstein* HRI showing the nucleus and jet (Schreier et al. 1982). The dotted line outlines the radio jet and lobes (adapted from Owen et al. 1980).

---

near-infrared (α_{ro} = 0.6-0.7) and steepens to α_o = 1.7 $\pm$ 0.1 in the optical band. *IUE* ultraviolet and the *Einstein* X-ray measurements are consistent with an extrapolation of this steepening spectrum if the difference between α_{ox} and α_o is not considered to be critical. The X-ray jet then requires that electron acceleration achieve energies up to $\Gamma = 2 \times 10^7 B_{eq}^{-1/2}$ with individual electrons decaying on timescales $\tau = 5B_{eq}^{-3/2} \sin^{-2}\Psi$yr, where $B_{eq} = 3 \times 10^{-4}$G is the equipartition magnetic field derived from radio maps and Ψ is the pitch angle. All of the theoretical difficulties presented by the optical synchrotron jet (e.g. Felten 1968) are accentuated by the need for electrons 20 times more energetic to account for X-ray synchrotron emission.

Thermal and inverse Compton explanations for the X-ray jet require rather special conditions. In a thermal model, the magnetic field of each knot must compress about $10^4 M_\odot$ of hot gas to densities 10^2 times higher than the ambient interstellar medium. The thermal and relativistic gases could not be mixed as the radio/optical emission is not depolarized. The inverse Compton X-ray emission under standard assumptions (equipartition, power law spectrum, etc.) is 10^4 times lower than observed. It can be increased only if the radio knots are smaller than measured with the VLA (which is not verified by recent VLBI measurements) or if one of the standard assumptions is radically changed. A model combining low-Ψ electron propagation from the nucleus with inverse Compton X-ray emission is proposed by Stocke et al. (1981) to explain the curious similarities of the infrared-optical spectra of the jet knots.

III. THE CENTAURUS A JET

The X-ray source in Centaurus A has been known since 1970 as a prototypical active galactic nucleus: a compact, highly variable source located at the nucleus with strong absorption of its soft X-rays. There was hope that inverse Compton emission from the inner lobes would be seen with *Einstein*. The lobes were not seen, but a dramatic of unexpected picture emerged (Figure 2a): a string of X-ray knots extending almost 6' from the nucleus into the NE inner radio lobe. About seven knots, all aligned along P.A.53°, are statistically significant above background. They are 10-20" in diameter and are separated by 20-40", (Feigelson et al. 1981). The knots closer to the nucleus are brighter ($L_x \approx 1 \times 10^{39}$ erg/s in the 0.5-4.5 keV band) than the outer knots. When the X-ray knots were discovered (Schreier et al. 1979), only a uniform radio bridge was known to connect the nucleus and lobe. Maps made with the VLA, however, revealed that at least the inner two knots are also distinct radio knots (Figure 2b; Schreier et al. 1981; Feigelson 1981). Maps produced with the VLA A and B configurations show that the radio and X-ray morphologies of the inner knots are very similar (Burns et al. 1983).

The interpretation of the Cen A knots is more complex than that of the M87 jet. The latter is too X-ray luminous for plausible thermal or inverse Compton models. The Cen A knots are ∿100 times less luminous and are ∿7 times larger than those of M87, making it more difficult to exclude these models. However, the new detailed radio data does favor a synchrotron model. The magnetic field in Knot B, for instance, must be 100 times lower than the equipartition value of 34 μG to produce sufficient inverse Compton emission. Although depolarization is present, it seems unlikely the jet can contain the 0.2 cm^{-3} density of hot plasma required for a thermal model. If the X-ray emission is synchrotron, optical synchrotron must also be present. Brodie and colleagues (this conference) have present preliminary evidence for optical continuum emission in the knots.

IV. LOBES OF VIR A AND CEN A

Further out from the nuclei of these same two galaxies, additional intriguing X-ray features are seen in *Einstein* images. Embedded within the gaseous envelope of M87, which extends more than a degree in all directions, distinctly asymmetrical structure is seen between 2' and 5' from the center (Schreier et al. 1982). About 0.04 HRI cts/s, or 1-2 x 10^{41} erg/s in the *Einstein* band, emission above the symmetrical component is seen to the E and SW. The SW component in particular is very distinct, appearing as a distinct radial stripe ∿1' wide. There is little doubt that these structures are related to the radio halo, which has recently been resolved into misaligned components E and SW of the nucleus. The inverse Compton emission expected from the halo assuming an equipartition magnetic field of 4μG corresponds to 0.02 HRI cts/s, very close to the observed excess. This quantitative agreement makes the radio halo of M87 the most likely case of detected inverse Compton

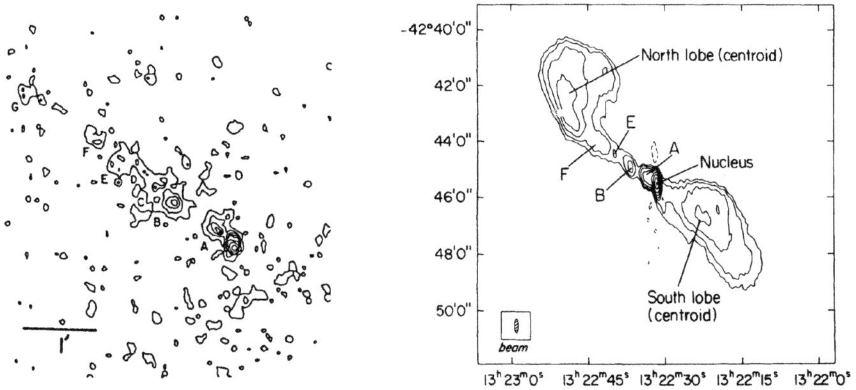

Figure 2. (a) *Einstein* HRI image of Centaurus A showing the nucleus and string of X-ray knots pointed along P.A.53° (Feigelson et al. 1981). (b) A low resolution VLA map at 6 cm showing the nucleus, inner lobes, and some of the X-ray jet knots (Burns et al. 1983).

X-ray emission in any radio jet or lobe in the sky. Unfortunately, the case is not yet airtight. First, the published radio maps have poor resolution ($\gtrsim$ 1' compared to 3" in the X-ray image) and dynamic range due to the bright radio core. Higher resolution VLA maps have recently been acquired and are being compared to the HRI image. A second problem is more fundamental. The excess X-ray emission could be thermal in origin if the ejected radio lobes have compressed the surrounding (or internal) gaseous medium.

The other reported case of X-ray lobe emission is the middle lobe of Centaurus A about 25' NE of the nucleus. The X-ray feature is weak, with a luminosity of $\sim 1 \times 10^{39}$ erg/s, and has a distinctly soft spectrum (Feigelson et al. 1981, Feigelson 1981). It is surprising that this relatively weak lobe ($\sim$10 Jy at 5GHz) is seen in X-rays while the stronger inner NE lobe ($\sim$70 Jy) and the giant outer lobe ($\sim$250 Jy) are not. A suggestive clue is the presence of turbulent gas, H II regions and young stars in the vicinity of the middle lobe (Graham and Price 1981). The X-ray emission could be thermal emission from the compressed or entrained interstellar medium. The soft spectrum implies the gas would rapidly cool to produce the optical emission line regions. Recent VLA observations, however, have complicated this picture by revealing a radio hot spot in the lobe coincident with the peak of the X-ray distribution. This suggests a non-thermal rather than thermal model.

V. OTHER POSSIBLE X-RAY JETS AND LOBES

3C273 (Feigelson 1980; Willingale 1981). A very deep HRI exposure of this nearby quasar was obtained during the first few weeks of the

Einstein mission with the express intention of detecting the jet. The principal difficulty is measuring a faint, resolved source within the wings of the extremely strong point source associated with the nucleus. After strenuous deconvolution of the point spread function, emission from the jet at a level of 1/400 that of the nucleus is seen at the 3-5 sigma confidence level. With $\alpha_{ro} \approx 0.9$ and $\alpha_{ox} \approx 1.3$, the electromagnetic spectrum of the jet is qualitatively similar to that of M87, suggesting a synchrotron interpretation. Modeling would be helped by the acquisition of accurate UBVRI photometry and polarimetry of the jet with modern solid state detectors.

Per A = NGC 1275 (Fabian and Kembhavi, 1982; G.C. Stewart et al., in preparation). NGC 1275 lies at the center of the Perseus cluster, which has a dense cooling core of intracluster gas with $\sim$300 $M_{\odot}$/yr accreting onto the galaxy. In addition to a compact core, the radio map includes jet knots several arc seconds from the core (Pedlar et al, 1983) and a larger structure about 2 1/2' to the SW. The inner jet is not seen in the HRI of the cluster ($\alpha_{rx} \geq 0.9$) although traces of X-ray emission are seen towards the outer structure. It is not yet clear if this reflects non-thermal emission or a compression of the intracluster medium.

Radio Galaxy in Abell 754 (D. Harris, C. Costain and P. Dewdney, in preparation). One member of a group of galaxies on the outskirts (2.4 Mpc) of Abell 754 is a radio galaxy with a strong core and weak tail or jet about 3' long. An HRI image shows an unresolved source at the nucleus with a very weak feature with a statistical synificance of 2σ at the same position angle as the jet. It is unclear whether the X-ray structure, if real, is generated in or around the jet.

PKS 0301-123 (E. Nelson, J. Burns, and R. White, in preparation). The nearby cluster of galaxies WP08 centered on NGC 1200 appears to contain two radio galaxies, one a head-tail and the other a nuclear source. A deep IPC image shows X-ray emission associated with several of the galaxies, with an X-ray morphology similar to that seen in the radio. Although a non-thermal mechanism is suggested, the spatial resolution of the IPC may not be sufficient to eliminate contributions from cluster gas or active nucleus X-ray emission.

VI. CONCLUSIONS

The current observational status of X-ray detections of jets and lobes is analogous to the status of all X-ray extragalactic studies in the mid-1970's. The Uhuru satellite detected the nearest quasar, the nearest clusters and the nearest active galaxies. Giacconi (1976) predicted that the more sensitive *Einstein* satellite would find many more such objects, and this prediction was fully borne out. We find that *Einstein* was able to detect X-rays from jets and lobes in the two nearest radio galaxies but generally not in more distant objects. It is a reasonable prediction that the next generation X-ray satellite, the Advanced X-ray Astrophysics Facility (AXAF), will find many more. AXAF will be

$\sim$100 times more sensitive with $\sim$7 times the resolution of the *Einstein* HRI. Since arc second resolution as well as high sensitivity is crucial to resolve the jet/lobe emission from nucleus and cluster X-ray emission, it does not seem likely that the imaging X-ray missions expected in the 1980's (ESA's *EXOSAT*, Germany's *ROSAT*, Japan's *Astro C*, or NASA's Explorer or Spacelab series) will revolutionize this field as *AXAF* can.

The theoretical interpretations and implications of X-ray emission in ejected material are important but not easily attained. There is a continual difficulty in distinguishing thermal and non-thermal mechanisms. However, the detailed morphological correspondence of X-ray and radio structures in M87 and Cen A point towards non-thermal processes. The two jets are probably synchrotron in origin, requiring incredibly efficient particle acceleration--with electron energies up to 100 TeV--in the rarefied environment of jet knots. This is a challenge for jet models that invoke a single macroscopic shock and those that entail turbulent acceleration. The M87 halo emission is a gratifying discovery. If further analysis confirms its inverse Compton origin, then various unproved assumptions (e.g. regarding the relative energy densities of electrons, proton, and magnetic fields) that pervade interpretation of radio structures will be verified. Even the failures to detect X-rays from certain radio structures may be of theoretical value. The hot spots of Cyg A, for instance, must produce X-rays far less efficiently ($\alpha_{rx} > 1.2$) than the jet knots of M87 and Cen A ($\alpha_{rx} \gtrsim 0.8$).

This work is supported in part by NASA grant NAS8-30752.

REFERENCES

Burns, J.O.: 1983, This volume.
Burns, J.O. and Balonek, T.J.: 1982, Ap.J. 263, 546.
Burns, J.O., Feigelson, E.D. and Schreier, E.J.: 1983, submitted to Ap.J.
Canizares, C.R., Clark, G.W., Jernigan, J.G. and Markert, T.H.: 1982, Ap.J. 241, 552.
Cooke, B.A., Lawrence, A. and Perola, G.C.: 1978, M.N.R.A.S. 182, 661.
Elvis, M., Schreier, E.J., Tonry, J., Davis, M. and Huchra, J.P.: 1981, Ap.J. 246, 20.
Fabricant, D., Lecar, M. and Gorenstein, P.: 1980, Ap.J. 241, 552.
Fabian, A.C. and Kembhavi, A.K.: 1982, In IAU Sympspium #97, Extragalactic Radio Sources (C. Wade and D. Heeschen, eds.), 453.
Feigelson, E.D.: 1980, Ph.D. Thesis, Harvard University.
Feigelson, E.D. and Schreier, E.J.: 1980, In Highlights of Astronomy (P.A. Wayman, ed.) 5, 695.
Feigelson, E.D.: 1981, In IAU Symposium #97, Extragalactic Radio Sources (C. Wade and D. Heeschen, eds.), 107.
Feigelson, E.D., Schreier, E.J., Delvaille, J.P., Giacconi, R., Grindlay, J.E. and Lightman, A.P.: 1981, Ap.J. 251, 31.
Feigelson, E.D. and Schreier, E.J.: 1983, Sky and Telescope, 65, 6.
Feigelson, E.D. and Berg, C.J.: 1983, Ap.J. in press.
Felten, J.E.: 1968, Ap.J. 151, 861.

Giacconi, R.: 1976, Ann. N.Y. Acad. Sci. 224, 149.
Graham, J.A. and Price, R.M.: 1981, Ap.J. 247, 813.
Harris, D.E.: 1982, In IAU Symposium #97 Extragalactic Radio Sources (C. Wade and D. Heeschen, eds.), 77.
Hayakawa, S., Matsuoka, M. and Sugimoto, D.: 1965, Space Sci. Rev. 5, 109.
Henry, J.P., et al.: 1979, Ap.J. Lett. 234, L115.
Kotanyi, C.: 1980, Astr. As. 83, 245.
Kriss, G.A., Canizares, C.R., McClintock, J.E. and Feigelson, E.D.: 1980, Ap.J. Lett. 235, L61.
Lampton, M., Margon, B., Bowyer, S., Makoney, W. and Anderson, K.: 1972, Ap.J. Lett. 171, L45.
Marshall, F.J. and Clark, G.W.: 1981, Ap.J. 245, 840.
Owen, F.N., Hardee, P.E. and Bignell, R.C.: 1980, Ap.J. 239, L11.
Pedlar, A., Booler, R.V. and Davies, R.D.: 1983, M.N.R.A.S. in press.
Schreier, E.J., Feigelson, E.D., Delvaille, J., Giacconi, R., Grindlay, J., Schwartz, D.A. and Fabian, A.C.: 1979, Ap.J. Lett. 234, L39.
Schreier, E.J., Burns, J.O. and Feigelson, E.D.: 1981, Ap.J. 251, 523.
Schreier, E.J., Gorenstein, P. and Feigelson, E.D.: 1982, Ap.J. 261, 42.
Shklovskii, I.S.: 1967, Sov. As. 11, 45.
Stocke, J., Rieke, G.H. and Lebovsky, M.J.: 1981, Nature 294, 319.
Willingale, R.: 1981, M.N.R.A.S. 539, L94.

X-RAY JETS IN BL LAC OBJECTS?

Peter Biermann
Max-Planck-Institut für Radioastronomie
Bonn, W.- Germany

ABSTRACT

A complete sample of flat spectrum radio sources selected from the 5 GHz survey of the north pole region is being studied at wavelengths in the radio, mm, infrared, optical, ultraviolet and X-ray regions. These objects and the fields around them are also being observed with spatial resolutions down to milli-arcseconds with VLBI and various conventional radio interferometers and imaging systems in the optical and X-ray range. We give a progress report on the work. The purpose of this study is the investigation of a complete sample of compact nuclei with extreme activity.

1. INTRODUCTION

A flux-density limited complete sample of flat spectrum radio sources was chosen from the 5 GHz survey carried out with the MPIfR 100 m radio telescope (Kühr et al. 1979, Kühr et al. 1981a, b). We used as criteria that a) the spectral index between 5 and 10.7 GHz be flatter than -0.3 ($S \propto \nu^{\alpha}$), and b) that the declination be larger than 70° (the S5-catalogue, Kühr et al. 1981a). A primary sample was defined by chosing a flux density limit of 1 Jy at the survey epoch at 5 GHz. This primary complete sample comprises ten sources. Secondary complete samples extend the flux density limit to below 1 Jy. The sources in these samples are being studied by a team of scientists at several institutions with a multitude of instruments. The principal partners are given in Table 1. Obviously we are also making use of the surveys of radio sources carried out at the VLA (Ulvestad et al. 1981, Perley 1982) and other published relevant work. A number of papers have appeared on our work (Biermann et al. 1981, 1982, Eckart et al. 1982, Biermann 1981, 1982, Johnston et al. 1982).

A. Ferrari and A. G. Pacholczyk (eds.), Astrophysical Jets, 173–182.
Copyright © 1983 by D. Reidel Publishing Company.

2. THE COMPACT SOURCES

As an example of a spectrum we show the source 0212+73 in Fig. 1. The results show typically the following (Biermann et al. 1981, 1982, Eckart et al. 1982, Geldzahler and Kühr 1982):

1) The spectrum of the sources is approximately flat ($\alpha \sim 0$) from 408 MHz to 300 GHz; from 300 GHz to 2μ in the infrared and to 2 keV in the X-ray range the spectral index is close to a -1 power law. This type of spectrum is suggestive of the Synchrotron-Self-Compton model. Within the radio range the sources are known to vary, consistent with their compact structure. Radio polarization measurements at three frequencies allow an intrinsic polarization angle to be determined.

2) With intercontinental VLBI of milli-arcsecond resolution the sources still show 80-100% of their total flux density in compact components; they are thus typically more compact than normal BL Lac objects (Weiler and Johnston 1980) which show only 50-60% of their total flux density in compact components. The compact structure defines a position angle if it has two components or is elongated.

3) The intrinsic radio polarization angle is close to being either parallel or perpendicular to the direction defined by the VLBI structure. Since for an optically thin Synchrotron source the radio polarization is perpendicular to the direction of the dominant magnetic field (and parallel for an optically thick source), we obtain information on the magnetic field direction.

4) The spectrum steepens from the infrared to the optical range, where it is often close to a powerlaw of slope -2 or even steeper. The slope in the infrared is marginally consistent with a straight powerlaw from the mm range. The extension of the optical spectrum lies below the X-ray flux density.

5) Over half of the sources in our primary sample have a steep optical spectrum, show no emission lines, are highly polarized and show evidence of variability in the optical wavelengths range. We conclude that we have identified BL Lac objects. In all cases, where the VLBI structure is well defined, the optical polarization angle is not far from the direction parallel to it; for an optically thin Synchrotron source, this would again suggest a magnetic field perpendicular to the VLBI axis. Scattering by an optically thin cloud flattened in a direction perpendicular to the VLBI axis could also explain this correlation (see Stockman et al. 1979, Angel and Stockman 1980).

6) Seven sources of the primary sample were observed in X-rays; all were detected. The optical to X-ray spectral index is typically flatter than for quasars (Zamorani et al. 1981).

The spectrum of the source 0212+73, Fig. 1, is especially interesting, since this BL Lac object (Biermann et al. 1981) shows a very hard X-ray spectrum. Adopting the HI-column density of Heiles (1975), we can attempt to fit a power-law spectrum to the data (800 photons obtained in two observations with a total observation time of 13000 seconds): Such a fit stronghly suggests a flat or inverted spectrum. A Synchrotron-Self-Compton model for this source can be constructed in the following manner (see, e.g., Blandford and Königl 1979): We adopt a -1 law from ca. 5 10^{10} Hz to 1.35 10^{14} Hz (2μ) from optically thin Synchrotron emission and then a sharp cutoff in the spectrum due to losses (see Rieke et al. 1982). The first Compton order peaks at energies at the upper end of the EINSTEIN-band (0.3-4.5 keV) or beyond, i.e. at 10^{18} Hz or higher. This requires then a value for the Lorentz-factor γ of the relativistic electrons of about 4.5 10^3 or higher, implying in turn a magnetic field of 6 10^{-4} Gauss or less. We can estimate the magnetic field also from the minimum energy argument (see e.g. Miley 1980).

Using the limiting Compton case (Kellermann and Pauliny-Toth 1969) of $T_b \approx 10^{12}$K, we obtain an estimate for the size of the source at 50 GHz (Kellermann and Pauliny-Toth 1981) of 40 μarc-sec. Putting that size into the minimum-energy argument yields a magnetic field of at least B = 0.6 Gauss; this is obtained for no energy at all in relativistic protons and is the value at z = 0.2 where the magnetic field estimated in this way is smaller than at any other redshift (spectral index -1, see above). Since bulk relativistic motion changes the flux density and the frequency, but does not change the size of a marginally resolved VLBI-component (the Lorentz contraction makes it unresolved in one dimension), the magnetic field obtained from the minimum energy argument like the boosting factor D_j^{-1}. However, the magnetic field obtained from the inverse Compton argument (relating mm and X-ray wavelengths) also varies as D_j^{-1} because of the frequency dependence on D_j. There are three ways to bring these two estimates of the magnetic field into agreement.

First, we could change our estimate of the Lorentz-factor of the relativistic electrons radiating from 4.5 10^3 down to 45 or less implying that we are observing several Compton orders in the various wavelength bands: then the mm-radiation would be the Synchrotron-Emission, the infrared emission would be first-order-Compton and the X-ray emission would have to be second- or third-order Compton emission.

Second, one could obviously abandon the assumption, that we are at a state of minimum energy. This would imply much higher particle energy densities and obviously much lower magnetic field strengths.

Third, one could attribute the infrared and/or X-ray emission to another component and then no easy conclusions are possible.

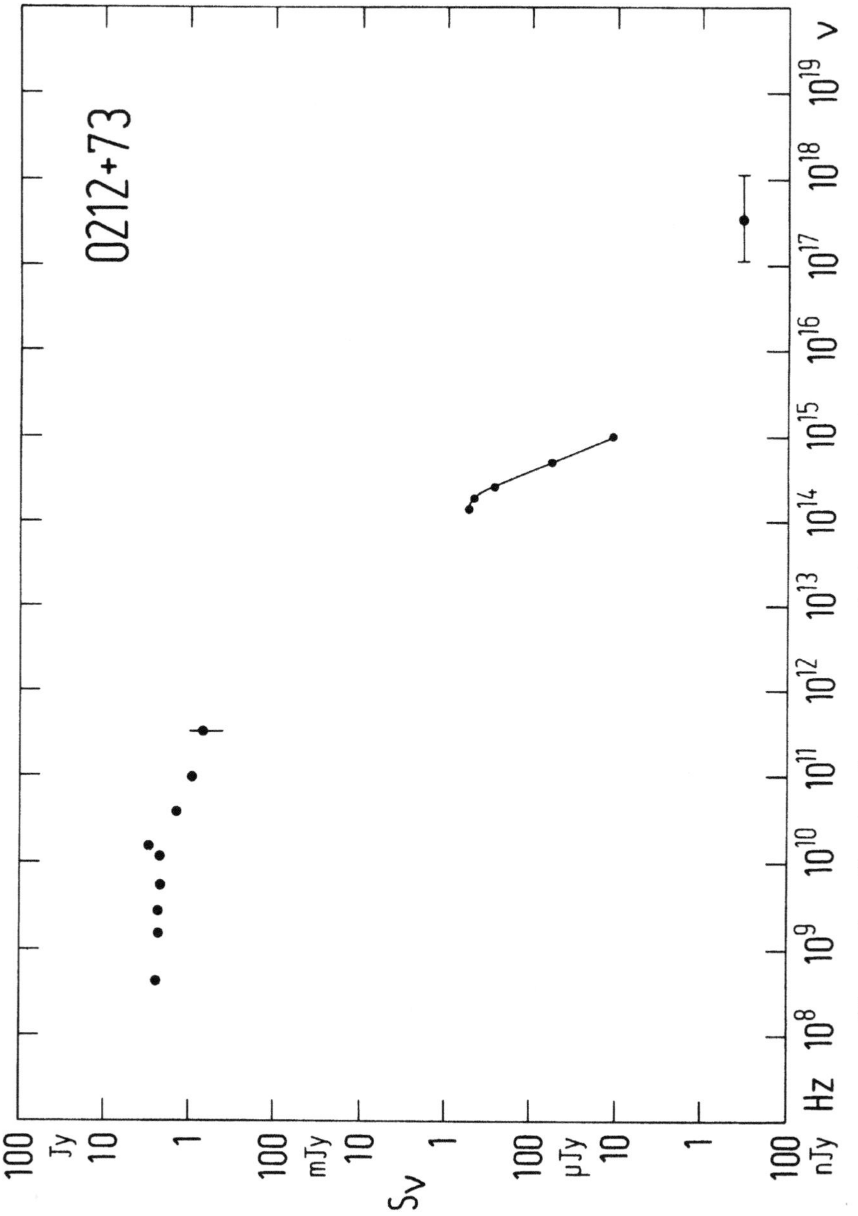

Figure 1: The total spectrum of the BL Lac object 0212+73. The X-ray data suggest a flat spectrum, as indicated by the horizontal line.

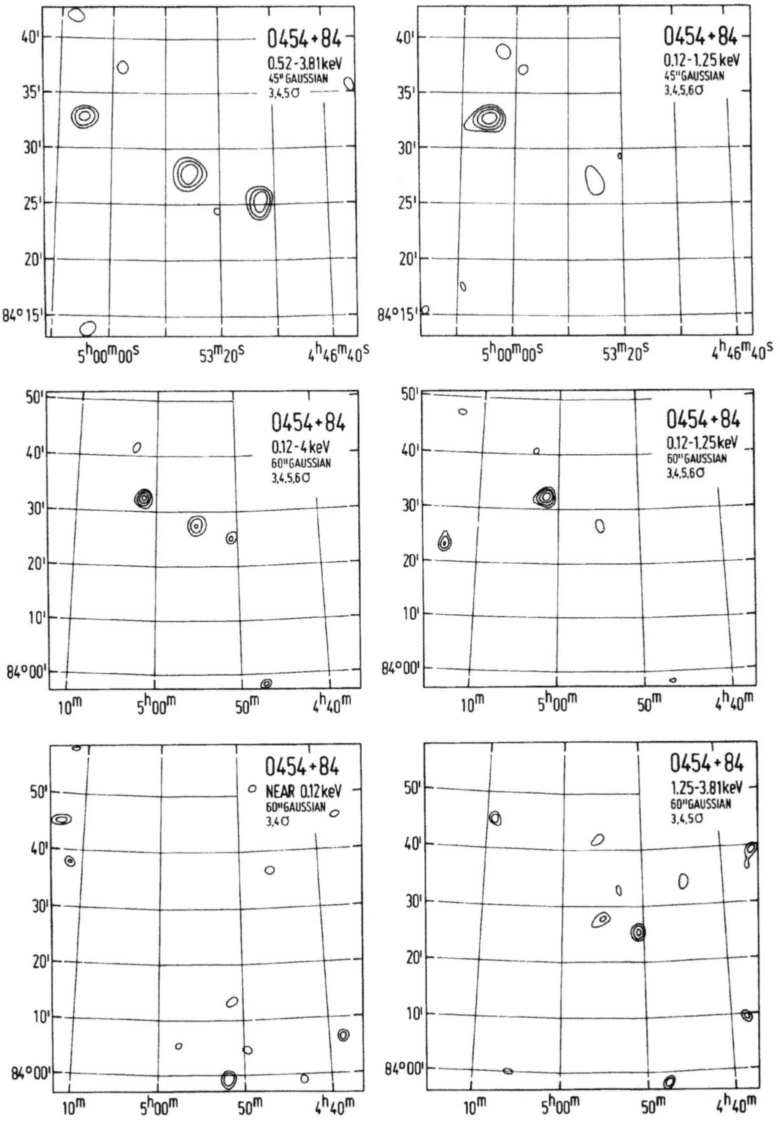

Figure 2: X-ray maps of the BL Lac object 0454+84 obtained with the IPC of the EINSTEIN-Observatory. The different maps have been formed with different smoothing gaussians and using different energy channels.

Further progress requires mm and submm data, mm-VLBI measurements and a better X-ray spectrum.

3. THE EXTENDED SOURCES

Here we discuss the fields around the compact sources (see Biermann et al. 1982), which we have studied with the X-ray maps, radio-maps obtained with the 100 m radiotelescope, with Westerbork, with the VLA, and with CCD on the MMT (see Table 1).

As example we give the X-ray maps of the sources 0454+84 (Fig. 2), 1803+78 (Fig. 3) and 1928+73 (Fig. 4), and discuss them (cf. Johnston et al. 1982).

We have selected these three fields because both in 0454+84 and 1803+78 there are three X-ray sources aligned in a triplet containing the central compact BL Lac object, while in 1928+73 there appears to be an X-ray extension of the central compact object, a quasar of redshift $z = 0.36$ (Biermann et al. 1981). At our sensitivity we do not expect more than at most one serendipitous X-ray source per field, and so the appearance of three aligned sources did suggest the possibility of emission from a large X-ray jet.

In 0454+84 we detected radio emission from the western X-ray knot (with evidence for hard X-rays). The radio spectrum is flat and with the VLA-position we have identified the source with a cluster of galaxies; presumably we are seeing the active dominant galaxy. The eastern X-ray knot (with evidence for very soft X-rays) has not been identified.

In 1803+78 we have detected radio emission from the northern X-ray knot. This radio emission shows structure at high resolution with a P.A. of $\sim -120°$. Even with positions good to better than $0\overset{''}{.}5$ and very deep CCD plates we have not been able to find any candidate identification. So we remain with various possibilities: a) If it were an active galaxy with extended radio emission (steep spectrum) at high redshift, the optical-X-ray spectral index would have to be flatter than usual. It would have to be a very red active nucleus. Also, the radio - X-ray spectral index is at the steep end of the distribution for serendipitous X-ray sources (Feigelson et al. 1982). b) If the X-ray emission were inverse Compton radiation off the microwave background from the radio photons seen (Biermann and Schlickeiser 1981), we again require an active nucleus to feed the radio lobes: This active nucleus feeding this emission through a channel might be an object not seen so far, or might also be the central BL Lac object. c) If the X-ray emission were unrelated to the radio emission, there could be no argument. The problem as to what the northern X-ray knot is, is unresolved.

In 1928+73, the magnetic field direction in the compact source, the optical polarization position angle, the VLBI structural angle and

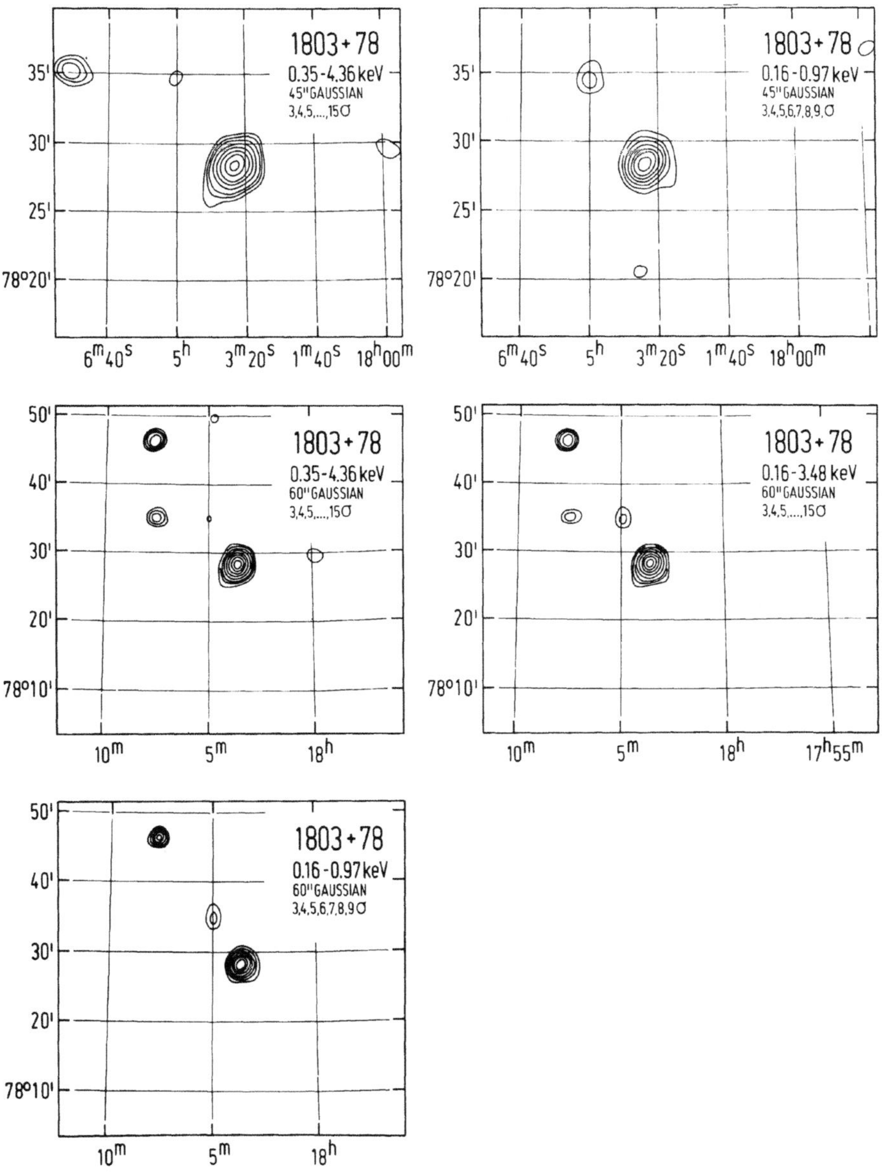

Figure 3: X-ray maps of the BL Lac object 1803+78.

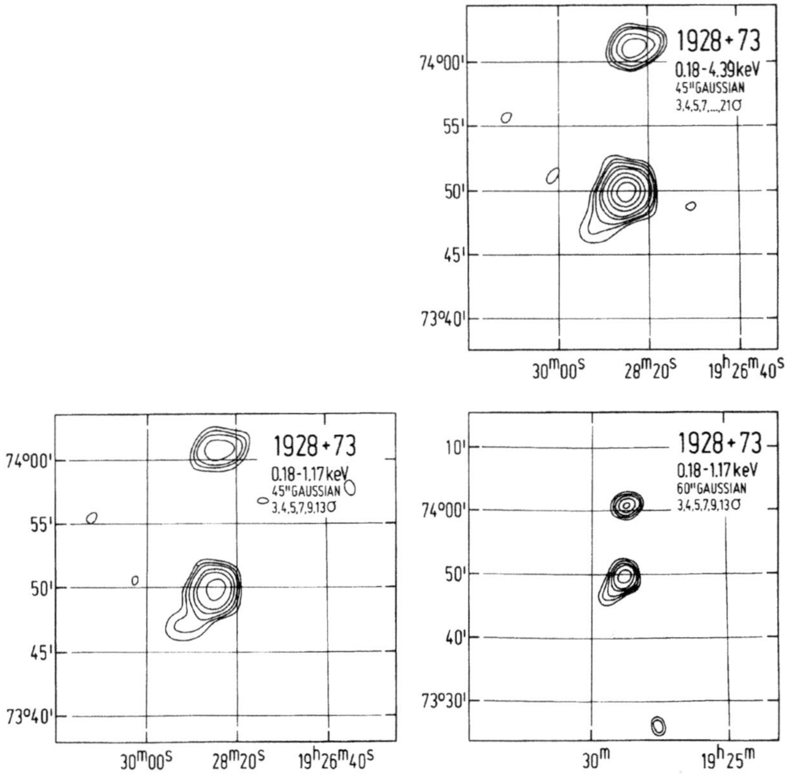

Figure 4: X-ray maps of the QSO 1928+73 (z = 0.36).

an apparent X-ray extension (Fig. 4) all agree at P.A.$\sim$160°. This might suggest, that we are seeing an X-ray jet; if so it would have a length of about 2 Mpc. Higher resolution X-ray studied are cleary required to check such a suggestion.

ACKNOWLEDGEMENT

The X-ray work of the author is supported by the Deutsche Forschungs-gemeinschaft (SFB131 and Bi 191/5). Helpful comments from L. Maraschi and E. Feigelson have been incorporated into the paper.

REFERENCES

Angel, J.R.P., Stockman, H.S.: 1980, Ann Rev. Astron. Astrophys. 8, 321
Biermann, P., Duerbeck, H., Eckart, A., Fricke, K., Johnston, K.J., Kühr, H., Liebert, J., Pauliny-Toth, I.I.K., Schleicher, H., Stockman, H., Strittmatter, P.A., Witzel, A.: 1981, Astrophys. J. Lett. 247, L53
Biermann, P., Schlickeiser, R.: 1981, Cosmic Ray Symposium Proceedings, OG3.2-11
Biermann, P: 1981, Space Science Rev. 30, 33
Biermann, P., Fricke, K., Johnston, K.J., Kühr, H., Pauliny-Toth, I.I.K., Strittmatter, P.A., Urbanick, M., Witzel, A.: 1982, Astrophys. J. Lett. 252, L1
Biermann, P.: 1982, Proceedings Nice workshop on hot astrophysical plasmas (in press)
Blandford, R.D., Königl, A.: 1979, Astrophys. J. 232, 34
Eckart, A., Hill, P., Johnston, K.J., Pauliny-Toth, I.I.K., Spencer, J.H., Witzel, A.: 1982, Astron. Astrophys. 108, 157
Feigelson, E.D., Maccacaro, T., Zamorani, G.: 1982, Astrophys. J. 255, 392
Geldzahler, B.J., Kühr, H.: 1982, Astron. J. (submitted)
Heiles, C.: 1975, Astron. Astrophys. Suppl. 20, 37
Johnston, K.J., Biermann, P., Eckart, A., Kühr, H., Strittmatter, P.A., Strom, R.G., Witzel, A., Zensus, A.: 1982, Astrophys. J. (submitted)
Kellermann, K.I., Pauliny-Toth, I.I.K.: 1969, Astrophys. J. 155, L71
Kellermann, K.I., Pauliny-Toth, I.I.K.: 1981, Ann. Rev. Astron. Astrophys. 19, 373
Kühr, H., Nauber, U., Pauliny-Toth, I.I.K., Witzel, A.: 1979, MPI-Preprint No. 55 "A catalogue of Radio Sources"
Kühr, H., Pauliny-Toth, I.I.K., Witzel, A., Schmidt, J.: 1981, Astron. J. 87, 854
Kühr, H., Witzel, A., Pauliny-Toth, I.I.K., Nauber, U.: 1981, Astron. Astrophys. Suppl. 45, 367
Miley, G.: 1980, Ann Rev. Astron. Astrophys. 18, 165
Perley, R.A.: 1982, Astron. J. 87, 859
Rieke, G.H., Lebofsky, M.J., Wisniewski, W.Z.: 1982, Astrophys. J. 263, 73

Stockman, H.S., Angel, J.R.P., Miley, G.K.: 1979, Astrophys. J. Lett. 227, L55
Ulvestad, J., Johnston, K.J., Perley, R., Fomalont, E.: 1981, Astron. J. 86, 1010
Weiler, K.W., Johnston, K.J.: 1980, Monthly Notices Roy. Astron. Soc. 190, 269
Zamorani, G., Henry, J.P., Maccacaro, T., Tananbaum, H., Soltant, A., Avni, Y., Liebert, J., Stocke, J., Strittmatter, P.A., Weymann, R.J., Smith, M.G., Condon, J.J.: 1981, Astrophys. J. 245, 357

Table 1

The Team working on flat spectrum radio sources from the S5-catalogue (Kühr et al. 1981a):

Radio flux density and polarization measurements:
408 MHz: Bologna cross: Gregorini, Mantovani
1.4-22 GHz: MPIfR 100 m telescope: Eckart, Pauliny-Toth, Witzel, Zensus et al.
Radio polarization: Beck, Forkert, Sieber, Turlo, Wielebinski
31.4 GHz: 11 m Kitt Peak: Geldzahler, Kühr
87 GHz: Hat Creek: Bieging
300 GHz: 6 m optical telescope USSR: Gemünd, Kreysa, Schultz, Sherwood

Radio structural observations:
606 MHz: Westerbork: Strom
1666 MHz: MERLIN: Eckart, Thomasson
1.4, 5, 15 GHz: VLA: Johnston, Kühr
1.666, 5 GHz: Transatlantic and European VLBI: Eckart, Johnston, Witzel

Infrared and optical measurements:
1.25, 1.6, 2.2 μ: MMT: Rieke-Lebofsky, Rieke
optical continuum: 5 m Palomar: Green
Spectroscopy: MMT, Steward 90": Kühr, Liebert, Strittmatter
Polarization: Steward 90": Stockman
CCD: Steward 90": Kühr, Strittmatter

Satellite observations:
IUE-satellite: Fricke, Kollatschny
EINSTEIN-satellite: Biermann

EXTENDED SOFT X-RAY EMISSION FROM NGC 4151

Ulrich G. Briel
Max-Planck-Institut für Extraterrestrische Physik

Martin Elvis
Harvard-Smithsonian Center for Astrophysics

J. Patrick Henry
University of Hawaii

INTRODUCTION

Extended X-ray emission associated with radio and optical jets has been reported in different types of active galaxies (see the review articles by E. Feigelson and by P. Biermann, both in this volume). The extended radio and optical features in the Seyfert galaxy NGC4151 are much weaker than in other active galaxies, and only careful optical (Ulrich, 1973; Fricke and Reinhardt, 1974) and radio (Wilson, 1981) measurements have revealed weak resolved emission, associated with the active nucleus. A program of long ($\sim 10^4$ sec) EINSTEIN- High Resolution Imager (Giacconi et al. 1979, Henry et al. 1977) observations of nearby Seyferts with resolved optical or radio features has been undertaken. NGC4151 seemed the best candidate for extended X-ray emission for the following reasons: Holt et al. (1980) reported that NGC4151 shows an excess of soft (< 1 keV) X-rays over that expected from an extrapolation from higher energies. This excess matches the peak response of the HRI which is around 0.7 keV. Besides that, the large and variable ($N_H \sim 10^{23}$ atoms cm^{-2}, Barr et al. 1977) low energy cut-off in the nucleus eases the dynamic range problem of searching for weak features near a strong source.

OBSERVATION AND SEARCH FOR EXTENT

NGC4151 was observed twice, in 1979 and 1980 for about 8000 sec in each case. To avoid any possible spurious effects introduced by the aspect solution we used only those parts of the observations when the star trackers were locked on their chosen guide stars. A first examination of the image using the standard procedures showed no significant deviation from the HRI-PSF.

A. Ferrari and A. G. Pacholczyk (eds.), Astrophysical Jets, 183–188.
Copyright © 1983 by D. Reidel Publishing Company.

We therefore chose a different approach of searching for smaller effects close to the nucleus, namely binning the image data in pie-shaped bins as in Figure 1. Important for this approach is the use of the correct center of the point source about which the binning would be performed, since a small displacement will cause photons from the central peak to be counted in outer bins resulting in an apparent asymmetry. We chose the position that gave the most symmetric distribution of counts in the central 2.5 arcsec radius. At this center we then chose a large radius of 10 arcsec and looked for asymmetries between the different quadrants outside the 2.5 arcsec inner circle. Figure 1a shows the result. A χ^2 fit gives a probability P (constant) $\leq 10^{-4}$ that the four outer quadrants are consistent with constant counts in each. The excess of 82 counts in the SW quadrants over the the mean of the other 3 quadrants is significant at the 5σ level.

This level of asymmetry is not a normal feature of HRI-images. To demonstrate this we have performed the same analysis on a segment of 3C273 data containing a similar number of counts (Figure 1b). In this case the counts are consistent with constant counts in each quadrant.

To estimate the position angle of the excess we have rotated the pie bins in 22.5° steps and performed the same χ^2 fit. The peak in the χ^2 distribution occurs at the initial binning chosen (Figure 2). Therefore our best estimate of the position angle of the excess is 225°.

Taking into account the scattering performance of the telescope this excess amounts to $\sim$ 15% of the total counts in the entire emitting region. The total number of background counts expected within this area is only about 5 and is neglected. Assuming a cut-off due only to our galaxy (2×10^{20} atoms cm^{-2}, Heiles, 1975) and a thermal spectra with 1 keV $\leq$ kT $\leq$ 10 keV, the excess flux (in the 0.5 - 4.5 keV band) lies in the range 6.5×10^{-13} erg cm^{-2} sec^{-1} to 1.3×10^{-12} ergs cm^{-2} sec^{-1}. With a distance to NGC 4151 of 20 Mpc these fluxes give luminosities of 3.1×10^{40} ergs sec^{-1} and 6.2×10^{40} ergs sec^{-1} respectively.

DISCUSSION

The probability of finding a source with a flux of 6.5×10^{-13} ergs cm^{-2} sec^{-1} in any given 10 arcsec radius circle is only $\sim 10^{-5}$ using the log N/log S of Maccacaro et al. (1982). Therefore it is very unlikely that the excess is due to a point source unassociated with NGC 4151. If it were a single point source associated with NGC4151 it would have a remarkable luminosity of $\sim 5 \times 10^{40}$ ergs sec^{-1}. Non-nuclear point sources with luminosities up to 10^{39} ergs sec^{-1} may have been seen (Long and van Speybroeck, 1982) and we cannot definitely exclude the possibility that we have detected a high luminosity point source near the nucleus of NGC4151. However, it seems more plausible that we are seeing extended emission at a position angle of 225° with a maximum approximately 5 arcsec from the nucleus.

The extended emission that we resolve to the SW amounts to about 20% of the soft component Holt et al. (1980) reported. Removal of this emission will modify somewhat their conclusions by decreasing the uncovered fraction of the central continuum source.

With what feature is this emission associated, assuming it to be extended? Figure 3 shows a composition of the extended optical (dotted areas; outer from Fricke and Reinhardt, 1974; inner from Ulrich, 1973), radio (lines of dots, Johnston et al., 1982) and our extended X-ray emission (shaded area to SW). Although the radio jet is close to the X-ray emission, it is at a different position angle and it does not appear that the two features are associated with each other.

More promising is an association with the forbidden line emitting gas. The position angle is in both cases the same and recent video camera data with large dynamic range (Heckman and Balick, 1983) have shown that the O III-emission is indeed asymmetric towards the SW, like the X-rays.

We will assume now, that the X-ray emission is associated with the forbidden line emitting region. We will discuss three possible mechanisms for producing it: thermal emission from a hot medium confining the forbidden emission line clouds; shock heating of the outflowing forbidden line clouds as they impact on the normal interstellar medium of the galaxy; enhanced star formation, stimulated by shocks originating from outflowing clouds leading to an unusually high concentration of SNR and massive X-ray binaries.

Krolik, McKee and Tarter (1981) proposed that the broad emission line clouds in AGN's are in pressure equilibrium with a hotter thermal plasma in order to confine them. A similar situation might exist for the narrow line clouds. For reasonable values of $n_e T$ (10^7 - 10^9 K cm^{-3}, Boksenberg et al. 1975) in the emission line clouds and making simple assumptions, the X-ray luminosity (Tucker 1975) would be 10^{45} - 10^{46} ergs sec^{-1}, far larger than observed. If the above assumptions are right, we can rule out this explanation, but it is, of course, very crucially dependent on the product $n_e T$.

The second possibility is that, as the narrow line clouds are ejected from the nucleus, possibly by interaction with the radio jet (Wilson, 1981), they should cause shocks in the normal interstellar medium of the galaxy. The line-of-sight velocities measured by Fricke and Reinhardt (1974) would produce a temperature of $\sim 2 \times 10^6$ K at the shock front. Assuming the interstellar density $n_e \simeq 1$ cm , then the emitting volume needed to give the observed X-ray luminosity corresponds to a sphere of radius 0.5 kpc which is similar to the observed size.

A factor two lower outflow velocity would reduce the X-ray emission, but could lead to an increase in star formation as in galactic spiral arms. To explain the luminosity we see would require

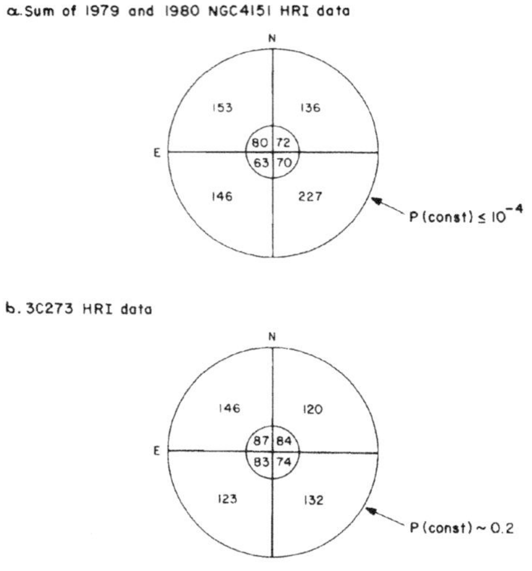

Figure 1 - Pie diagrams centered on the nuclear X-ray source in NGC4151 (a) and 3C273 (b). There is an excess of counts in the SW quadrant in the NGC4151 diagram. The probability that the counts in the outer quadrant are independent of azimuth, P (const), is given for each diagram.

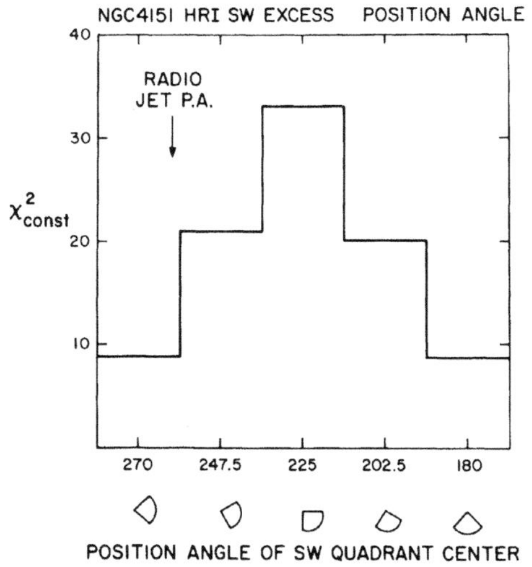

Figure 2 - NGC4151 SW excess as function of the position angle. Also indicated is the position angle of the radio jet.

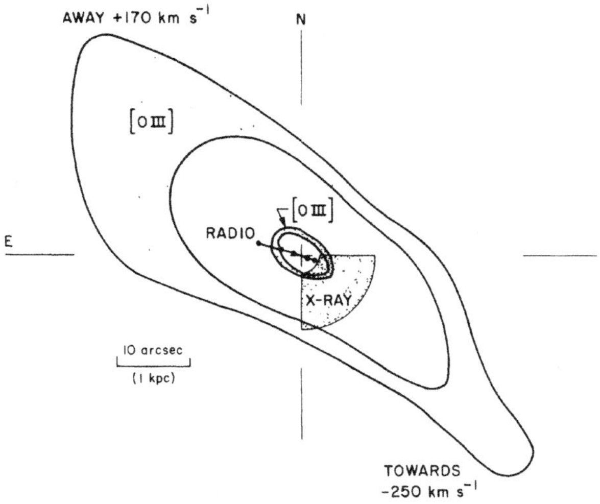

Figure 3 - Composite of extended optical (dotted areas; outer from Fricke and Reinhardt, 1974; inner from Ulrich, 1973), radio (line of dots; from VLA map of Johnston et al., 1982) and X-ray (shaded area to SW; this paper). The optical data come from deep (outer) and short (inner) long slit, image tube spectra at various position angles. For each area the inner contour shows the saturation level and the outer shows the detection limit. Also shown are the velocities of the O III lines in the NE and SW quadrants relative to the nucleus.

$\sim$ 5 x 10^4 SNR (at 10^{36} ergs sec^{-1} each), or $\sim$ 5 x 10^3 X-ray binaries (at 10^{37} ergs sec^{-1} each), or $\sim$ 5 x 10^2 luminous, SMC-X-1-like binaries (at 10^{38} ergs sec^{-1} each). Footprints of such a "starburst" hypothesis would be a strong UV flux from the sites of star formation regions. A preliminary examination of short U-band photographs of NGC4151 shows no clumping to indicate the presence of large numbers of H II regions (Malkan, M. 1982; private communication).

We prefer shock heating of the normal interstellar medium of NGC 4151 by the outflowing forbidden line clouds since our simple numerical estimates give reasonable, consistent results, and especially since we know there must be a shock front as the clouds impact on the interstellar medium.

This paper is a condensed version of a paper accepted for publication in The Astrophysical Journal (1983). This work was performed under NASA contract NAS 8-30751.

REFERENCES

Barr, P., White, N.W., Sanford, P.W., and Ives, J.C. 1977, Mon. Not. R. Astron. Soc., 181, 43
Fricke, K.J. and Reinhardt, M. 1974, Astron. Astrophys., 37, 349
Giacconi, R., et al. 1979, Ap. J., 230, 540
Heckman, T. and Balick, B. 1983 Ap. J., in press
Heiles, C. 1975, Astron. Astrophys. Suppl., 20, 37
Henry, J.P., Kellogg, E.M., Briel, U.G., Murray, S.S., Van Speybroeck, L.P. and Bjorkholm, P.J. 1977, Proceedings of the SPI, 106, 196
Holt, S.S. Mushotzky, R.F., Becker, R.H., Boldt, E.A., Serlemitsos, P.J., Symkoviak, A.E., and White, N.E. 1980, Ap. J., 241, L13
Johnston, K.J., Elvis, M., Kjer, D., and Shen, B.S.P. 1982, Ap. J., in press
Krolik, J.H., McKee, C.F., and Tarter, C. 1981, Ap. J., 249, 422
Long, K.S. and Van Speybroeck, L.P. 1982, "Accretion Driven Stellar X-ray Sources", eds. Lewin and van den Heuvel
Maccacaro, T., et al. 1982, Ap. J., 253, 504
Tucker, W.H. 1975, "Radiation Processes in Astrophysics", MIT, Cambridge, Mass., USA
Ulrich, M.-H. 1973, Ap. J., 181, 51
Wilson, A.S. 1981, IAU Symposium no. 97, "Extragalactic Radio Sources", pp. 179-188, (D. Reidel)

JETS FROM SUPERCRITICAL ACCRETION DISKS

M. Calvani
Institute of Astronomy, University of Padova, Italy
L. Nobili
Institute of Physics, University of Padova, Italy

INTRODUCTION

A large variety of astrophysical objects like extended (Miley 1980, Perola 1981) and compact (Kellermann and Pauliny-Toth 1981) radio-sources, quasars and BL Lac objects (O'Dell 1979), etc. have given rise to the question on the nature of their central powerhouse. Several observational evidences like enormous energy output, short time-scale variability in the flux, high and variable optical polarization (Angel and Stockman 1980) suggest that the central core hides a large concentrated mass, probably a black-hole of $\sim 10^8 M_\odot$.

The above objects are frequently associated with, or imply the existence of beams of gas squirting from the core. These beams may extend up to hundreds of kiloparsecs and their orientation, at least for some sources, must have remained steady for $\sim 10^8$ yrs. Recently moreover similar beams have been observed (on a much smaller scale) also in galactic objects like SS433 (Mammano et al., 1980) and R Aquarii (Sopka et al., 1982).

A model which in many respects can account for the above phenomenology is a geometrically thick accretion disk orbiting the central black-hole. In fact the gas originating from stellar collisions, tidal disruption and stellar mass loss is likely to form an accretion disk whose thickness will depend on the mass inflow rate and the efficiency of radiative processes relative to the local rate of energy dissipation.

Two kinds of geometrically thick accretion disks (called also tori) have been recently considered: 1) radiation pressure supported disks which form when the accretion rate is supercritical (Paczynski 1982, Wiita 1982); 2) ion-supported disks which may form at very subcritical rates (Rees et al., 1982). Both these models share the common feature of having two narrow funnels free of material around the rotational axis of the hole, a configuration which suggests the possibility of producing beams as pointed out by Lynden-Bell (1978).

A. Ferrari and A. G. Pacholczyk (eds.), Astrophysical Jets, 189–199.
Copyright © 1983 by D. Reidel Publishing Company.

However the two models differ in the way in which beams are accelerated: for radiation supported disks the acceleration mechanism could be a hydrodynamical one, while for ion-supported disks it is the extraction of rotational energy of the hole (via magnetic fields anchored to the disk) which powers the beams.

In the following we shall consider in more details radiation pressure supported disks and we shall discuss the possibility of producing fast jets.

SUPERCRITICAL ACCRETION DISKS

When the accretion rate onto the black-hole becomes supercritical, the disk must become geometrically thick otherwise instabilities will develop and the disk will puff up in a dynamical time-scale. The critical accretion rate is

$$\dot{M}_{cr} = \frac{64\pi GM}{ck} \simeq 5\times10^{-8} \left(\frac{M}{M_\odot}\right) \quad M_\odot \; yr^{-1}$$

(G is the gravitational constant, c the velocity of light, k the opacity per unit mass and M the mass of the hole) and corresponds to the rate which gives rise to a total luminosity equal to the Eddington limit $L_{Edd} \simeq 10^{38}(M/M_\odot)$ erg/s of a spherical star of mass M.

If the disk is thick, so that its height h is comparable with the distance r from the axis, the approximation used in the standard theory of thin accretion disk (Pringle 1981, Hoshi 1981) breaks down and it must be replaced with a theory of thick disks.

On the basis of a simplified Newtonian description of thick disks (Abramowicz et al., 1980) (which is really not a strong limitation as one can easily model those relativistic effects whose influence is important by using a modified Newtonian potential (Paczynski and Wiita 1980, Jaroszynski et al. 1980)

$$V(r,z) = -GM \; \{(r^2+z^2)^{\frac{1}{2}} - r_g\}^{-1}$$

where (r,z) are cylindrical coordinates and $r_g = 2GM/c^2$ is the gravitational radius) it was shown that all the relevant properties of the disk like total luminosity, shape, etc. depend only on the mass M and the geometrical ratio r_{out}/r_{in} of outer to inner edge of the disk, when the disk has the maximal luminosity.

The shape of a thick disk with $r_{out}/r_{in} = 10^5$ is shown in Fig. 1; note the steep and narrow funnels along the rotational axis. The surface temperature of the disk, also shown in the figure, is evaluated assuming a blackbody relationship $T = (F/\sigma)^{1/4}$ (where F is the flux of radiation) and scales as $M^{-1/4}$.

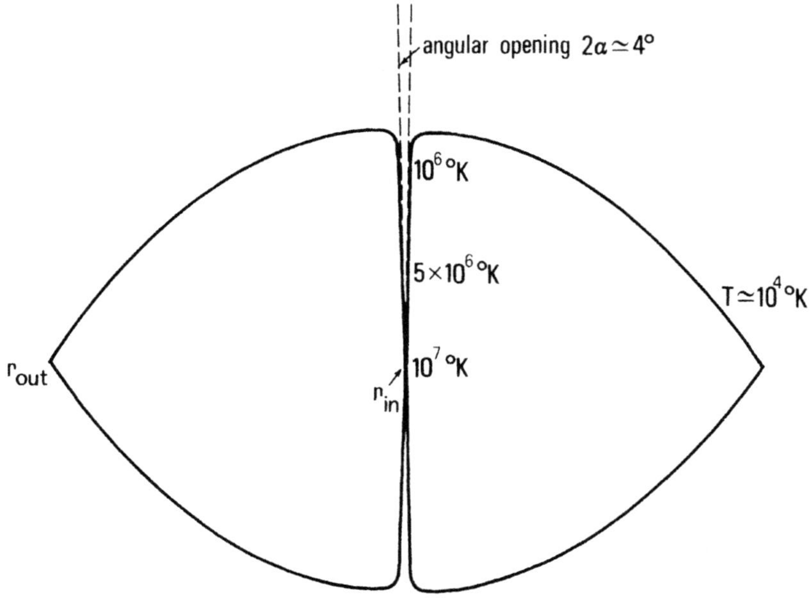

Fig. 1

The shape of a maximal luminosity disk with $r_{out}/r_{in} = 10^5$. The temperature distribution is also shown for $M = 1\ M_\odot$ and it scales as $M^{-1/4}$.

A relevant property of thick disks is that their total luminosity, obtained integrating the flux over the whole surface, may be well in excess of the Eddington limit for a spherical star of mass M (Paczynski 1982, Abramowicz et al., 1980). The flux moreover is strongly anisotropic, a major fraction of the luminosity being radiated from the surface of the funnels (Sikora 1981) and therefore the apparent luminosity along the axis of the funnels can be as high as $\sim 100\ L_{Edd}$. In fact it is this last result, together with the shape of the funnels, which suggest that radiative acceleration may be relevant in explaining cosmic jets.

JETS FROM SUPERCRITICAL ACCRETION DISKS

The several papers (Abramowicz and Piran 1980, Sikora and Wilson 1981, Piran 1982, Abramowicz and Sharp 1982) which were up to now devoted to the study of acceleration and collimation of jets from thick disks considered the acceleration of a single particle under the action of the radiation flux of the funnels. It was shown that good collimation (beams of $\sim 5°$ half-opening angle) and γ Lorentz factors of ~ 2 for ordinary plasma, could in such a way be achieved. However none of them allowed for the gas in the funnels to be optically thick nor a correct hydrodynamical treatment was made.

In the following we shall discuss a model in which a continuous mass loss from the walls of the funnels is considered and we shall present a hydrodynamical description of the problem.

The strong flux of radiation F(z) from the surface of the disk will cause a mass loss which we assume to be described by a law similar to the one used in stellar wind theory (Chiosi 1981, Lamers 1981), that is

$$\phi = \phi_o F(z)^n \qquad \text{g cm}^{-2}\text{s}^{-1} \tag{1}$$

where ϕ_o and n are constants ($1 < n < 1.5$). The above relation is a good choice for an accretion disk around a black hole of few solar masses because we expect such a disk to resemble the atmosphere of an early type supergiant(+). For more massive systems the law of mass loss can be different; nevertheless in the lack of more informations we shall use the same relation for all configurations.

Schematically we can outline the flow in the funnel as shown in Fig. 2, in which z_o gives the location of the "stagnation point" where the instantaneous gas velocity is zero: below this region the gravitational forces of the central black hole prevail over gas and radiation pressure forces; the reverse is true for $z > z_o$.

For simplicity we have neglected gas viscosity and, as the opening angle of the funnel is small, we have reduced the problem to a one dimensional flow. On doing this semplification we must introduce in the hydrodynamical equations some "creation terms" which take into account the inflow of matter with zero momentum but non zero internal energy, and which are related to the function ϕ.

With this assumption the stress-energy tensor takes the form:

$$T_i^j = (\rho c^2 + p)u_i u^j + p\delta_i^j + Q_i^j + \Pi_i^j \tag{2}$$

where u^i is the four-velocity of the fluid, $p = p_G + p_R$ is the isotropic (gas + radiation) pressure, ρ the gas density. The two terms Q_i^j and Π_i^j give the heat flow and the mass inflow contributions respectively (their explicit derivation will be given elsewhere in Calvani et al. (1982)). Taking the divergence of T_i^j in a Schwarzschild background and using the standard procedure (Novikov and Thorne 1972) we can derive the explicit form of the Navier-Stokes equations which, with the

(+) In the funnels however the tangent component of the radiation field on the surface could somehow change the law (1).

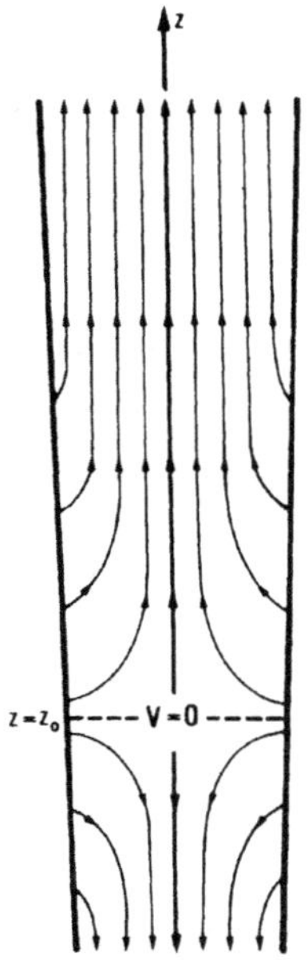

Fig. 2

Schematic view of the gas flow in the funnel.

rest mass conservation law, provide the full set of the relativistic hydrodynamical equations. They read:

$$\frac{d(\gamma\rho\beta)}{dz} = \frac{\Phi}{\sqrt{1-\chi}} - 2\gamma\rho\beta\left[\Lambda + \frac{1}{z}\left(\frac{d\ln r}{d\ln z} + \sqrt{1-\chi} - 1\right)\right] \qquad (3)$$

$$\gamma^2\beta\,\frac{d\beta}{dz} = -\left(\frac{P'}{\rho c^2} + \Lambda + \frac{\beta\Phi}{\rho\sqrt{1-\chi}}\right) \qquad (4)$$

$$\gamma\rho\beta\sqrt{1-\chi}\,\frac{de}{dz} = -p\,\theta + (\gamma-1)\,\Phi c^2 + Q/c \qquad (5)$$

where β is the gas velocity in units of c and

$e = \frac{3k}{m_p}T$ = specific internal energy of the gas

$\chi = \frac{2GM}{c^2 z} = \frac{r_g}{z}$ = gravitational "potential"

$$\Lambda = \frac{r_g}{z^2\ (1-\chi)} \quad = \text{gravitational "force"}$$

$$\theta = (\Phi - \gamma\beta\sqrt{1-\chi}\ \frac{d\rho}{dz}) / \rho \quad = \text{expansion}$$

$$\Phi = \frac{2\phi}{r} \quad = \text{rate of mass accretion per unit volume}$$

$$P' = \begin{cases} \dfrac{dp}{dz} & \text{for optically thick gas} \\ \dfrac{dp_G}{dz} + \dfrac{k\rho}{c} H_z & \text{for optically thin gas} \end{cases}$$

H_z = z-component of the radiative flux

$r = r(z)$ = radius of the funnel

The first term in the r.h.s. of eq. (5) describes the expansion work, the second one the dissipation of mechanical energy of the material captured by the flow, whilst Q is the rate of heat exchanged by the gas (per unit volume) whose explicit form depends on the physical conditions of the jet. When the gas is optically thick, and the diffusion approximation holds, Q is a cumbersome function of the first and second derivatives of temperature, density and velocity according to the relativistic radiative transfer equation. Considering only the leading terms, the explicit form of Q is, in this case:

$$Q = \frac{p_R}{k\rho} \left\{ \frac{1}{p_R} \frac{d^2 p_R}{dz^2} + 4\gamma^2\beta \frac{d^2\beta}{dz^2} + 7\gamma^2\beta \frac{d\beta}{dz} \frac{d\ln p_R}{dz} + 4\gamma^4 (1+4\beta^2) (\frac{d\beta}{dz})^2 + \right.$$
$$\left. - (\frac{d\ln p_R}{dz} + 4\gamma^2\beta \frac{d\beta}{dz}) \frac{d\ln\rho}{dz} \right\} + Q_w$$

where $Q_w = 2F(z)/r$ is the heating due to the flux from the walls.

When the gas is optically thin, Q is merely a balance between radiative gains and losses and it takes into account all absorption and emission processes (Shapiro 1973), including the Compton heating and cooling mechanisms which in our model can be a very efficient way of exchanging energy (Calvani and Nobili 1982).

The mathematical solution of this hydrodynamical problem is obtained by solving the set of Eqs. (3), (4), (5) and (1), with the following boundary conditions:

$$\text{at} \quad z = z_o \quad \text{(stagnation point)} \qquad \begin{cases} \beta = 0 \\ \rho = \rho_o \\ T = T_o \\ \dfrac{dT}{dz} = T_o' \end{cases}$$

However the choice of the parameters z_o, ρ_o, T_o and T_o' is not arbitrary because when the flow is supersonic we must add to the equations some special conditions in order to obtain non-singular (and then physical) solutions. In fact when the velocity of the gas approaches the isothermal sound velocity $\beta_c = \frac{1}{c}(\partial p/\partial \rho)_T$, the derivatives of velocity and density diverge unless z_o, ρ_o etc. are related in some particular way. Moreover the energy equation (5) is badly behaved near the adiabatic sound velocity $\beta_S = \frac{1}{c}(\partial p/\partial \rho)_S$ and this provides a further condition on the parameters (for a more detailed discussion see Flammang (1982)). Because in our configuration the sound velocities β_c and β_S can be attained by the gas either below or above the stagnation point, four indipendent conditions must be satisfied, so that the problem has only one degree of freedom: once we fix the value of ϕ_o in Eq. (1) we obtain the complete behaviour of the physical quantities which describe the gas flow. In particular we obtain values for the mass loss rate, velocity, density and temperature of the gas at the top of the funnel, which is the region of greatest interest as we can compare these values with those inferred from the observational data.

From a numerical point of view the problem is rather complicated and it needs much computer time. In order to overcome the difficulty connected with the energy equation, we have used the reasonable assumption that in the regime of moderately low velocity up to $\beta \simeq \beta_S$ there is a good thermal contact between the beam and the walls of the funnel, so that in that region $T_{gas} \simeq T_{wall}$.

Figs. 3a and 3b show the behaviour of $\beta(z)$, $\rho(z)$ and $T(z)$ which were obtained using the above approximation for two relevant values of the mass of the central black hole, i.e. $M = 5M_\odot$ and $M = 10^8 M_\odot$ respectively. In the first case the value of the accretion parameter ϕ_o was choosen in such a way to obtain $\beta \simeq 0.26$ as the final velocity (model for SS433).

It is worth noting the steep increase in the gas temperature in the outer part of the funnel. This particular behaviour is due to two concomitant physical effects which occur when the gas velocity reaches very high values: (a) the gas becomes optically thin and this reduces its efficiency to irradiate internal energy; (b) the interaction of the jet with the matter coming from the walls causes a strong dissipation of energy. This interaction acts as a friction so that on one hand the gas velocity tends to decrease (in absence of any other forces), on the other hand the conversion of mechanical into thermal energy produces the increase of temperature shown in the figures. In this region the main cooling mechanism is inverse Compton scattering of electrons with the photons outflowing from the disk surface. Note that in the region of high temperature the shape of the disk is likely to be unaffected by the jet owing to its low density.

The mass loss rate in the jet ($\dot{M} = 10^{-8} M_{\odot}$/yr for $M = 5M_{\odot}$ and $\dot{M} = 0.1 M_{\odot}$/yr for $M = 10^{8} M_{\odot}$) as well as the final density deduced from our model are in good agreement with the observational data on SS433 and on several quasars.

DISCUSSION

Aim of this paper was to stress the relevance of geometrically thick disks in explaining the origin of astrophysical jets. Although we have addressed ourselves mainly to radiation pressure supported disks, which form when the accretion rate is supercritical, also ion supported tori could help in explaining the jet phenomenology, depending on the supply of gas which "feeds the monster". As suggested by Rees and his collaborators (1982) the majority of radio-quiet quasars could in fact be explained in terms of a $\sim 10^{8} M_{\odot}$ black hole surrounded by radiation supported tori without dominant magnetic fields, while radio-loud quasars, BL Lac objects and superluminal sources could be associated with electromagnetically powered tori.

There are however still many open problems which should be solved before thick disks may be considered a viable explanation of the activity of galactic nuclei and we shall now briefly comment on those which we believe relevant in the context of astrophysical jets.

The evidence on the velocity of extragalactic jets is rather conflicting. There are arguments, based on observations of steep spectrum cores of double radio-galaxies (Saikia 1981) or on radio variability of central components (Hine and Scheuer 1980), which seem to imply typical values of $\gamma \simeq 2$; on the other hand, a scheme which attempts to unify flat-spectrum with steep-spectrum quasars is in agreement with observations if $\gamma \simeq 5$ (Orr and Browne 1982). Moreover extreme values of $\gamma \simeq 10$ are needed to explain apparent superluminal velocities on the basis of relativistic bulk motion (Cowen and Unwin 1982). The model which we have discussed in the previous section can easily account for jets of ordinary plasma with $\gamma \simeq 2$; somewhat higher values could be obtained changing the law (1) and assuming for instance that mass loss occurs mainly in the deepest part of the funnels. High values of γ seem however hard to obtain with hydrodynamical acceleration of the gas because either Compton drag prevents the

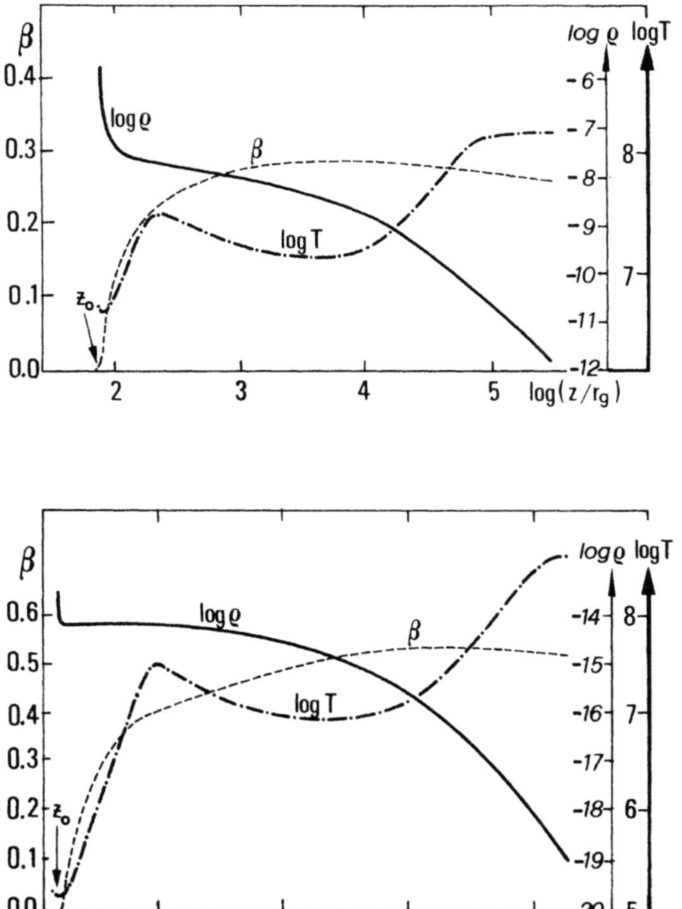

Fig. 3

Velocity, density and temperature distribution in the jet along the funnel for the two cases $M = 5\ M_\odot$ and $M = 10^8 M_\odot$. The top of the funnel is at $\log(z/r_g) \simeq 5.3$.

attainment of relativistic speeds (Rees 1981) or the radiation is seen more and more red-shifted by the gas.

At last we shall mention the question of stability of thick disks which is still an open problem. One can envisage this matter in the light either of thermal and dynamical instabilities which are typical of rotating configurations (Hacyan 1982, Abramowicz et al. 1982) or of instabilities which are peculiar of the system disk + black hole, like for instance catastrophic mass exchange (Abramowicz et al. 1983).

ACKNOWLEDGEMENTS

We wish to thank M.A. Abramowicz for several helpful discussions. Work supported by the Italian National Council for Research (C.N.R.).

REFERENCES

Abramowicz, M.A., Calvani, M. and Nobili, L., Astrophys. J., 242, 772 (1980)
Abramowicz, M.A., Calvani, M. and Nobili, L., Nature, submitted (1983)
Abramowicz, M.A., Livio, M., Piran, T. and Wiita, P., (1982) preprint
Abramowicz, M.A. and Piran, T., Astrophys. J., 241, L7 (1980)
Abramowicz, M.A. and Sharp, N., (1982) preprint
Angel, J.R.P. and Stockman, H.S., Ann. Rev. Astron. Astrophys., 18, 321, (1980)
Calvani, M. and Nobili, L., Lett. Nuovo Cimento, 35, 7 (1982)
Calvani, M., Nobili, L. and Turolla, R., (1982) in preparation
Chiosi, C., Astron. Astrophys., 93, 163 (1981)
Cowen, M.H. and Unwin, S.C., in "Extragalactic radio sources" IAU Symp. n. 97, p. 345 (1982) Heeschen and Wade eds., Reidel, Dordrecht, Holland
Flammang, R.A., M.N.R.A.S., 199, 833 (1982)
Haycan, S., Astrophys. J., (1982) to appear
Hine, R.G. and Scheuer, P.A.G., M.N.R.A.S., 193, 285 (1980)
Hoshi, R., Suppl. Prog. Theor. Phys., 70, 181 (1981)
Kellermann, K.I. and Pauliny-Toth, I.I.K., Ann. Rev. Astron. Astrophys., 19, 373 (1981)
Jaroszynski, M., Abramowicz, M.A. and Paczynsky, B., Acta Astron., 30, 1 (1980)
Lamers, H.J., Astrophys. J., 245, 593 (1981)
Lynden-Bell, D., Physica Scripta, 17, 185 (1978)
Mammano, A., Ciatti, F. and Vittone, A., Astron. Astrophys., 85, 14 (1980)
Miley, G., Ann. Rev. Astron. Astrophys., 18, 165 (1980)
Novikov, I. and Thorne, K.S., in "Black holes", p. 343 (1972), De Witt ed.
O'Dell, S.L., in "Active galactic nuclei", C. Hazard and S. Mitton eds., Cambridge Univ. Press (1979)
Orr, M.J.L. and Browne, I.W.A., M.N.R.A.S., 200, 1067 (1982)

Paczynski, B., Karl Schwarzschild, Lecture 1981, to appear in Astronomische Gesellschaft Mitteilungen (1982)
Paczynski, B. and Wiita, P.J., Astron. Astrophys., 88, 23 (1980)
Perola, G.C., Foundamentals of Cosmic Physics, 7, 59 (1981)
Piran, T., Astrophys. J., 257, L23 (1982)
Pringle, J.E., Ann. Rev. Astron. Astrophys., 19, 137 (1981)
Rees, M.J., in "Origin of Cosmic Rays" IAU Symp. n. 94, p. 139 (1981), Setti, Spada and Wolfendale eds.
Rees, M.J., Begelman, M.C., Blandford, R.D. and Phynney, E.S., Nature, 295, 17 (1982)
Saikia, D.J., M.N.R.A.S., 197, 1097 (1981)
Shapiro, S.L., Astrophys. J., 180, 531 (1973)
Sikora, M., M.N.R.A.S., 196, 257 (1981)
Sikora, M. and Wilson, D.B., M.N.R.A.S., 197, 529 (1981)
Sopka, R.J., Herbig, G., Kafatos, M. and Michialitsianos, A.G., Astrophys. J., 258, L35 (1982)
Wiita, P.J., Comments in Astrophysics (1982) to appear

BLACK HOLE - DRIVEN HYDROMAGNETIC FLOWS
flywheels vs. fuel

E.S. Phinney
Institute of Astronomy, Madingley Road, Cambridge CB3 OHA, England

SUMMARY

A Kerr black hole in a magnetic field shares many of the properties of a conductor rotating in a magnetic field. In particular it can act as a dynamo. Physical processes in the black hole magnetosphere suggest several astrophysically plausible "loads". One of the most likely is a hydromagnetic wind. This leads to a nearly perfect impedance match with the black hole dynamo and extracts energy from the hole at a rate $\sim$ 64% of the maximum rate allowed by the second law of (black hole) thermodynamics. The emergent wind may drive the jets seen in radio sources.

I. INTRODUCTION

Ever since it was realized that energy could be extracted from a rotating black hole (Penrose 1969) people have been searching for applications, from garbage disposal (Misner et al. 1973) to the construction of bombs (Press and Teukolsky 1972). The idea that the rotational energy of black holes (flywheels) may power active galactic nuclei has been made more compelling by the observation (Phinney 1981, Rees et al. 1982) that the large ratio of L_{jet} to $L_{nucleus}$ in some radio galaxies may preclude accretion (fuel) as a source of power. A rotating black hole may also act as the fire under the pot in the cauldron model of jet formation (Begelman, these proceedings) which is relevant at the other extreme of supercritical, highly trapped accretion flow.

All magnetospheres are fundamentally the same: one has a rotating conductor, a magnetic field and a current-carrying plasma. In what follows, the reader is introduced to the relevant concepts in the simple context of a conducting sphere in an externally imposed magnetic field. The analysis is then repeated with a Kerr black hole replacing the conducting sphere, and the close analogies are pointed out. The elements of black hole circuit theory are introduced, and the problem

A. Ferrari and A. G. Pacholczyk (eds.), Astrophysical Jets, 201–213.
Copyright © 1983 by D. Reidel Publishing Company.

of determining the external load discussed. It is shown that the most likely load, a relativistic MHD wind, gives a nearly perfect impedance match to the black hole, and thus extracts energy at a rate near the maximum possible for a given magnetic flux imposed and confined by surrounding matter (say in an accretion disc or torus). The magnetically dominated relativistic wind can drive a bipolar jet flow and accelerate entrained or injected particles.

II. THE ANALOGY OF A CONDUCTOR

A perfectly conducting sphere of charge rotating in an aligned uniform magnetic field $B_o \hat{z}$ (Fig. 1) has induced on it a surface charge density

$$\sigma_s = - \frac{\Omega R}{8\pi c} B_o (3\cos^2\theta - 1) + Q/_{4\pi R^2} \tag{1}$$

and thus an electric field with only monopole and quadrupole ($C_2 = -\frac{1}{3} \frac{\Omega R}{c} R^4 B_o$) components in the vacuum above its surface. The poloidal electric field gives an azimuthally circulating Poynting flux, but no net radiation to infinity, and thus no energy loss.

In vacuum, if the rotation axis is inclined at a nonzero angle γ to B_o, currents will circulate over the conductor, and if it has a surface impedance Z_s, Joule heating will dissipate energy on the surface at a rate

$$\dot{Q}_s = \frac{1}{Z_s} \frac{2\pi}{3} \left(\frac{\Omega R}{c}\right)^2 B_o^2 R^2 \sin^2\gamma \tag{2}$$

and the sphere will feel an aligning torque.

If a plasma surrounds the aligned rotating sphere, it can carry an axial current, generating an azimuthal magnetic field and allowing a net Poynting flux, even in an axisymmetric steady state (the unipolar induction dynamo). This Poynting flux must be present to replenish energy which in turn must be dissipated at large distances. Suppose the surrounding plasma creates a force-free magnetosphere, so that currents can there flow only along magnetic field surfaces. The vacuum quadrupole electric field (cf. Fig. 1) would try to drive surface currents from the equator towards the poles. In steady state these currents must be carried out and returned along field lines after passing through a region where the force-free approximation breaks down sufficiently for the currents to cross field lines (Fig. 2).

The power dissipated on the "sphere at infinity" depends crucially on its surface impedance Z_∞. The total voltage drop around the circuit is essentially fixed at

$$V \sim \Omega_s R^2 B_o/c = \frac{\Omega_s}{c} \Phi , \tag{3}$$

where Φ is the flux threading the sphere. If Z_∞ is too low ($Z_\infty << Z_s$) there is a large current, but most of the voltage drop (and hence Joule dissipation) occurs across the rotating sphere. If Z_∞ is too large ($Z_\infty >> Z_s$), very little current flows and the joule dissipation is correspondingly low. The power dissipated on the sphere at infinity is maximized when $Z_\infty = Z_s$:

$$L_\infty \text{ (Max)} = \frac{V^2}{4Z_s}$$

III. THE ROTATING BLACK HOLE DYNAMO

A rotating black hole has many properties in common with our rotating conductor. The effective surface charge density σ_H induced on a black hole of total charge Q in an asymptotically uniform vacuum magnetic field B_o is

$$4\pi\sigma_H = \frac{1}{(r_+^2+a^2\cos^2\theta)^2}\left[(r_+^2 - a^2\cos^2\theta)(Q+ aB_o(r_+-M)(1+\cos^2\theta)) - 2aB_o r_+(r_+^2 - a^2)\cos^2\theta\right] . \tag{4}$$

For $a/M << 1$, this simplifies to

$$\sigma_H \simeq -\frac{\Omega_H r_+}{8\pi c} B_o (3\cos^2\theta - 1) + Q/4\pi r_+^2 \tag{5}$$

which is identical in form to Eqn. (1). The multipole structure has associated far-field tangential electric fields which will drive currents when we add a magnetosphere. For $r >> r_g$, the electric field induced by the black hole is (as seen by non-rotating observers)

$$\vec{E}_{ZAMO} = \frac{a}{M}\frac{B_o r_g^2}{r^2}\left[(Q/aB_oM + (1-3\cos^2\theta))\,\hat{r} - \sin(2\theta)\hat{\theta}\right] \tag{6}$$

It may be surprising that even an uncharged black hole gives rise to an electric field dropping as r^{-2}. This can be understood in an intuitive (though only half correct) way as an effect of the dragging of inertial frames. These have angular velocities with respect to Boyer-Lindquist observers of

$$\omega_{ZAMO} = \frac{2aM}{r^3} \rightarrow V_{ZAMO} \sim \frac{aM}{r^2} .$$

The reader is hereby warned that in most of the following G = c = 1. The notation for quantities in the Kerr metric and sign conventions are those of Misner et al. (1973): M is the mass of the hole ($M = GM/c^2 \equiv r_g$ also has dimensions of length), a is the hole's angular momentum per unit mass, r_+ = Boyer-Lindquist radius of the horizon ($r_+ \rightarrow 2M$ for $a \rightarrow 0$, $r_+ \rightarrow M$ for $a \rightarrow M$), etc.

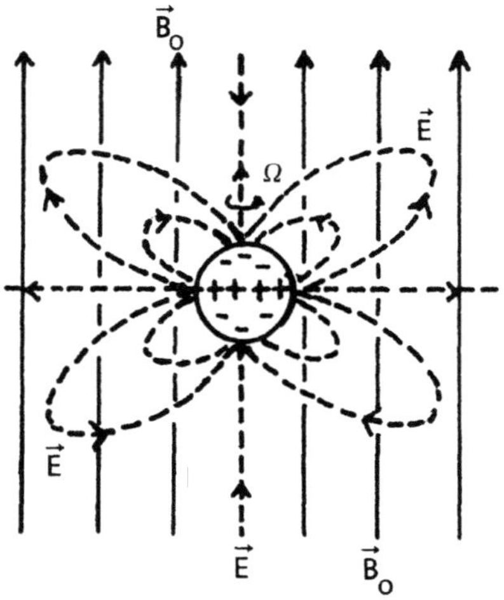

Fig. 1: An uncharged conducting sphere rotating in vacuum in an imposed magnetic field $B_o\hat{z}$.

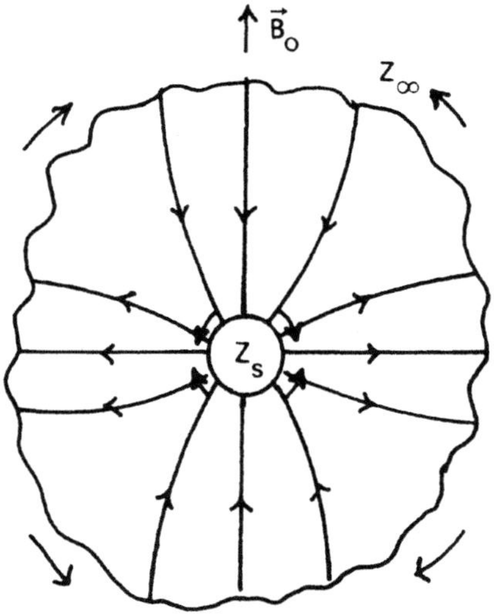

Fig. 2: A conducting sphere rotating in a force-free magnetosphere with an imposed magnetic field. Arrows indicate the direction of current flow. Compare with the vacuum electric fields of Fig.1.

hence

$$E_{\text{FRAME DRAGGING}} \sim V_{\text{ZAMO}} \times B \sim \frac{aMB_o}{r^2} = \frac{a}{M}\frac{B_o r_g^2}{r^2}$$

As before the induced poloidal electric field gives rise to an azimuthally circulating Poynting flux, but not to any net radiation to infinity.

A black hole in a slowly rotating ($\sim \cos(m\phi - \omega t)$) vacuum external field will superradiantly scatter the incident waves, losing energy at a rate (Znajek 1977)

$$L_{\text{rot,vac}} \sim \left(\frac{\omega r_g}{c}\right)\left(\frac{\Omega_H r_g}{c}\right) c\, B_\perp^2\, r_g^2$$

while one with rotation axis misaligned by an angle γ with $\vec{B}_o(\infty)$ (steady and uniform) feels an aligning torque and surface currents dissipate energy on the horizon at a rate (Damour 1978)

$$\dot{Q}_H = 1/6 \left(\frac{\Omega_H r_+}{c}\right)^2 c\, B_o^2\, r_+^2 \sin^2\gamma \left[\frac{(r_+^2 + a^2)(2M)^2}{r_+^4}\right] \quad (7)$$

which is identical to Eqn.(2) for $a \ll M$ if we identify the surface impedance of the horizon $Z_H = 4\pi/c = 377$ ohms.

If we put the hole in a "circuit" with currents flowing into and out of the horizon, some of this energy becomes available to do work at large distances. Thus if the hole is surrounded by plasma, it can, like the conducting sphere, drive a DC Poynting flux (Blandford and Znajek 1977, Carter 1979, Macdonald and Thorne 1982). The dissipation in the circuit of Fig. 2 is determined as before by the impedances of the hole (Z_H) and the sphere at infinity (Z_∞). Remarkably, we can calculate Z_H in the limit of degenerate fields: the fields are kept force-free ($F_{\mu\nu} J^\nu = 0$) by particles sliding along field lines. But as "seen" from infinity (Boyer-Lindquist coordinates) particles never cross the horizon (time dilation) but orbit it forever at $d\phi/dt \to \Omega_H$. For a given poloidal magnetic field, the azimuthal magnetic field must tend to infinity fast enough that particles can slide around the horizon along the field lines:

$$F_{\mu\nu} U^\nu = 0 \rightarrow F_{\mu\nu}\, dx^\nu = 0$$

which implies (Ω_F is the angular velocity of field lines):

$$B_\phi = \frac{dt}{dr}\left(\frac{d\phi}{dt} - \Omega_F\right) A_{\phi,\theta}\,.$$

But as $r \to r_+$, $\Delta \to 0$, $\frac{d\phi}{dt} \to \Omega_H$, $\frac{dt}{dr} \to -\frac{r_+^2 + a^2}{\Delta} \to \infty$

$$\underbrace{B_T}_{\text{finite}} \equiv \frac{\Delta \sin\theta}{\underbrace{\rho_+^2}_{\to 0}} \underbrace{B_\phi}_{\to\infty} \to \frac{(r_+^2 + a^2)\sin\theta}{\rho_+^2} (\Omega_F - \Omega_H) \underbrace{A_{\phi,\theta}}_{\text{poloidal field}} \tag{8}$$

at the horizon. This boundary condition (in fact only a consequence of $\vec{E} \cdot \vec{B} = 0$), fixes the azimuthal magnetic field (or equivalently, the axial current) in terms of Ω_F (a measure of Z_∞, as we shall see below) and the imposed poloidal magnetic field, and therefore determines $Z_H \sim 1/c = 30$ ohm. We can even construct a little table on black hole circuit theory for engineers:

Quantity	Value in: Geometric units	SI units
V	$\Phi\Omega_H$	$\sim 10^{19}$ volts$(B/10^4 G)(M/10^8 M_\odot)^2(a/M)$
Z_H	1	30 ohms
Z_∞	$\frac{\Omega_F}{\Omega_H - \Omega_F}$	??? (but see section V)

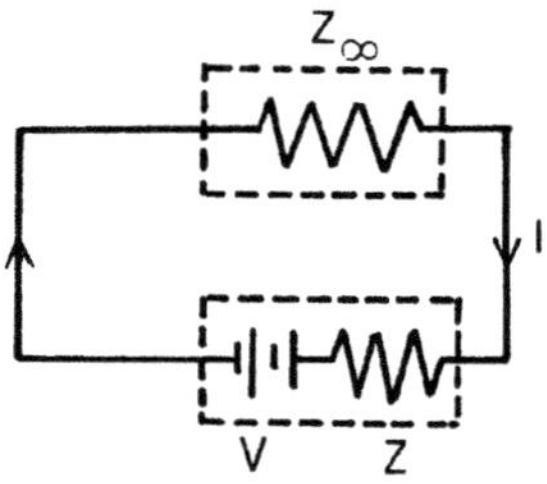

particle accel^n region

force-free magnetosphere

black hole

IV. MAGNETOSPHERIC PHYSICS

In making the approximation of a globally force-free magnetosphere, we have thrown away all information about the inertia of the particles in the magnetosphere. Since it is the acceleration of these particles and the associated transfer of energy from electromagnetic to kinetic form in the region of Joule-heating which determines Z_∞, we must re-introduce inertia if we are to predict Z_∞ and the extracted power. The force-free approximation can break down in three ways:

1) A vacuum region may allow acceleration of particles by electrostatic fields (appropriate to gap models of pulsars (Arons 1979), whose strong-wave wind can exclude all external sources of neutralizing particles and whose internally created particles are so relativistic that the Debye length is of the order of the neutron star radius).

2) A region may become unstable to plasma modes which can accelerate ("heat") particles in countless complicated ways (familiar from auroral physics and reconnection in the earth's magnetotail).

3) On sufficiently large scales, the guiding centre (perfect MHD) approximation becomes useful, and energy is transferred from fields to bulk kinetic energy only by drift forces. The equations are simple only if plasma heating is negligible or has a simple form (appropriate to the solar wind, winds from Ap stars, and parts of Jupiter's magnetosphere).

(Only a cynic would notice that in the applications mentioned, the physics has gotten most complicated in the areas most accessible to observational tests).

It was initially thought (Lovelace, Mac Auslan, Burns 1979) that a scenario of type (1) provided an accurate picture of black hole magnetospheres. This is unlikely to be the case, however, for if the hole is surrounded by an accretion disc with luminosity L_c in photons of energy > 1 MeV (and observations indicate that most of the energy radiated by Seyfert galaxies and quasars <u>is</u> carried off by such photons), then photon-photon collisions in the "vacuum" above the disc will create $e^+ e^-$ pairs at such a rate that the pair density in the magnetosphere

$$n_p \gtrsim (r_g/c) \ \dot{n}_{\gamma\gamma} \sim (m_p/m_e) \ (L_c/L_{Edd})^2 \ n_{Edd}$$

where $n_{Edd} = 10^{13} M_8^{-1} \ cm^{-3}$. This is orders of magnitude higher than the density required to maintain a force-free magnetosphere, and in fact is so high that if we tried to accelerate a relativistic charged beam in such a plasma, bunching instabilities would grow (Godfrey et al. 1975) on timescales $\sim \omega_{pe}^{-1}$ which are much shorter than the dynamical times:

$$\omega_{pe}^{-1}/(r_g/c) \simeq \frac{\lambda_{Debye}}{r_g} \simeq 10^{-16} \ M_g^{-1/2} \ (L_c/L_{Edd})^{-1} \qquad (9)$$

This suggest that scenarios (2) or (3) are more appropriate. The huge difference in scales motivates us to consider (3), and assume that there are enough charges and collisionless pitch-angle scattering that we can everywhere find a moving reference frame ("fluid frame") in which the electric field is so small that Joule heating can be neglected on the dynamical time. The pair creation plasma density is a bare minimum value: flares from the disc may inject additional particles, and in the quasi-spherical "cauldron" scenario (Begelman, these proceedings) the distinction between disc and magnetosphere disappears, and the density is $\gtrsim n_{Edd}$ everywhere.

V. HYDROMAGNETIC FLOWS

The equations of stationary, axisymmetric MHD in a curved background spacetime can be integrated to give 4 constants of the motion. These fix all properties of the flow except the poloidal magnetic field structure which is determined by A_ϕ, the azimuthal component of the vector potential. A_ϕ is in turn the solution of an unpleasantly nonlinear second order partial differential equation in two variables. Fortunately, one can extract a great deal of useful information about possible flows from the integrals of motion alone without knowing much about the precise solutions for A_ϕ (other than their existence, a pious hope). The integrals of the motion are:

$\mathcal{e}$, the energy-at-infinity per unit rest mass
ℓ , the angular momentum-at-infinity per unit rest mass
k , the ratio of rest mass flux to magnetic flux in a flux tube
Ω_F, the angular velocity of the magnetic field lines.

It is a straightforward exercise to write down the full general relativistic wind equation along an arbitrary poloidal field line. This is an algebraic equation (too lengthy to reproduce here) which determines all the flow parameters (density, four-velocity and azimuthal magnetic field) as a function of position along the field line, in terms of the four integrals of the motion (which enter as constant parameters). In the limit of a cold (pressureless) equation of state, this is simply a quartic for $u^r(s)$, the radial Boyer-Lindquist component of the four-velocity.

The MHD equations have two boundary conditions at the black hole horizon. One is the electromagnetic boundary condition discussed in Section III. The other is essentially that the fluid 4-velocity become asymptotically null ingoing at the horizon. It can be shown that these two boundary conditions guarantee that the second law of black hole thermodynamics is satisfied (the area of the horizon always increases, though the hole's mass may decrease if energy is extracted).

Because of the horizon boundary condition, and because we are interested in solutions with outflow to infinity, since flow cannot reverse direction on field lines when particles are conserved, there must be a region between the horizon and infinity in which particles are created, some to fall into the hole, others to be driven off to infinity. This particle creation is provided naturally by the photon-photon collisions and disc flares already discussed (bubbling of the cauldron is another mechanism). The flow is thus divided in two - an ingoing half and an outgoing half, separated by a surface of zero velocity in the particle creation region (see Fig. 3). The wind equation has three critical points at which pairs of solutions cross: the slow magnetosonic (which disappears in the limit of pressureless flow), the Alfvén, and the fast magnetosonic. A solution which passes through a sonic point makes a smooth transition from subsonic to supersonic flow. Solutions which do not pass through a sonic point are either everywhere subsonic or everywhere supersonic with respect to the

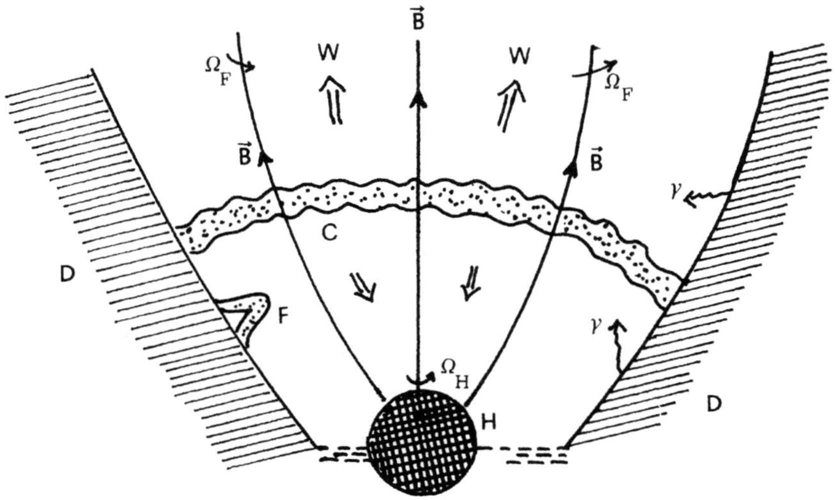

Fig. 3

A Kerr black hole in a magnetosphere, driving a hydromagnetic wind W. Particles appear on open magnetic field lines in the particle creation region C (through $\gamma\gamma \rightarrow e^+ e^-$, or from flares F on the accretion disc D). The disc D is necessary to confine the magnetic field B which threads the hole's horizon H. Magnetic flux will be carried onto the hole by the accretion flow until $B^2/8\pi \sim p_{disc}$. After that, accretion can proceed only by a 'drip' process followed by reconnection, which avoids adding any more flux to the hole. The magnetic field may also be generated by dynamo action in a convectively unstable disc. If the disc is radiation-supported, we can build up $B^2/8\pi \sim p_{rad}$, provided that the field gradients occur on scales much longer that the photon mean free path.

relevant wave. The four solutions of the quartic correspond to physical inflowing and outflowing solutions, and to two unphysical solutions which originate on, and have infinite velocity gradients at the light cylinder (which is not a critical point).

The horizon boundary condition forces the ingoing half of the flow to pass through all three critical points and cross the horizon as a supersonic flow (if it didn't, MHD waves could propagate out of the hole!). The condition that it does so replaces the horizon boundary conditions by another set, since the flow between the last sonic point and the horizon can no longer communicate with the outside world. If this occurs close enough to the horizon, this should be immaterially different from the force-free condition

$$\mathcal{k}\mathcal{e} \simeq \Omega_F (\Omega_H - \Omega_F) \Phi \mathcal{k}_H \qquad (10)$$

where $\mathcal{k}_H \simeq 1/4\pi$ (and depends weakly on the poloidal field structure).

The outflow is more interesting. Consider first the cold wind limit. In the case, the fast magnetosonic point moves out to infinity (the Alfvén point always remains inside the light cylinder). Define $\sigma_F = (\Omega_F)^2 \Phi/\mathcal{k}\mathcal{k}_F$, where $\mathcal{k}_F \simeq \pi$. As is well known (Michel 1969, Goldreich and Julian 1970, Okamoto 1978, Fujimura and Kennel 1982) the condition that the flow pass through the critical point is

$$\mathcal{e} = \sigma_F (1 + \sigma_F^{-2/3})^{3/2} \qquad (11)$$

Defining

$$\sigma_H = (\Omega_H)^2 \Phi/\mathcal{k}\mathcal{k}_F \ ; \ \alpha = \Omega_H/\Omega_F \ ; \ K = \mathcal{k}_F \ \mathcal{k}_H \simeq 1/4 \ , \qquad (12)$$

the condition that the inner and outer energy fluxes match is

$$(1-\alpha)\ K = -[\ 1 + (\alpha^2/\sigma_H)^{2/3}\ ]^{3/2}, \qquad (13)$$

with solution α_c. In the situations of interest $\sigma_H >> 1$, so that $\alpha_c \simeq 1 + 1/K \simeq 5$ and $\Omega_{Fc} \simeq 1/5\ \Omega_H$, which corresponds to an extracted power 64% of the maximum conceivable (when $\Omega_F = \frac{1}{2}\Omega_H$).

Cold flows which do not pass through the critical point have poloidal four-velocities u_p at infinity given by the roots of

$$(1-\alpha)^2\ K^2\ u_p^{\ 2} - (1+u_p^{\ 2})[\ 1 + \alpha^2\ \frac{u_p}{\sigma_H}\ ] = 0 \qquad (14)$$

These are shown in Fig. 4 for $\sigma_H = 1{,}000$, $K = 1/4$. For $\alpha < \alpha_c = 5.61$ ($\Omega_F > .18\ \Omega_H$) there are no outgoing wind solutions. For $\alpha = \alpha_c$, the wind passes through the critical point (and the equation has a double root). For $\alpha > \alpha_c$ ($\Omega_F < .18\ \Omega_H$), the equation has two real roots: a high velocity one, belonging to the unphysical solutions which originate at the light cylinder, and a low velocity one belonging to the

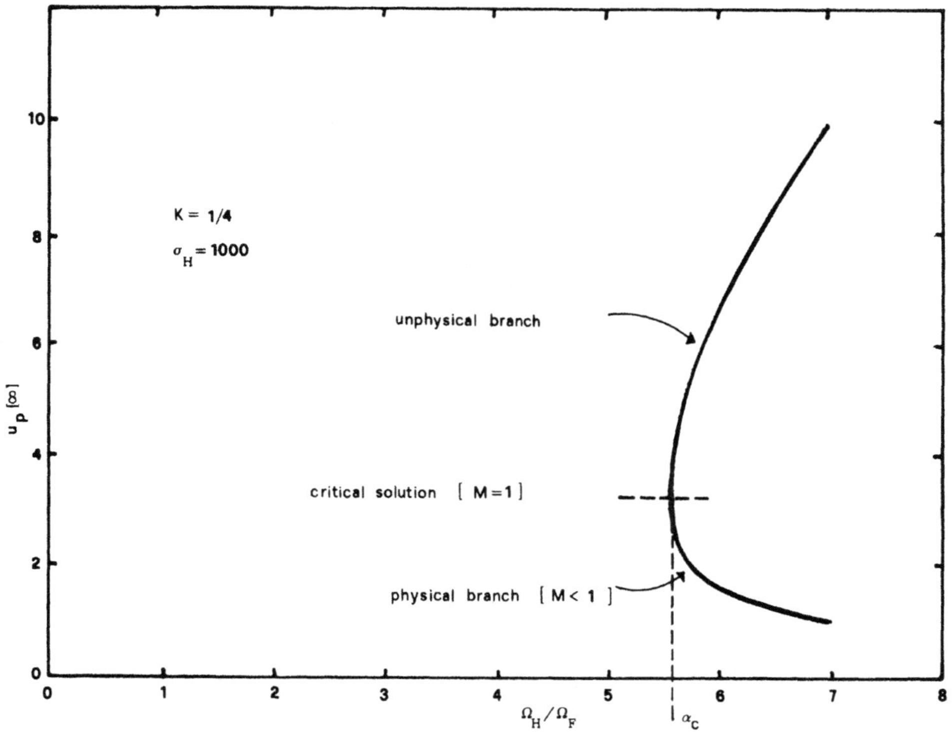

Fig. 4

Cold black hole-driven wind solutions.

physically accessible solutions originating within the light cylinder and everywhere subsonic. These seem to have finite velocity and zero pressure at infinity. This is an artifact of the cold plasma approximation, which has moved the critical point to infinity. If the plasma is allowed to have finite pressure, the critical point moves in from infinity, and one finds that the subsonic solutions decelerate beyond it to zero velocity and finite pressure. The only solution starting inside the light cylinder with a fast magnetosonic Mach number $M < 1$ and extending smoothly to a zero-pressure flow (with $M > 1$) at infinity is the critical one. For sufficiently large σ_H, the inclusion of finite pressure does not affect the critical condition significantly, and one still has $\Omega_{Fc} \simeq (1/5)\,\Omega_H$.

What happens to the flow beyond the fast critical point cannot affect the inner part. Interesting things that can happen include the injection and acceleration of particles in the electromagnetically dominated flow, until it becomes a jet whose energy is carried mainly by particles, rather than a Poynting flux.

REFERENCES

Arons, J., 1979, Space Sci. Rev., 24, 437

Blandford, R.D. & R.L., Znajek, 1977, MNRAS, 179, 433

Carter, B., 1979 in General Relativity, S.W., Hawking & W., Israel, eds. (Cambridge U. Press, Cambridge), 294

Damour, T., 1978, Phys. Rev. D, 18, 3598

Fujimura, F.S. & C.F., Kennel, 1982, preprint

Godfrey, BOB, W.R., Shanahan & L.E., Thode, 1975, Phys. Fluids, 18, 346

Goldreich, P. & W.H., Julian, 1970, Ap. J., 160, 971

Lovelace, R.V.E., J., MacAuslan, & M.L., Burns, 1979 in Particle Acceleration Mechanisms in Astrophysics, J. Arons, C. McKee & C. Max, eds. (AIP, N.Y.)

Macdonald, D. & K.S., Thorne, 1982, MNRAS, 198, 345

Michel, F.C., 1969, Ap. J., 157, 1183

Misner, C.W., K.S., Thorne, & J.A., Wheeler, 1973, Gravitation (W.H. Freeman, San Francisco), 908

Okamoto, I. 1978, MNRAS, 185, 69

Penrose, R., 1969, Nuovo Cimento, 1, 252

Phinney, E.S., 1981 in Plasma Astrophysics, T. Guyenne ed. ESA SP-161, 337

Press, W.H. & S.A. Teukolsky, 1972, Nature, 238, 211

Rees, M.J., M.C. Begelman, R.D., Blandford & E.S., Phinney, 1982, Nature, 295, 17

Znajek, R.L., 1977, MNRAS, 179, 457

Znajek, R.L., 1978, MNRAS, 185, 833

DISCUSSION

R. RUFFINI: In your model is the magnetic field an assumption or is it the consequence of other physical ingredients?

S. PHINNEY: The magnetic field is something which is almost certainly present. If the accreting matter has some embedded magnetic field, in the perfect MHD approximation every open magnetic line, which is carried onto the horizon by the inflow, must remain stuck on the horizon. The flux through the horizon will therefore build up until the associated magnetic field pressure becomes large enough to support the exterior matter against gravity. Accretion would then shut off, except that such a situation is probably Rayleigh-Taylor unstable, and matter will accrete by "dripping", bringing in closed magnetic field loops (after reconnection in a small region at the neck of the drop), without adding flux to the horizon. If the matter around the hole is in a convectively unstable disk, magnetic fields may also be built up by dynamo action, as R. Pudritz has shown in a preliminary attack on a complicated problem.

G. BENFORD: How does the vacuum break down in your magnetosphere, and how are particles created on the field lines?

S. PHINNEY: I did not have time to go into this aspect in the talk. The minimum particle creation rate is that provided by γ-rays from the surrounding accretion disk via $\gamma + \gamma \rightarrow e^- + e^+$ reactions. Even for tiny γ-ray luminosities, this is more than adequate to break down the vacuum. Another possibility is that in the non-stationary drip accretion which may occur when the magnetic flux through the horizon becomes strong enough to support the exterior matter against gravity, particles are injected impulsively on the temporarily non-steady field lines.

G.C. PEROLA: For the benefit of those of us who make observations, can you estimate how much collimation your model can produce?

S. PHINNEY: The answer depends very much on the nature of the matter confining the magnetic field. If the matter is in an accretion torus, the flow can be collimated to an angle equal to the opening angle of the paraboloidal funnel at the height where the jet pressure becomes equal to the pressure in the torus. Of course, if the flow becomes magnetically confined, the exterior medium plays no role in the confinment, and the collimation may be arbitrarily good. If the energy extraction is the "fire" in Begelman's "cauldron", the jet must be collimated in a nozzle in the settling matter.

SUPERCRITICAL JETS FROM A "CAULDRON"

M.C. Begelman* and M.J. Rees
Institute of Astronomy, Madingley Road
Cambridge CB3 OHA, England

ABSTRACT

We consider various schemes whereby a supercritical power level can be generated around a compact object, and discuss how optically thick radiation-driven jets (which may become relativistic) can be channeled through nozzles confined by an external cloud supported by radiation pressure.

1. INTRODUCTION

Our instruments have inadequate angular resolution to reveal directly the places where jets are produced and collimated. Thus, we know that many extragalactic jets are well on their way by $\sim$ 1 pc, yet we infer theoretically that the energy source probably occupies a region only 10^{-5} times this size. Likewise, we see unambiguous evidence for well-collimated jets in SS433 on scales $\sim$ a few a.u.; yet the "black box" at the center may be only a few kilometers across. This observational barrier allows ample scope for speculating on the origins of jets, and two schools of thought have arisen, distinguished by whether the primary collimation process is assumed to be intrinsic or extrinsic to the source of energy.

The exemplar of the "extrinsic" school is the "twin-exhaust" model proposed by Blandford and Rees (1974). Here, no directionality is required in the initial release of energy; bifurcation and collimation into a pair of oppositely directed jets can occur much further out, and the buoyant fluid rises into a gas cloud whose isobars are flattened by rotation. Numerical modellers such as Wiita (1978) and Norman et al. (1981), have treated the energy sources as a spherical symmetric wind emerging from a point source and have found that the de Laval nozzles envisaged by Blandford and Rees do form, but are stable only if the

* Present address: Joint Institute for Laboratory Astrophysics
University of Colorado, Boulder, CO 80309, USA

A. Ferrari and A. G. Pacholczyk (eds.), Astrophysical Jets, 215–226.
Copyright © 1983 by D. Reidel Publishing Company.

nozzles are not too narrow or wide compared with the scale of the cavity carved by the wind. This, in turn, places constraints on the relationship between the luminosity of the jet, its velocity (assumed to be predetermined by the energy injection process) and the properties of the confining cloud (Smith et al. 1981).

Most "intrinsic" models involve a geometrically thick accretion disk. The energy supply is intimately tied up with the properties of the disk, either through the accretion itself or the extraction of rotational energy from the central compact object by a disk-confined magnetic field (Rees et al. 1982). Collimation is provided by the inertia of material which surrounds a funnel or vortex, but the jets are not part of a flow pattern which relates directly to the accretion.

In this paper, we examine a class of models where jets form via "nozzles" which can be analysed by gas-dynamical methods, but where the interdependence of the jets and the surrounding cloud places extra constraints on a self-consistent model. We envisage a class of flows where material accretes towards a central mass quasi-spherically. The inflowing gas may be nearly in free fall, slowly settling, or have its inflow regulated by viscous transfer of angular momentum (provided the pattern is still quasi-spherical). Over some characteristic cavity radius, r_{cav}, energy associated with the presence of the central object is deposited in the inflowing gas, which leaves it able to escape. It is within this "cauldron" that some of the gas is turned around and eventually passes through a pair of nozzles of width $\lesssim r_{cav}$. Because of slight asphericity in the accreting gas cloud, the heated has can escape through nozzles, rather than escaping convectively or blowing the cloud apart.

"Cauldrons" may support stationary flow patterns dynamically dominated by the pressure of trapped radiation, where the jets carry away power at a rate than can <u>far exceed</u> the Eddington limit L_E.

2. THE CAVITY AND NOZZLES

Suppose that sufficient anthalpy to escape is introduced into the inflowing gas on a scale $\sim r_c$, and that the inflow "turns around" on this scale. Denoting conditions just outside the cavity by subscript "e", the pressure near the boundary of the cavity is

$$P_c \sim \rho_e v_{fl}^2 (r_c) \equiv \rho_e \frac{GM}{r_{cav}} , \qquad (2.1)$$

regardless of whether the gas flows in at nearly the free fall speed v_{ff} or settles in subsonically. Nozzles will form where the pressure in the hot outflowing gas has dropped to $\sim p_c/2$, typically at a radius, $r_{cav} < r_{nozzle} \lesssim 2\, r_{cav}$. Each nozzle has a cross-sectional area $A_{nozzle} = 2\pi \xi r_{cav}^2$; stability considerations (Smith et al. 1981) may rule out nozzles with $\xi \ll 1$. If a fraction η of the inflowing material escapes through the jets then the flux density of rest mass through each nozzle is

$$\rho_{cav} v_{nozzle} \simeq \eta \xi \rho_e v_e, \tag{2.2}$$

v_e being the inward drift speed just outside the cavity boundary.

The total energy flux L_j carried by the mass flux $\eta \dot{M}$ along the jets is then of order $P_c v_{nozzle} A_{nozzle}$, where v_{nozzle} is close to the sound speed within the cavity (and is also of order the final velocity v_j of the jet). L_j, v_{nozzle} are then related to the properties of the flow:

$$v_{nozzle} \sim \min \left[\xi \eta^{-1} \left(\frac{GM}{r_{cav}}\right)^{\frac{1}{2}} \left(\frac{v_{ff}}{v_e}\right), c \right] \tag{2.3}$$

$$\gamma_{cav} \sim \max \left[\xi \eta^{-1} \left(\frac{GM}{r_{cav} c^2}\right)^{\frac{1}{2}} \left(\frac{v_{ff}}{v_e}\right), 1 \right] \tag{2.4}$$

We define γ_{cav} as the ratio of the total mass-energy density to the rest mass density of matter in the cavity. The energy flux carried through the nozzles by the jets is then

$$L_{nozzle} \sim \eta \xi \dot{M} \gamma_{cav} v_{nozzle}^2 \tag{2.5}$$

In the relativistic case ($\gamma_{cav} \gg 1$) most of the energy carried through the nozzle still resides in internal degrees of freedom and may be radiated away as the jet propagates beyond the nozzle. If this happens, the asymptotic kinetic energy flux L_j may be much smaller than L_{nozzle}.

The further specification of cavity and nozzle parameters depends on the particular physical processes involved in powering the jet, e.g. gas pressure or radiation pressure. Therefore, we will now specialize the subsequent discussion to systems dominated by radiation pressure, where the photons interact with the gas via Thompson scattering. One can easily check that, for the parameter range where $L \gtrsim L_E$, the densities are generally high enough that $t_{cool} \ll t_{dynamical}$, implying that the gas and radiation temperatures will be equal; but low enough that the radiation pressure $1/3\ a\ T^4$ much exceeds the gas pressure nkT. The factor γ_{cav} is then $\sim a\ T^4/nm_pc^2$; it can be large because photons greatly outnumber particles, though the temperatures are never high enough for even the electrons to become relativistic.

Two requirements for powering jets with radiation pressure are: (i) that the available energy flux, L_{nozzle}, exceed the Eddington limit L_E (= 1.3×10^{38} $M/M_\odot$ erg/s for ordinary plasma and Thompson scattering); and (ii) that the radiation at the boundary of the cavity be trapped by the inflow. The first is necessary in order to get radiation-driven outflow at all; while the second must apply, given the first, in order that the surrounding gas cloud not be blown away by radiation escaping from the cavity. Photons are trapped when their diffusion speed outward through a region characterized by optical depth $\tau, v_{diff} \sim c/\tau$, is exceeded by the bulk speed which advects them inward. For spherical inflow at a given rate $\dot{M}$, there is a "trapping radius" within which any photons present in the flow are dragged inward, given by

$$r_{tr} \sim (\dot{M}/\dot{M}_E)(r_s/2) \tag{2.6}$$

(Begelman 1978), where $\dot{M}_E \equiv L_E/c^2 = 1.4 \times 10^{17}$ $(M/M_\odot)$ gm s^{-1} and r_s is the Schwarzschild radius. Thus, condition (ii) translates to

$$r_{cav} \lesssim r_{tr} \tag{2.7}$$

A given value of r_{tr} ($\propto \dot{M}$) does not specify the value of P_c: one can have free fall ($v_e \simeq v_{ff}$); alternatively, the flow for $r_{cav} \lesssim r \lesssim r_{tr}$ may be a "settling solution" with $p \propto r^{-4}$, $\rho \propto r^{-3}$, $v \propto r$. The latter option allows a much higher P_c for a given $\dot{M}$.

The relations (2.1) - (2.5) show that, with $\dot{M}$, η and ξ fixed, L_j and v_j depend only on the combination $\left(\frac{v_e}{v_{ff}}\right)\left(\frac{r_{cav}}{r_{tr}}\right)^{\frac{1}{2}}$. For certain values of v_j, one may obtain the same values of this product by setting $v_e \sim v_{ff}$, $r_c \ll r_{tr}$; or, alternatively, $r_c \sim r_{tr}$, $v_e \ll v_{ff}$. However, from (2.3) it is apparent that the solution with $v_e \sim v_{ff}$ is consistent only with sub-relativistic outflow, unless $\eta \ll 1$.

A third type of inflow, in which the cavity occupies a region within the pressure maximum of a thick disk, is also possible. In this case v_e is regulated by viscosity and the properties of the jet channels which form depend in part upon the distribution of angular momentum in the funnel. Since the collimation may be determined largely by the inertia of the funnel walls, there need not even be nozzles per se - a flow which is accelerated subsonically by trapped radiation will become supersonic when the radiation becomes untrapped. While this configuration may be operating in some jet-producing clouds, we will not discuss it here.

3. ENERGY SOURCES

The injection of enthalpy into a cavity filled with inflowing gas is a highly complex, possible chaotic process. When the inflow at r_{cav} is able to trap radiation (relation (2.7)), the energy injection cannot occur by a diffusion of radiation, but rather must involve bulk motions, convection, or shocks. Below we describe four energy injection processes which may play a role in powering supercritical jets in this way.

(i) Accretion

Because a black hole can absorb arbitrary amounts of thermal and kinetic energy, there is in principle no limit to the rate at which it can accrete (Begelman 1978, Gilden & Wheeler 1980, Flammang 1982). Even if the inflowing gas is highly dissipative, the energy released at $r < r_{tr}$ can simply be dragged into the hole. This is not the case for accretion onto a neutron star, however, where the stellar surface at r_* (or possible the magnetopause at the Alfvèn radius r_A) can act like a brick wall, forcing the infalling gas to give up its energy to the gas further out, at a rate $GM\dot{M}/r_*$ (or $GM\dot{M}/r_A$). If the accretion flow were absolutely spherically symmetric, release of energy at a rate exceeding L_E would ultimately limit the accretion to a rate not exceeding $\dot{M}_E(r_*/r_s)$ (or $\dot{M}_E(r_A/r_s)$). However, if jets provide an alternate escape route for the released energy, then we might envisage a situation in which a luminosity $\sim (r_{tr}/r_{cav})L_E$ is radiated inward from the cavity walls, and blobs rain down towards the stellar surface, releasing energy which both supports the cavity and powers the jets.

Suppose a fraction $(1-\eta)$ of the inflowing material produces energy at a rate $(1-\eta)\,\varepsilon\dot{M}c^2$, where $\varepsilon \simeq (r_s/r_*,\ r_s/r_A) \lesssim 0.1$ is the efficiency of mass- to-energy conversion, and that nearly all of this (except at most $\sim L_E$) goes into powering a radiation-driven jet with mass flux $\eta\,\dot{M}$. From energy conservation, it is obvious that we can consider two regimes:

$$\begin{aligned} \eta &> \varepsilon \qquad &&\text{(subrelativistic)} \\ \eta &\ll \varepsilon \qquad &&\text{(ultrarelativistic)} \end{aligned} \tag{3.1}$$

If η is much larger than $\sim \frac{1}{2}$, the material in the cavity at $r \ll r_c$ will still be gravitationally bound to the central object; hence it is plausible to further restrict the subrelativistic regime to $\varepsilon \lesssim \eta \lesssim \frac{1}{2}$.

Now consider the constraints imposed by the two solutions for quasi-spherical inflow, described in section 2. If the gas is nearly in free fall, then

$$r_{cav}/r_s \simeq \{\max\,[\,\eta\varepsilon,\ \varepsilon^2\,]\}^{-1} \tag{3.2}$$

while we have, for a settling solution,

$$\frac{v_e}{v_{ff}(r_{tr})} \simeq \left\{\left(\frac{\dot{M}}{\dot{M}_E}\right)^{\frac{1}{2}} \max\,[\,\sqrt{\eta\varepsilon},\ \varepsilon]\right\}^{-1} \tag{3.3}$$

These solutions are self-consistent if $r_{cav}/r_s \ll \frac{\dot{M}}{\dot{M}_E}$ and $\frac{v_e}{v_{ff}} \ll 1$, respectively - i.e., if

$$\dot{M}/\dot{M}_E \gg 1/\max\,[\,\eta\varepsilon,\ \varepsilon^2\,] \tag{3.4}$$

The presence of a strong magnetic dipole anchored in the accreting body places further constraints on the properties of the cavity, but it also loosens the restrictions on η. Since material cannot penetrate the magnetosphere except in blobs which eventually make their way to the surface (cf, e.g., Arons & Lea 1976, Michel 1977), we require that $r_{cav} > r_A$, but we also allow η to range upward to $(1-\eta) > (r_s/\varepsilon r_A)$. Here the relevant value of the Alfven radius is

$$r_A = \left(\frac{\mu^2}{8\pi P_c}\right)^{1/6} \sim \left(\frac{\mu^2 r_{cav}^3 v_e}{2GM\dot{M}}\right)^{1/6} \sim \qquad (3.5)$$

$$\sim 3.2 \times 10^8 \ \mu_{30}^{4/7} \left(\frac{M}{M_\odot}\right)^{-3/7} \left(\frac{\dot{M}}{\dot{M}_E}\right)^{-2/7} \left(\frac{v_e}{v_{ff}}\right)^{2/7} \text{cm},$$

for a given magnetic moment $\mu = 10^{30}\ \mu_{30}$ G x cm^3. From (3.2) and (3.3), the condition $r_c > r_A$ becomes

$$\max\ [\eta\ (1-\eta)\varepsilon,\ \varepsilon^2] \le 10^{-3}\ \mu_{30}^{4/7} \left(\frac{M}{M_\odot}\right)^{10/7} \left(\frac{\dot{M}}{\dot{M}_E}\right)^{2/7} \qquad \text{(free fall)} \qquad (3.6)$$

$$\frac{\dot{M}}{\dot{M}_E} > 127\ (\max[\eta(1-\eta)\varepsilon,\ \varepsilon^2\ ])^{-10/7}\ \mu_{30}^{2/5} \left(\frac{M_\odot}{M}\right) \qquad \text{(settling)}$$

If we further imposed the conditions $r_{cav} \lesssim r_{tr}$ and $v_e \lesssim v_{ff}$, we obtain

$$\frac{\dot{M}}{\dot{M}_E} > 213\ \mu_{30}^{4/7} \left(\frac{M}{M_\odot}\right)^{-10/9} \qquad (3.7)$$

for the free fall solution.

If (3.6) is marginally satisfied, then the cavity extends just outside the magnetosphere. In general, this marginal case will be consistent only with free fall inflow, and will result in a subrelativistic jet with $(1-\eta)$ determined by $\dot{M}$.

(ii) "Egg-beaters"

If the central neutron star is spinning, its rotational energy provides another heat input which may be more important than the energy derived from infall. Perhaps it would be more appropriate to refer to a rapidly spinning magnetized neutron star with an oblique magnetic axis as a "magnetic stirrer". It is still controversial whether the rotating magnetopause would act as a "propeller" attempting to spin the surrounding material into co-rotation (Illarionov and Sunyaev 1975, van den Heuvel 1977) or whether the field lines themselves would be swept back sufficiently to minimize the torque (Davies and Pringle 1981). The latter seems likely if the dipole moment is aligned with the rotation axis, but if there is substantial misalignment then the "equilibrium" Alfven radius at any given point will pulsate with the frequency of rotation Ω . The efficiency with which this process feeds energy into the cavity is uncertain, but an upper limit for a highly inclined dipole is

$$L_{max} \sim p(r_A)\, r_A^2 \times \begin{cases} \Omega\, r_A & \rightarrow \quad \Omega\, r_p < v(r_A) \\ (\Omega\, r_A)^3/v^2(r_A) & \rightarrow \quad \Omega\, r_A > v(r_A) \end{cases} \tag{3.8}$$

where $p(r_A)$ is the pressure at the magnetopause, $v(r_A)$ is the corresponding free fall speed, and we have assumed that r_A lies within the light cylinder. These quantities may differ from the cavity pressure and nozzle velocity, respectively, if some of the gas in the cavity is bound to the neutron star (in the latter case $v(r_A) \sim v_{ff}(r_A)$).

We still require $\dot{M}/\dot{M}_E > r_{cav}/r_s$ in order to prevent the energy generated by the "egg-beater" from escaping isotropically. However there is no reason why the luminosity should not be $\gg(\dot{M}/\dot{M}_E)L_E$, provided that interactions with the "flailing" magnetosphere can have the cumulative effect of dissipating more than the rest-mass energy of the material and that the required high pressure at the cavity boundary can be maintained (by $v_e \ll v_{ff}$ if necessary).

(iii) "Paddle-wheels"

There are other astrophysical contexts - apart from those involving magnetised neutron stars - where a non-axisymmetric rotating body could generate a supercritical luminosity if it were "smothered" by a dense cloud. For instance, a close binary in a common envelope would generate a heat input with an effective cross section related to the accretion radius GM/v_{orb}^2 of each component. In each orbital period, the matter interacting with the system would be given $\sim v_{orb}^2$ of energy per unit mass. Non-steady-state applications might include bipolar nebulae and some planetary nebulae. On a larger scale, massive black hole binaries in galactic nuclei, or a bar-mode instability in the core of a supermassive object, or even sudden increases in viscosity in a disc, could lead to the same phenomena.

(iv) Smothered black hole "flywheel"

As discussed by Phinney (these proceedings), a spinning black hole can generate power at the expense of its rotational energy if it is embedded in a magnetic field. If the magnetic stresses are comparable with the cavity pressure, the maximum luminosity is

$$L_{max} \simeq p_{cav}\, r_s^2 \left(\frac{J}{J_{max}}\right)^2 c \tag{3.9}$$

i.e.

$$\frac{L_{max}}{L_E} \simeq \frac{\dot{M}}{\dot{M}_E} \left(\frac{GM}{r_{cav}c^2}\right)^{1/2} \left(\frac{r_s}{r_{cav}}\right)^2 \left(\frac{v_{ff}}{v_e}\right) \left(\frac{J}{J_{max}}\right)^2 \tag{3.10}$$

If the "fluid" approach adopted in this paper were appropriate, the spin energy of a hole surrounded by a radiation-pressure-supported cloud with $r_{cav} \sim r_s$ could be converted into a radiation driven jet with γ_{cav} of order (v_{ff}/v_e). If the hole is surrounded by an accretion torus, there is the possibility that a small fraction of the matter achieves even higher γ's.

4. TRAPPING AND COLLIMATION

Insofar as the jet and the (lower-entropy) surrounding cloud can be treated as two separate fluids - i.e. insofar as diffusion and viscosity can be neglected - it is straightforward to calculate how the collimation angle θ varies with r after the jet has passed through the nozzle. If the external pressure falls off as r^{-4} (the case of an isentropic "settling solution"), and radiation dominates the internal pressure of the jet, then we have:

(i) for a non-relativistic jet,

$$\theta = \theta_{nozzle} \left(\frac{r}{r_{nozzle}}\right)^{\frac{1}{2}} \tag{5.1}$$

provided that pressure balance can be maintained ($\theta_{nozzle} \simeq \xi^{\frac{1}{2}}$). However, this balance is only possible so long as the jet's Mach number is $M_j < \theta^{-1}$. So, insofar as (5.1) is obeyed, $M_j \propto r^{1/2}$. Therefore the jet angle saturates when $\theta = \sqrt{\theta_{nozzle}}$; at larger distances, the jet would be overpressured with respect to its surroundings. The jet's terminal velocity exceeds v_{nozzle} by a numerical factor of order unity.

(ii) for an ultrarelativistic jet, the pressure balance condition yields

$$\theta \simeq \text{constant} \tag{5.2}$$

$$\gamma_j \simeq (r/r_{nozzle}) \; . \tag{5.3}$$

(5.3) holds until γ_j has built up to the initial γ_{cav} - at that stage all the internal energy has been converted into bulk kinetic energy of relativistic motion.

Let us now check under what conditions the radiation is well enough trapped to justify a fluid treatment. The optical depth across the cavity is

$$\tau_{cav} \simeq \eta \left(\frac{\dot{M}}{\dot{M}_E}\right) \left(\frac{r_{cav}}{r_s}\right)^{-1} \left(\frac{v_j}{c}\right)^{-1} \tag{5.4}$$

Trapping requires that the optical depth across the jet $\tau_j(r)$ exceed $\theta^{-1}(v_j/c)^{-1} (\gamma_j(r))^{-1}$, where

$$\tau_j \simeq \tau_{cav}\ \xi^{1/2} \left(\frac{\theta}{\theta_{nozzle}}\right)^{-1} \left(\frac{r}{r_{cav}}\right)^{-1} (\gamma_j(r))^{-1} \tag{5.5}$$

i.e.

$$\left(\frac{r}{r_{cav}}\right) < \left(\frac{v_j}{c}\right) \xi\ \tau_{cav} \tag{5.6}$$

(note the cancellation of the relativistic factor γ_j, from this expression).

The luminosity carried by the jets (assuming adequate trapping) can be written

$$L_j = L_E\ \xi\ \tau_{cav} \left(\frac{r_{cav}}{r_s}\right) \left(\frac{v_j}{c}\right) \gamma_{cav}. \tag{5.7}$$

Comparison of (5.6) and (5.7) then shows that the trapping condition can best be satisfied when $L_j >> L_E$.

Non-relativistic jets

Provided that the radiation from a non-relativistic jet does not leak out until the jet has got to a distance r_{esc} well outside r_{nozzle}, its energy will largely have been converted via adiabatic expansion into bulk kinetic energy, leaving a photon energy flux of

$$L_{esc} \simeq L_j \left(\frac{\theta_{esc}}{\theta_{nozzle}} \frac{r_{esc}}{r_{nozzle}}\right)^{-2/3} \tag{5.8}$$

Though a small fraction of Lj (cf (5.7)), L_{esc} may still be $>> L_E$. If the release occurred below the photosphere of the surrounding cloud, the resulting radiation pressure would eject a sheath of cloud material from around the jet.

Radiative diffusion will transport momentum out of the jet - i.e. there will be an associated radiative viscosity. However, the timescale for slowing down the jet is longer than the diffusion time by a factor $M_j^2 (v_j/c)^{-1}$.

Relativistic jets

For relativistic jets, the effects are more interesting, because until the jet has propagated so far out that γ_j "saturates" at a value $\sim \gamma_{cav}$, all the energy and momentum flux is in the form of photons. Thus, if the trapping condition (5.6) breaks down for $r_{esc} < \gamma_{cav}\, r_{nozzle}$, the entire luminosity will be radiated as beamed photons, the cone angle being max $[\theta_{esc}, (r_{esc}/r_{nozzle})^{-1}]$.

5. ASTROPHYSICAL RELEVANCE, ETC.

We have discussed some schemes whereby high-entropy material can be expelled as jets with energy fluxes >> L_E, driven by radiation pressure. The power may be supplied by supercritical accretion, a fraction $(1-\eta)$ of the infalling material falling onto the compact object and the remainder being ejected in a directed fashion; alternatively, the power may come mainly from a central spinning object ("paddle-wheel") which augments the entropy of accreted material so that all can be expelled along jets.

In all cases, the surrounding cloud provides the pressure which constrains the outflow along an axis. This does not require substantial flattening - only enough asymmetry to initiate the directivity. The cloud itself will radiate at $\sim L_E$; if the super-critical jets are accretion-powered, it must have a sufficiently large optical depth that most of the radiation released is trapped, and transported into the cavity by bulk flow rather than diffusing outwards. The details of the flow in the cloud are complex. The flow outside r_{tr}, and the properties of a photosphere (at radius $>(r_{tr}^2/r_s)$) depends on how the material is supplied; the effects of other opacities and the pressure of the gas itself may be crucial. However, for the purposes of our main discussion, the key features are the cavity radius and the pressure p_c at the cavity boundary. In the various cases considered in Section 3, one can test the stability of a particular pattern, by checking that the outflowing luminosity and mass depend more sensitively on the various parameters than the inflow does. Angular momentum in the cloud, leading to a pressure maximum <u>outside</u> r_{cav}, may allow better collimation than that derived in Section 4, allowing narrow jets to form without the risk of the instabilities discussed by Smith et al. (1981). Note also that we have assumed that gravitational deceleration is unimportant in the jets. For "slow" jets, or cases where entrainment is important, the critical points in the flow may not occur at "nozzle" as in the "twin exhaust" picture (Blandford and Rees 1974).

We mention in conclusion three specific applications.
(i) <u>SS433</u>. The supercritical energy flux along the jets is hard to reconcile with a conventional accretion model. If the compact object were a magnetised neutron star, the jet production mechanism would resemble that described in Section 3. The cloud's photosphere would be at $\sim 10^9$ cm, emitting thermal radiation ($\sim L_E$) in the soft X-ray band. The 164-day precession would result from the cloud being "slaved" to the companion star that supplies the mass (this requires the storage time in the cloud to be << 164 days, a condition that is easily fulfilled). If the jet material is ultrarelativistic, then, as discussed in Section 4, a supercritical flux of <u>beamed photons</u> can be generated, just as is required to provide an energy input into the observed thermal material in the jets. We discuss this particular application in fuller detail elsewhere.

(ii) Bipolar ejection in protostars. These are much less violent phenomena than SS433, but could involve a similar flow pattern, where non-axisymmetry or high viscosity in the core provides the energy and entropy input. In this (relatively cool) flow, the opacity is much higher than that from Thomson scattering, rendering radiation pressure effects important even far below the conventional Eddington luminosity.

(iii) Ultraluminous quasars. As already mentioned in Section 4, these models could yield supercritical jets, and supercritical fluxes of beamed photons, if a central black hole were smothered by plasma with a magnetic field, or if a galactic nucleus contained a supermassive binary acting as a "paddle-wheel". This ultraluminous phase might occupy only a brief fraction of the active lifetime of a galactic nucleus; though the flow near the cavity may be quasi-stationary (in the sense that it persists for much longer than the characteristic flow timescale), the surrounding cloud itself may not be in a steady state.

REFERENCES

Arons, J. and Lea, S., 1976, Astrophys. J., 207, 914
Begelman, M.C., 1978, M.N.R.A.S., 184, 53
Blandford, R.D. and Rees, M.J., 1974, M.N.R.A.S., 169, 395
Davies, R.E. and Pringle, J.E., 1981, M.N.R.A.S., 196, 209
Flammang, R.A., 1982, M.N.R.A.S., 199, 833
Gilden, D.E. and Wheeler, J.C., 1980, Astrophys. 232, 705
Illarionov, A.F. and Sunyaev, R.A., 1975, Astr. Ap., 39, 185
Michel, F.L., 1977, Astrophys., 213, 836
Norman, M.L., Smarr, L., Wilson, J.R., and Smith, M.D., 1981, Astrophys. J., 247, 52
Rees, M.J., Begelman, M.C., Blandford, R.D. and Phinney, E.S., 1982, Nature, 295, 17
Smith, M.D., Smarr, L., Norman, M.L. and Wilson, J.R., 1981, Nature, 293, 277
van den Heuvel, E.P.J., 1977, Ann. N.Y. Acad. Sci., 302, 15
Wiita, P., 1978, Astrophys. J., 221, 431

DISCUSSION

R.J. Davis. Do you have the same problems as the Blandford and Rees model? Do you have to match the energy flow to the shape of the particular cauldron?

M. Begelman. We deal with some of the stability problems by assuming that the width of each nozzle is a significant fraction of the radius of the cauldron. If the external material flows in nearly at the free-fall speed, then Rayleigh-Taylor instabilities will not develop to the point of disruption of the time-averaged structure of the cauldron. If the external gas is settling in slowly, then Rayleigh-Taylor instabilities may be important unless they are suppressed by "fire-polishing" at the cavity walls or some other sort of diffusive effect. Whether the former occurs would depend on the way in which energy is injected into the gas; the latter would require r_{cav} to be near the trapping radius.

T.P. Ray. In your paddle-wheel model, consisting of two black holes, collimation would occur on the scale of the paddle. In radio galaxies and quasars collimation can occur on a scale < 0.1 pc; but two black holes (masses about $10^8\ M_\odot$) so close would coalesce, due to gravitational dipole radiation, very quickly. Would this affect the time scale of such sources?

M. Begelman. The gravitational radiation lifetime of such a binary would be of order of 10^{12} yrs. However, if the paddle-wheel effect powered jets with energy $x\ L_E$, $x > 1$, the holes would spiral together in only in $10^4/x$ yrs. Thus it appears that such a wide paddle-wheel would rapidly spiral in as a result of energy loss through the jets, not via gravitational radiation. The two timescales would become comparable when the separation had shrunk to $\sim 10^{16}\ x^{-1/5}$ cm, at which point the binary would evolve on a time-scale $\sim 10^6\ x^{-4/5}$ yrs.

P. Biermann. The observations of superluminal velocity suggest a very well defined range of v/c of about 2 to 10. Any model of compact jets should produce the corresponding bulk Lorentz factor "naturally".

S. Phinney. Energy extraction from a black hole is one way to do that "naturally": if $v_{infall}/v_{ff} \sim 1/2$ and about 1/2 of the infalling matter is blown out in the jet, the bulk Lorentz factor, v_{app}/c, is ~ 4.

PROPAGATION AND MORPHOLOGY OF PRESSURE-CONFINED SUPERSONIC JETS

Michael L. Norman and Karl-Heinz A. Winkler
Max-Planck-Institut für Physik und Astrophysik
Institut für Astrophysik
D-8046 Garching b. München, FRG
and
Larry Smarr
Department of Astronomy
University of Illinois
Urbana, Illinois 61801

1. NUMERICAL JET LABORATORY

The beautiful, high resolution radio maps of jets, lobes, and hot-spots coming off the VLA challenge theorists to explain these phenomena in terms of simple, well-understood physical laws. The difficulty is often not one of finding a reasonable set of physical laws, but of finding their solutions. Analytic solutions of jet structure are limited to steady, isentropic, 1-dimensional flow in channels of slowly-varying cross section; analytic studies of jet stability are limited to modes of small amplitude which, in this regime, assume equal importance as well as unimportance for the gross structure of the flow. Numerical simulation provides a means for investigating those time-dependent, multidimensional, nonlinear solutions that are beyond the reach of analytic tools - solutions that must be relevant to an understanding of astrophysical jets. In an age when computing hardware and software are advancing so rapidly compared to analytic techniques, it is natural to turn to the computer for answers, approximate though they may be.

Numerical solutions can be approximate in several respects, chief of which are:

1) physics: The model system is based on a simplified, incomplete, or inappropriate set of physical laws, either through a desire to study simple cases first, or through ignorance of the correct set of physical laws appropriate to the phenomenon in question. Approximations of this sort are not limited to numerical model-builders.

2) dimensionality: The model system is constructed in a space of reduced dimensionality, either through a desire to study simple cases first, or because the necessary hardware and/or software is not available for the

A. Ferrari and A. G. Pacholczyk (eds.), Astrophysical Jets, 227–251.
Copyright © 1983 by D. Reidel Publishing Company.

higher-dimensional studies. Historically, hardware advances have been accompanied by important innovations in software design (algorithms, etc.). The next generation supercomputer (e.g. CRAY X-MP + SSD) will largely remove the dimensionality limitations of the past in modelling "simple" physics like gas-dynamics, and drive the development of new numerical techniques for more complicated physical systems (e.g. radiation hydrodynamics).

3) accuracy: The model system is further approximated as a discrete system so that it can fit into a finite-sized computer memory. Solutions of the discrete system only approximate to within a certain level of accuracy solutions of the model system. This level of accuracy usually improves as one attempts a finer discretization of the problem (i.e. using more grid points, zones, particles, modes, nodes, elements etc.), but need not improve uniformly over all parts of the solution. Accuracy is a major concern to numerical model builders.

In addition to being approximate, numerical solutions are also incomplete in at least two respects:

1) domain: The numerical solution is found for only a limited region in space-time (you have to cut it off somewhere.)

2) scope: A single numerical solution is a particular solution to the governing equations, and, unlike an analytic solution, does not show dependencies on the fundamental parameters. A family of numerical solutions comprising a parameter search do show dependencies on the basic parameters, however in practice one cannot search parameter space completely.

It is therefore the task of the numerical modeller to extract new qualitative and quantitative information (if any) from approximate and incomplete numerical solutions.

Blandford, Begelman, and Rees, in their 1982 Scientific American article entitled Cosmic Jets, suggest that laboratory investigations of supersonic jets may yield fruitful insights into the cosmic phenomena. They would indeed help, but even laboratory experiments are limited. The advantages of a numerical rather than an experimental approach for problems in fluid dynamics are encapsulated in the 1965 Scientific American article of Harlow and Fromm. They point out that even in cases where laboratory experiments are possible, the numerical approach may be preferred because of cost, flexibility, control of experiment, and accessibility of the data which are free of experimental uncertainties of measurement. Perhaps the most important advantage for understanding astrophysical jets is the ability to control the input physics at will. Here is a case where the appropriate and sufficient set of physical laws is not known, and is in fact one of the objects of inquiry.

Harlow and Fromm's arguments become compelling when one considers the new physical regimes which may exist in jets. Just consider the cost involved in studying relativistic jets in the laboratory! Indeed, the comprehensive experimental study of axisymmetric, supersonic jets by Love et al. (1959) encompasses only flow Mach numbers between 1 and 3, and describes only jets that are far out of pressure equilibrium with their surroundings. One advantage that laboratory experiments currently do have is that they are 3-dimensional, but rapid advances in computing hardware and software are eroding that advantage as well. Thus, we have undertaken a program to study basic jet physics in a numerical laboratory.

2. THEORETICAL PLANE

The problem we consider is the time-dependent penetration of a collimated, supersonic beam of gas into a constant ambient gas. The numerical experiments described here extend earlier work of this nature we reported in Norman, Smarr, Winkler, and Smith (1982; hereafter NSWS). The reader is referred to NSWS for a more complete description of our physical and numerical assumptions, as well as a comparison of our computer code against a flow measured in the laboratory. Here we shall only review our program and assumptions against the framework established in section 1.

physics: We assume ideal, nonrelativistic gas dynamics; adiabatic flow; $\gamma = 5/3$ ideal gas EOS for both jet and ambient medium. Beam is introduced with zero opening angle and with equal pressure to the undisturbed medium. Jets thus produced are members of a 2-parameter family, which we take to be 1) ratio of beam density to medium density $\rho b/\rho m$, and 2) beam Mach number M_b. By choosing a constant, static ambient medium, we have eliminated "environmental effects".

dimensionality: Flow is assumed to be symmetric about the jet axis, thus we deal in 2 space + 1 time dimensions. Potentially important non-axisymmetric structures and modes of instability are therefore not modelled.

accuracy: The numerical method we use is described in Norman and Winkler (1983), and is shown to compare favourably in accuracy with the BBC scheme on a series of standard test problems devised by Woodward and Colella (1983). Our method and the BBC scheme are generically similar; both are explicit, artificial viscosity methods for computing 2-dimensional flows with strong shocks. We compute with an Eulerian mesh of $75 \times 640 = 48{,}000$ zones, which allows adequate resolution of shocks and surface instabilities that develop in the flow. Sensitivity of our results to 1) zoning resolution, 2) order of accuracy, and 3) use of a material interface on the jet boundary are also discussed in Norman and Winkler (1983).

domain: Spatial domain comprises $0 \leq Z \leq 80$ (640 uniform zones), $0 \leq R \leq 15$ (60 uniform zones out to $R = 7.5$, 15 ratioed zones beyond). We evolve forward in time until the jet has propagated to $Z = 80$. The beam is admitted with a radius $R_b = 1.$, thus jets up to 40 times their diameter can be studied.

scope: Our numerical survey consists of 12 evolutions parameterized by the various combinations of $\rho_b/\rho_m = 0.01, 0.1, 1.0, 10.$ and $M_b = 1.5, 3., 6., 12.$ shown in Figure 1.

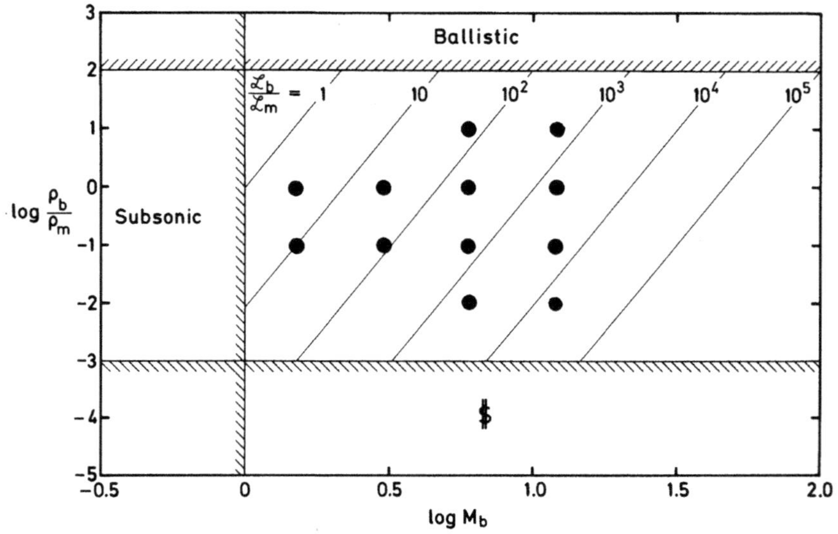

Figure 1. Density ratio - Mach number plane, and the location of the 12 evolutions presented here (large dots). Lines of constant beam kinetic luminosity $L_b = \frac{\pi}{2} R_b^2 \rho_b V_b^3$ in units of the ambient medium $L_m = \frac{\pi}{2} R_b^2 \rho_m C_m^3$ are shown in light. Hatched lines bound regions not considered for reasons given in the text.

Lines with hatchmarks indicate regions of the theoretical plane which have not been considered. Although an investigation of subsonic jets is probably necessary to understand wide-angle-tails and low luminosity jet morphologies, we do not consider $M_b < 1$ because 3D effects dominate in this regime (e.g. turbulence). We arbitrarily define $\rho_b/\rho_m > 10^2$ to be the ballistic regime, and do not consider it because there are better ways to do ballistics than with a hydro code (e.g. Hjellming and Johnston 1981). The region $\rho_b/\rho_m < 10^{-3}$ has not been investigated because of the prohibitive computational costs involved, despite the fact that astrophysical jets may be highly diffuse.

3. ARCHIVE OF PROPAGATING JETS

We have surveyed the "allowed" region of the theoretical plane (Figure 1) to the extent where we believe the essential dependencies of jet behavior on the beam parameters emerge. The 12 evolutions comprising this survey are archived here and analyzed in the following sections. We include this archive because 1) there is a general need for more complete documentation of numerical experiments than is now customary, and 2) each of the 12 evolutions is rather different from the others. By thus providing the data base from which our understanding of the flows is drawn, the interested reader is invited to make his own interpretations which can be compared with our own.

There remains, however, the problem of extracting useful information from the mountains of numbers the computer has produced, and presenting this information in a form that is transparent and unambiguous to the numerical modeller and interested reader alike. The problems of extracting and displaying information are often one and the same as the amount of data becomes large. One then speaks of image processing instead of plotting.

The traditional mode of display for a 2-dimensional flow field is to plot isocontour lines of density, temperature etc. with or without flow vectors superposed. We find this medium to be of limited use for clarifying very complicated flow fields involving many extrema, or simply to be an inappropriate medium for asking certain questions of the flow. The questions we ask here are 1) what is the kinematic structure of the jet?, 2) how does this depend on location in the theoretical plane?, and 3) how does the jet propagate?, i.e. rate of advance, flow stability etc. Questions of internal shockwave structure, working surface dynamics, and entrainment will be considered in separate future papers.

A mode of display which we find useful for elucidating kinematic structure and which is used in this archive is illustrated in Figure 2. As axisymmetry is assumed, the image has been mirrored through the jet axis.

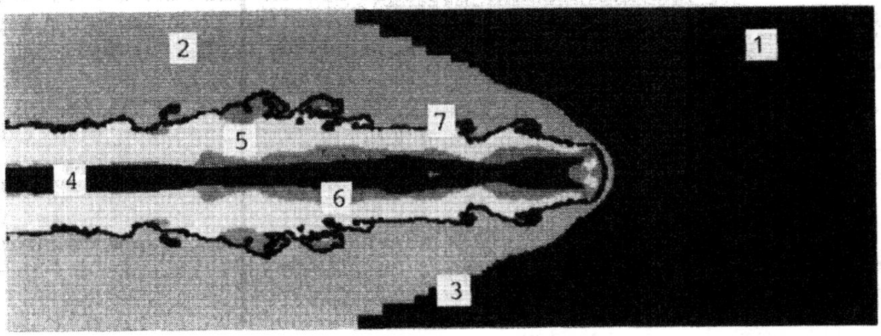

Figure 2. Kinematic structure of a supersonic jet ($\rho_b/\rho_m = 0.01$, $M_b = 12$). Seven distinct flow entities are labeled and defined in the text.

Different shades of grey represent parts of the flow having distinct identities or being in particular states of motion. These are labeled in figure 2 and have the following meanings: 1) undisturbed confining medium, 2) disturbed confining medium, 3) bow shockwave, 4) beam (defined where $\rho v_{\parallel}^3 > 5\%$ of the largest $\rho v_{\parallel}^3$ in the jet), 5) backward-flowing cocoon gas ($v_{\parallel} < 0$), 6) forward-flowing cocoon gas ($v_{\parallel} > 0$), and 7) contact discontinuity.

The images themselves are produced with a RAMTEK Color Display System, and recorded with a MATRIX INSTRUMENTS Color Graphic Camera. The display monitor has a raster field of 512 × 512 pixels, each 16 bits deep. Operationally, the flowfield is analysed zone by computational zone, each of which is assigned an integer between 1 and 7 according to the above criteria. The zone is then mapped into the raster field where its associated pixel(s) are illuminated with the color or grey tone assigned to its integer. Since the number of computational zones along the jet axis (640), exceeds the number of pixels (512), the images slightly under-resolve the data behind them.

The 12 evolutions presented in Figures 3-14 are grouped according to Mach number, in order of increasing beam kinetic luminosity. The dimensionless luminosity parameter $(\rho_b/\rho_m)^{-1/2} M_b^3$ varies by almost 4 orders of magnitude in this survey (cf. Figure 1). Each figure shows the jet at 4 stages of development chosen at regular length intervals. Each image covers the entire computational domain ($0 \leq Z \leq 80$, $0 \leq R \leq 15$) plus its mirror image. The beam is continuously admitted at left with radius $R_b = 1$ with the parameters indicated, and with the pressure of the undisturbed ambient gas. Boundary conditions are described in NSWS.

4. PROPAGATION EFFICIENCY

Define propagation efficiency η as the ratio of the actual velocity U of the head of the jet and usual ram-pressure estimate $U_{rp}=V_b(1+\sqrt{\rho_m/\rho_b})^{-1}$, where V_b is the beam velocity and ρ_b/ρ_m is the density ratio. η will in general be less than unity because U_{rp} is derived assuming a steady, head-on momentum transfer between the beam and the ambient medium over the entire area of the working surface, whereas the flow there is manifestly multidimensional(Blandford and Rees, 1974; NSWS), as well as time-dependent (NSWS).

In Table 1 we list 2 sets of propagation velocities and efficiencies for the 12 jets in our survey. The first set U_{av} and η_{av} are average values, where U_{av} is computed by dividing the total distance the jet has travelled by its evolutionary age in Figures 3d-14d. The second set U_e and η_e are end values, where U_e is similarly computed between Figures 3c-14c and 3d-14d. We further define a deceleration parameter $\delta \equiv \rho_e - \rho_{av}$.

An examination of these simply-defined quantities in Table 1 yields the following basic dependencies on the beam parameters:

a)
b)
c)
d)

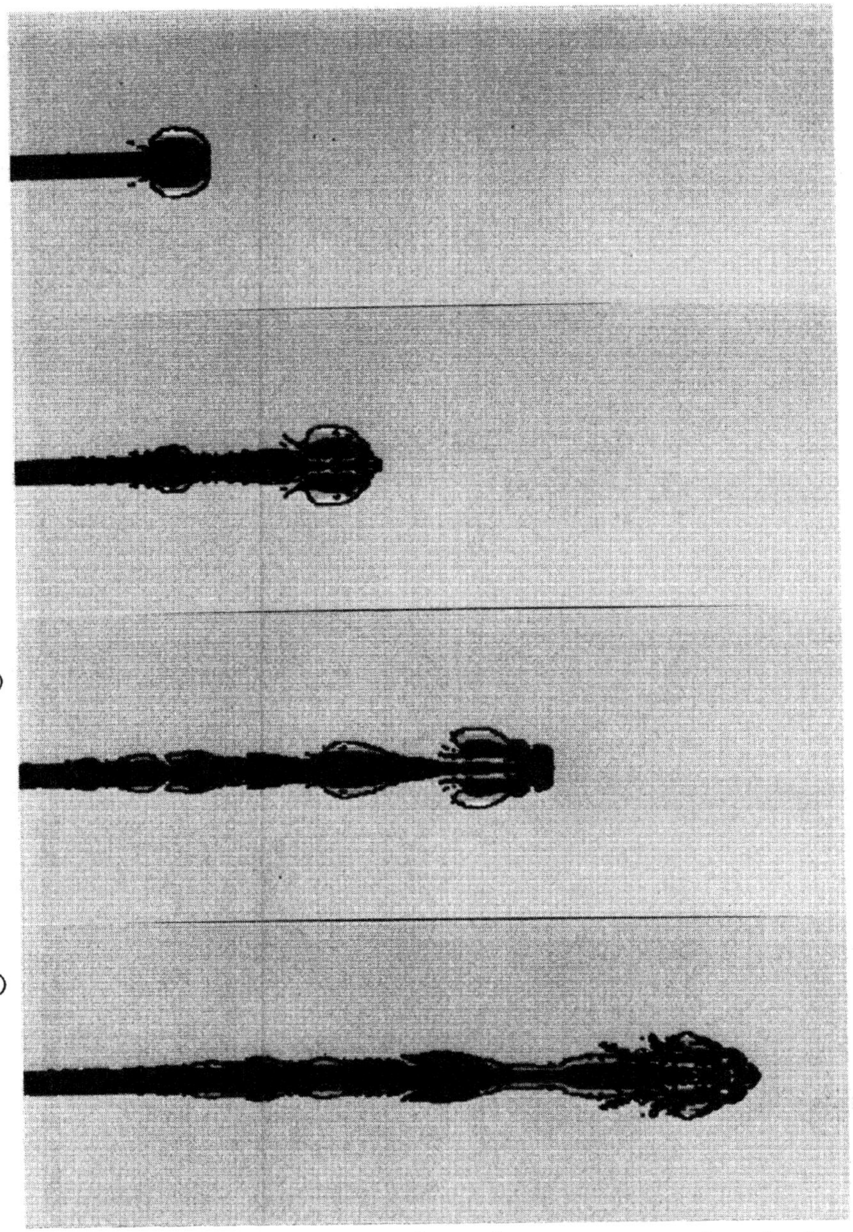

Figure 3. (ρ_b/ρ_m = 1, M_b = 1.5) Kinematic structure at 4 evolutionary times a) t = 30, b) t = 54, c) t = 78, d) t = 108.
Jet advances subsonically, therefore no bowshock is present.

a)
b)
c)
d)

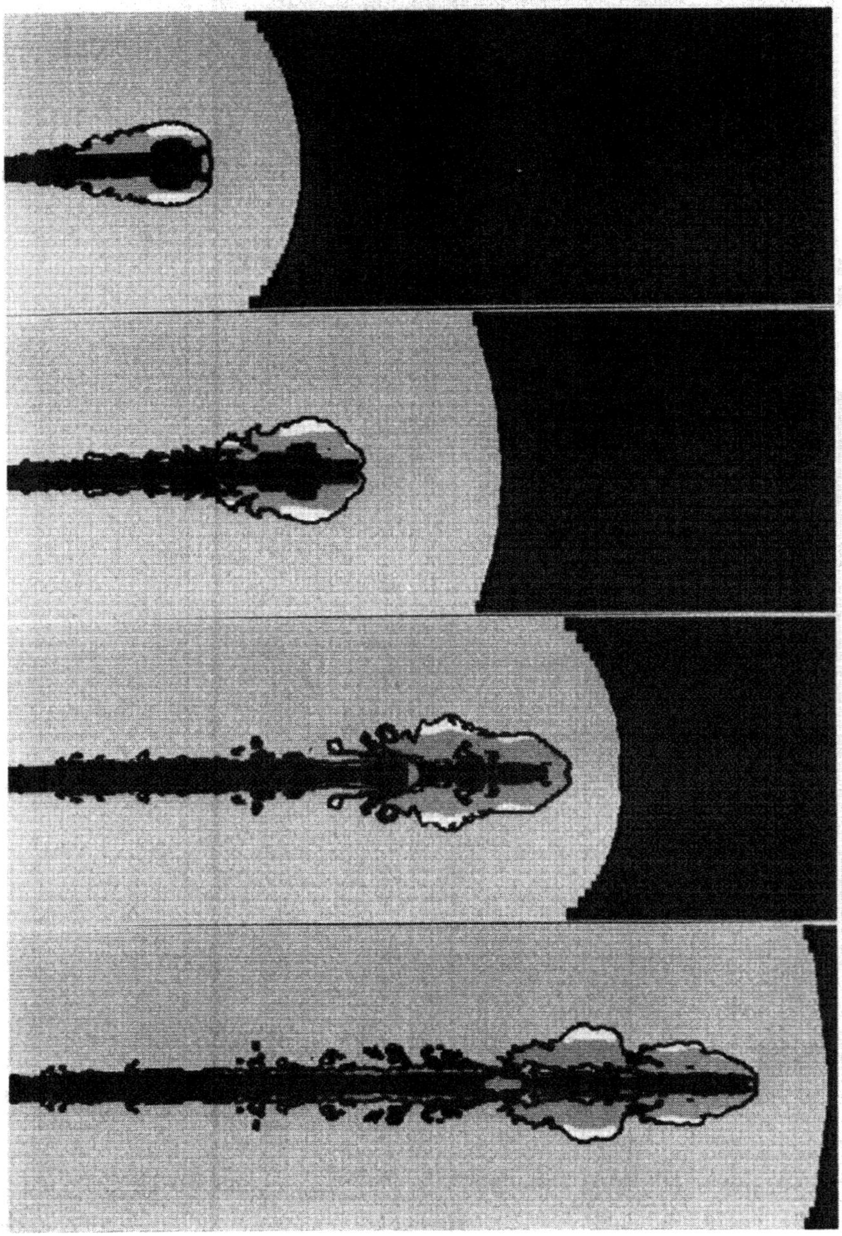

Figure 4. (ρ_b = 0.1, M_b = 1.5) Kinematic structure at 4 evolutionary times: a) t = 24, b) t = 42, c) t = 63, d) t = 81.

a)
b)
c)
d)

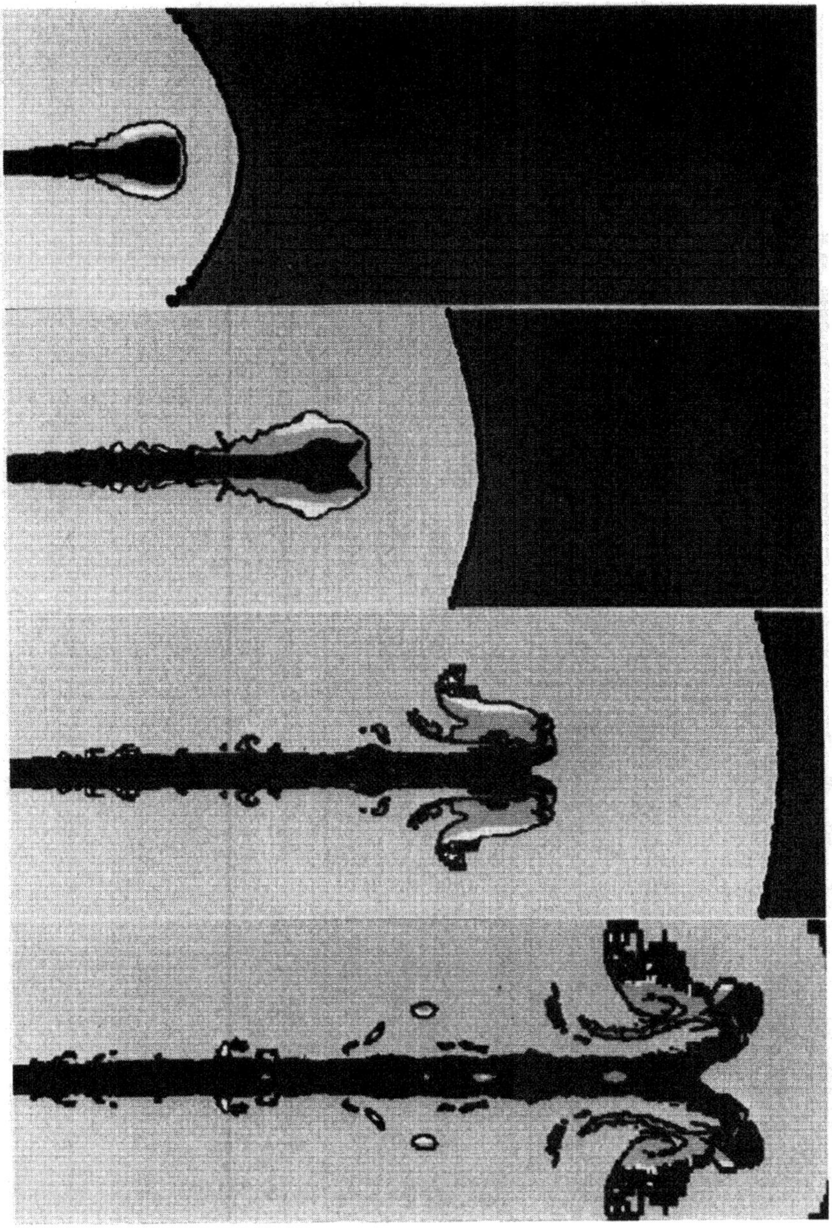

Figure 5. (ρ_b = 1, M_b = 3) Kinematic structure at 4 evolutionary times. a) t = 16, b) t = 36.5, c) t = 64, d) t = 100.

a)
b)
c)
d)

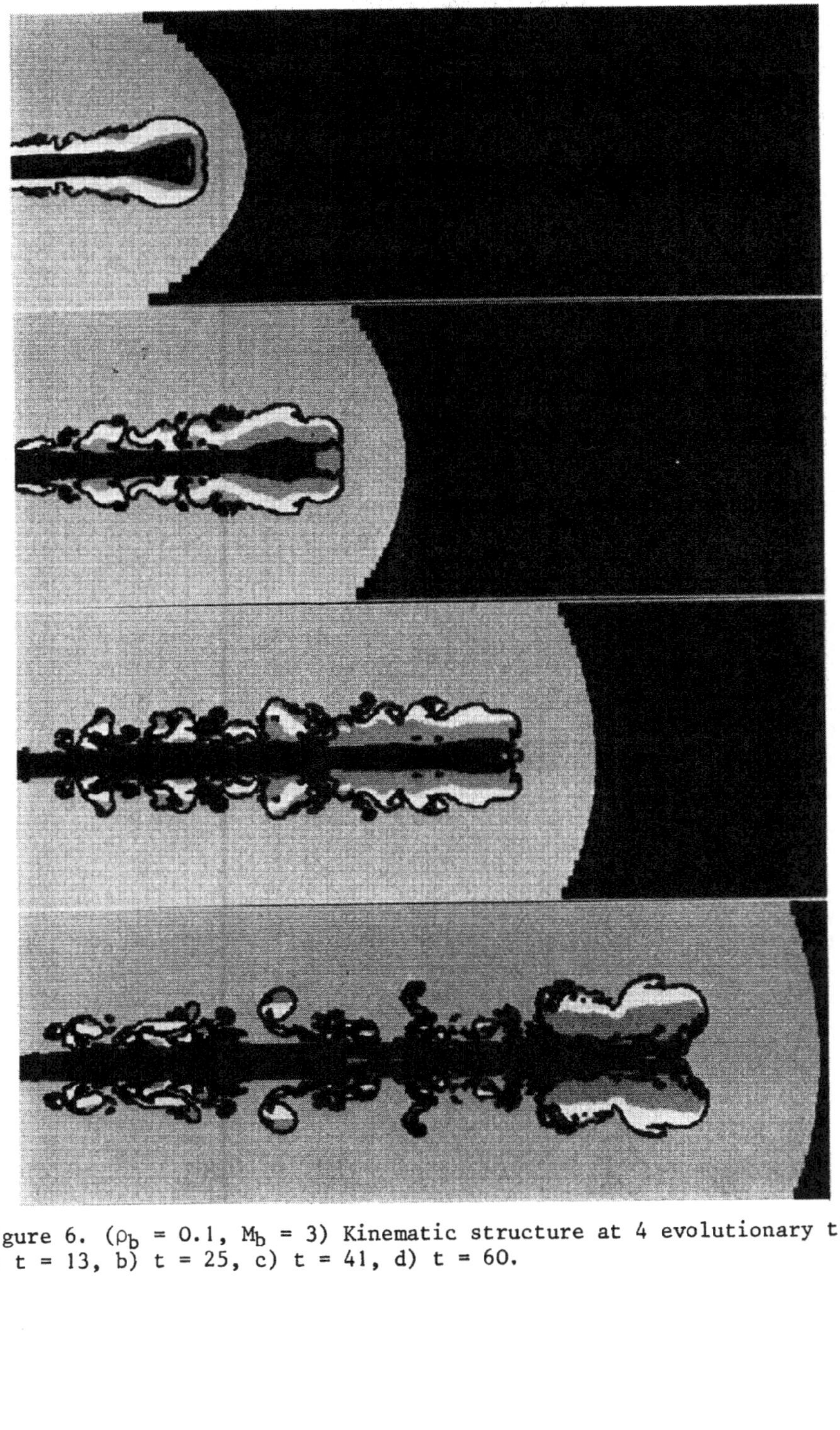

Figure 6. (ρ_b = 0.1, M_b = 3) Kinematic structure at 4 evolutionary times. a) t = 13, b) t = 25, c) t = 41, d) t = 60.

a)

b)

c)

d)

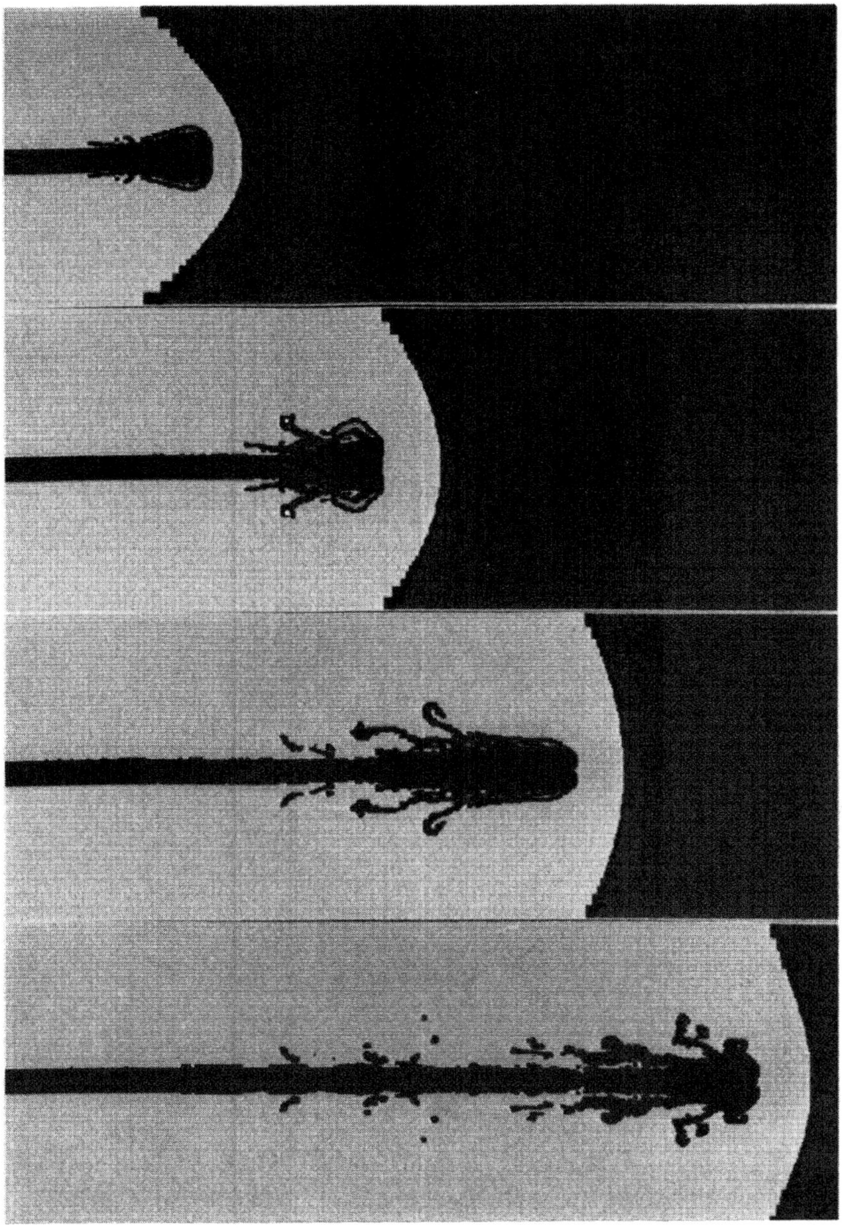

Figure 7. (ρ_b = 10, M_b = 6) Kinematic structure at 4 evolutionary times: a) t = 15, b) t = 30, c) t = 45, d) t = 60.

a)

b)

c)

d)

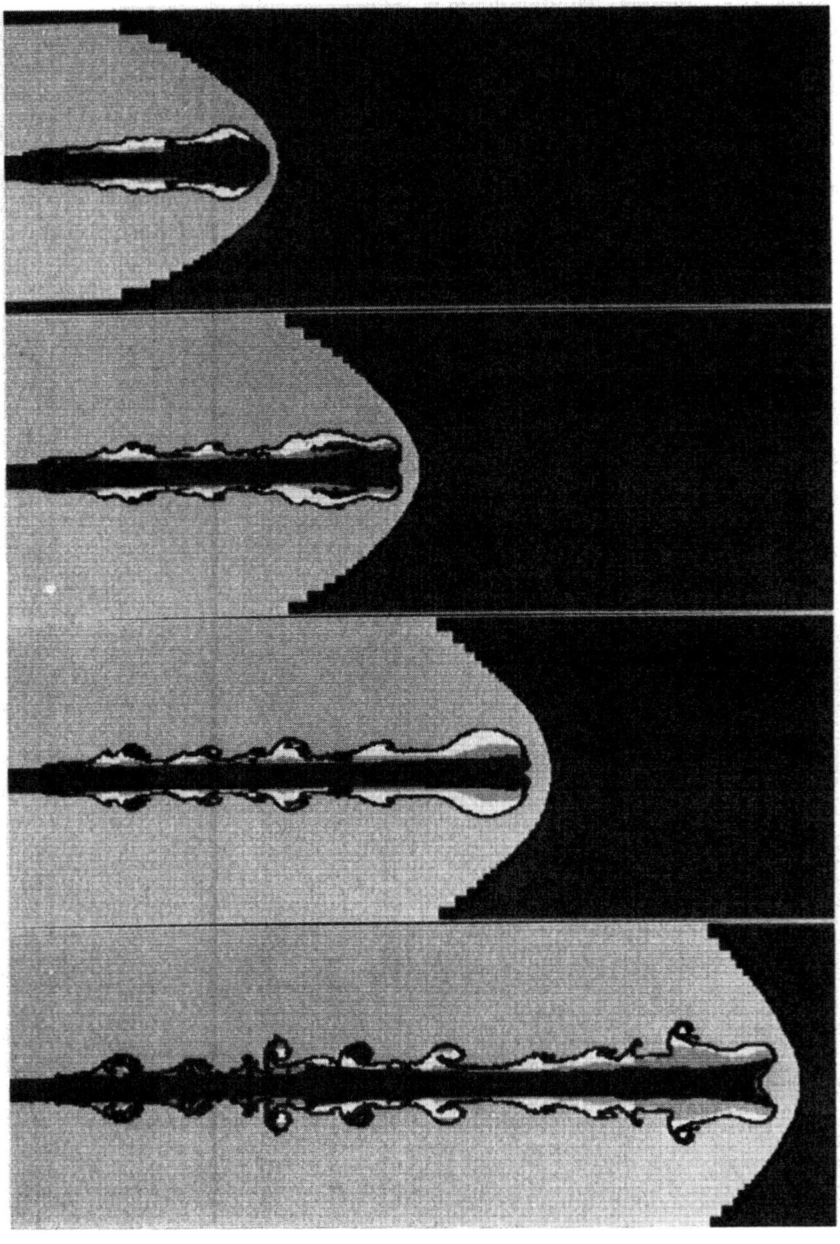

Figure 8. (ρ_b = 1, M_b = 6) Kinematic structure at 4 evolutionary times. a) t = 10, b) t = 16, c) t = 22, d) t = 34.

a)

b)

c)

d)

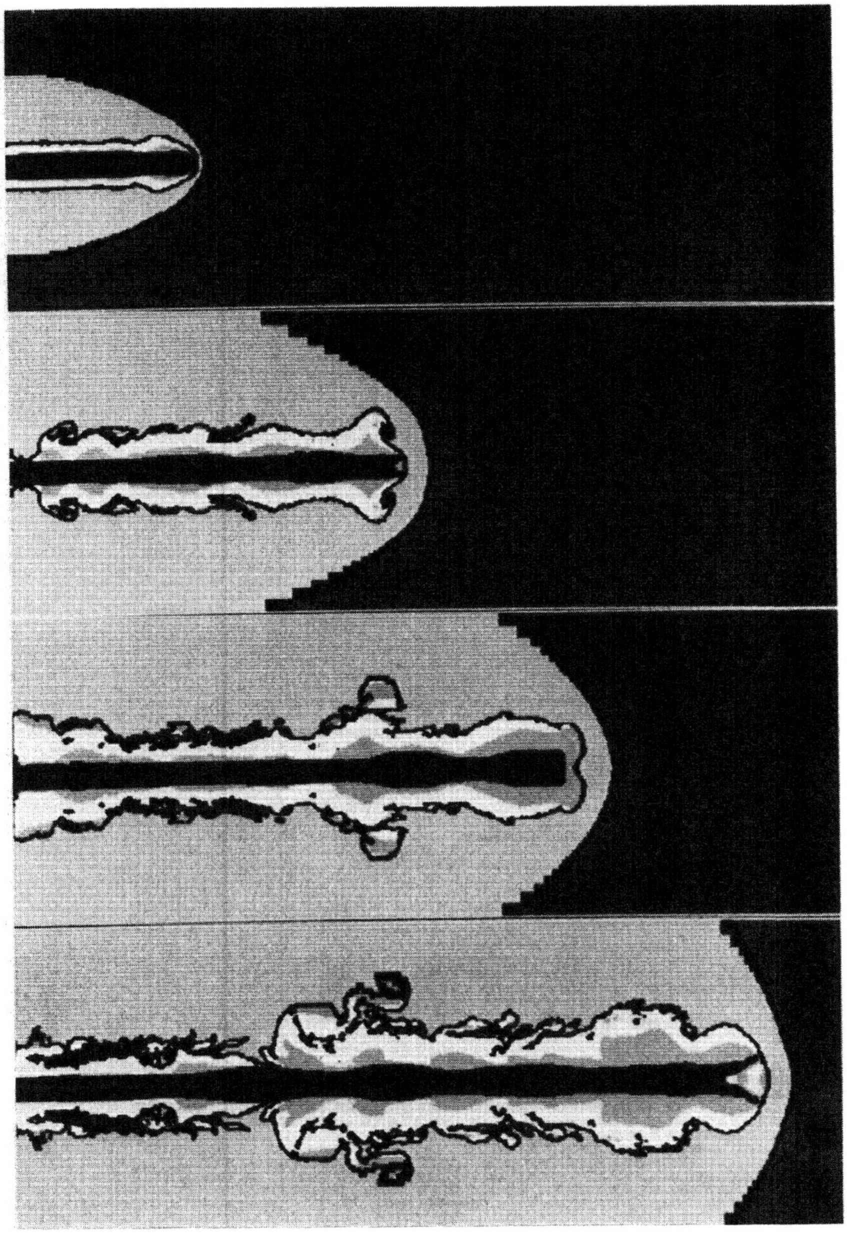

Figure 9. (ρ_b = 0.1, M_b = 6) Kinematic structure at 4 evolutionary times. a) t = 4, b) t = 11, c) t = 18, d) t = 27.

a)
b)
c)
d)

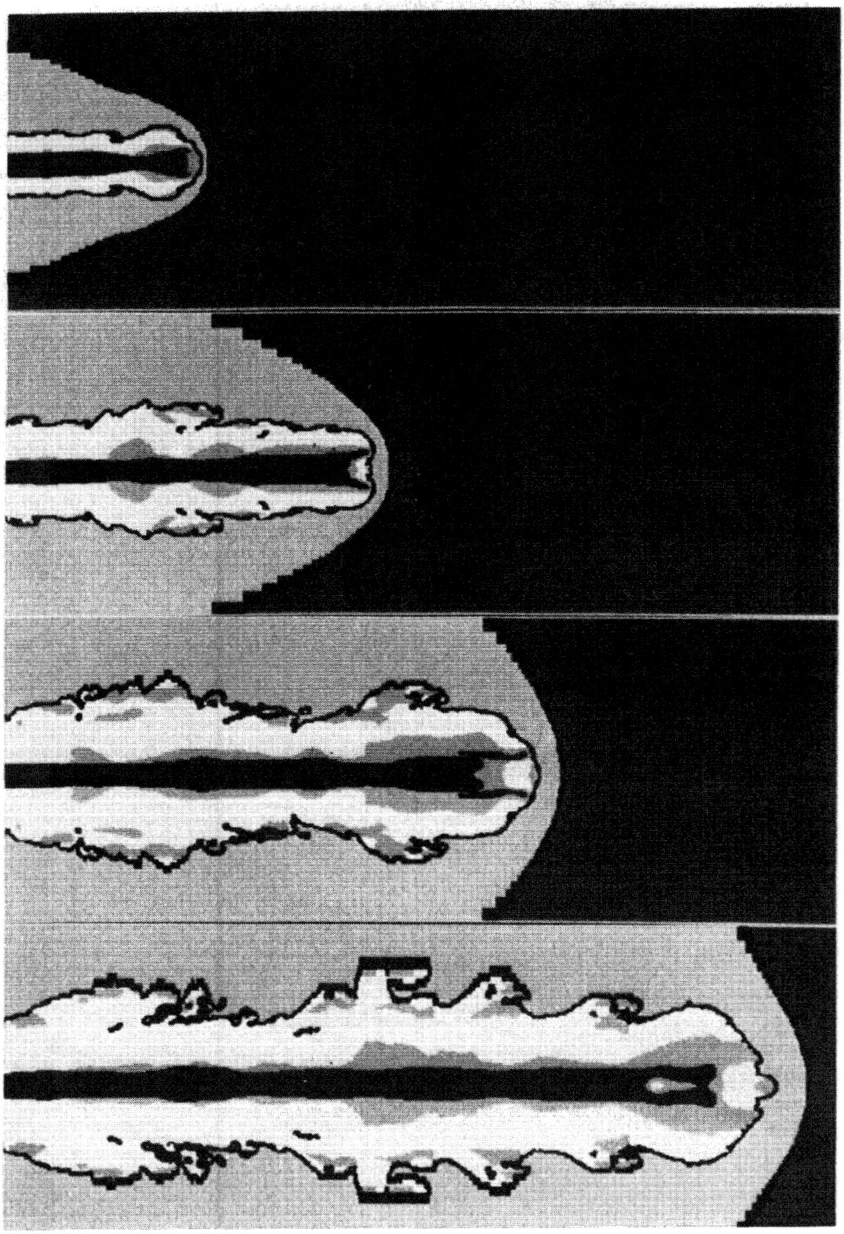

Figure 10. (ρ_b = 0.01, M_b = 6) Kinematic structure at 4 evolutionary times. a) t = 3, b) t = 8, c) t = 15.3, d) t = 27.

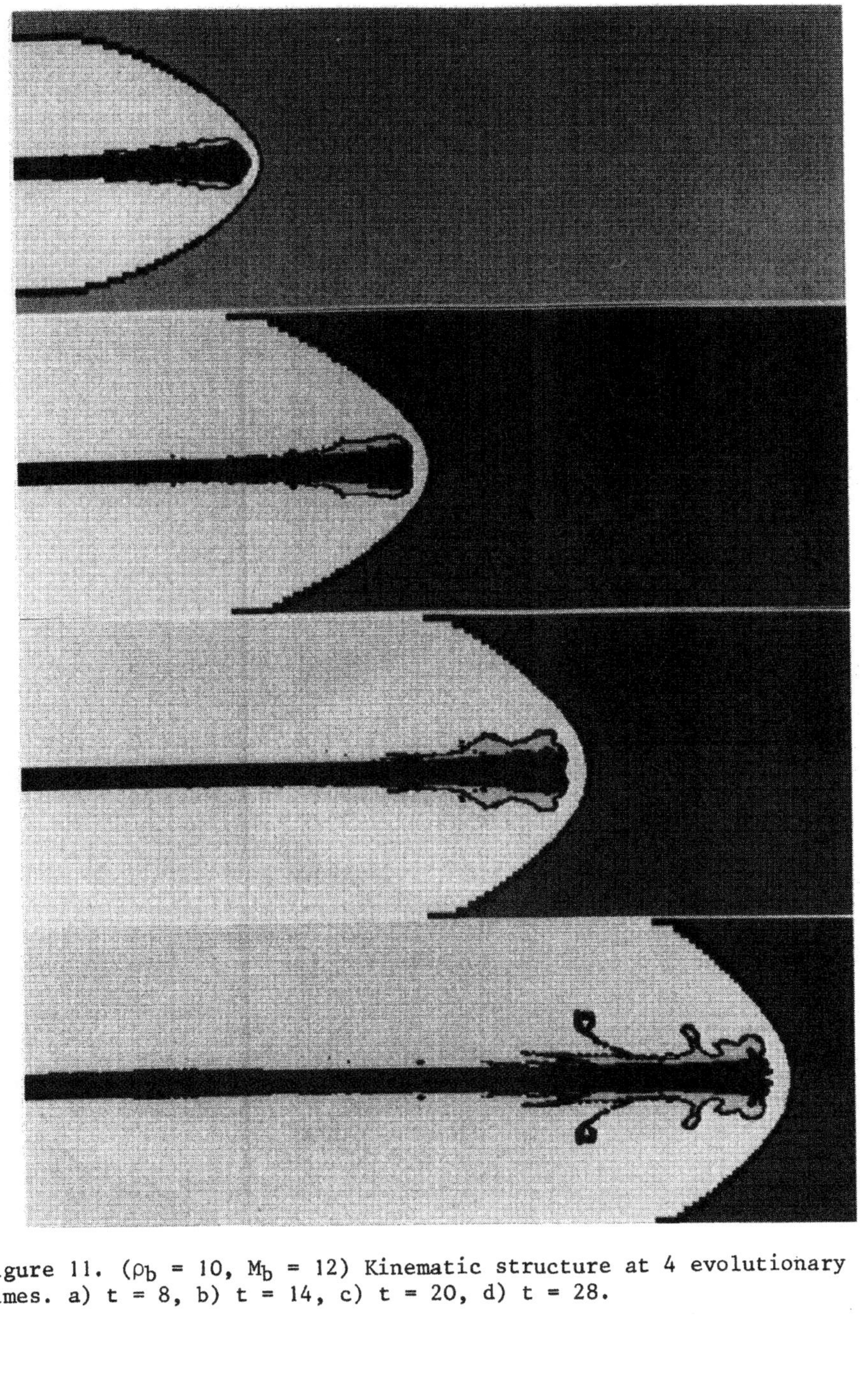

Figure 11. (ρ_b = 10, M_b = 12) Kinematic structure at 4 evolutionary times. a) t = 8, b) t = 14, c) t = 20, d) t = 28.

a)

b)

c)

d)

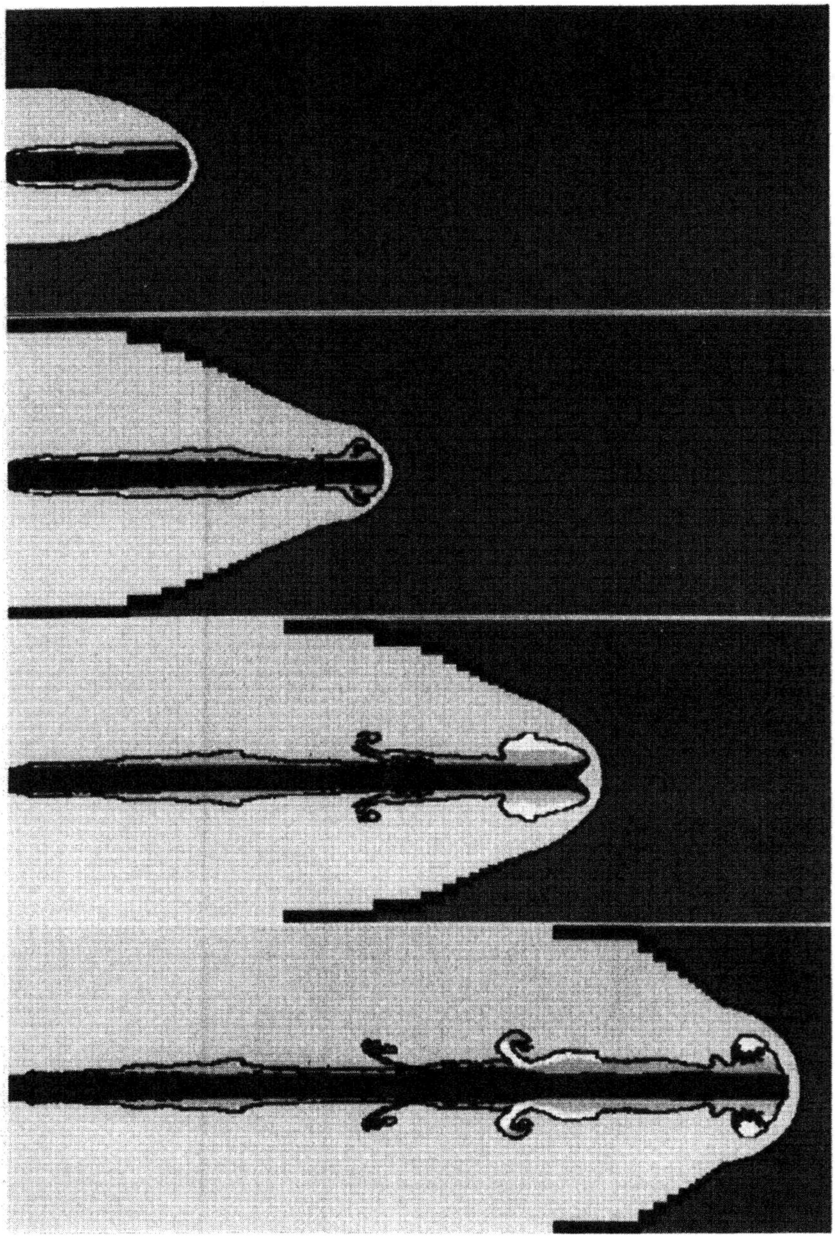

Figure 12. (ρ_b = 1, M_b = 12) Kinematic structure at 4 evolutionary times. a) t = 3, b) t = 6, c) t = 10, d) t = 14.

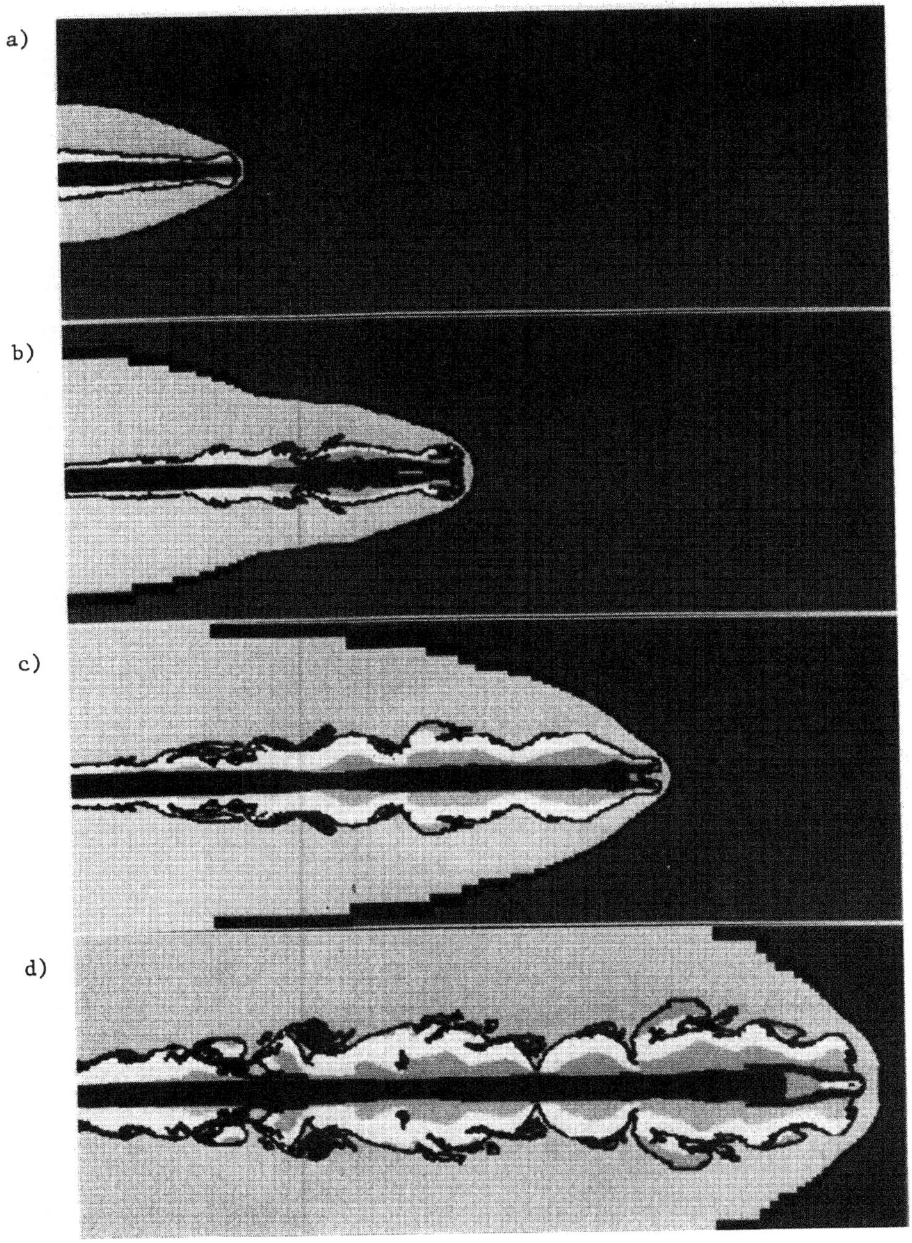

Figure 13. (ρ_b = 0.1, M_b = 12) Kinematic structure at 4 evolutionary times. a) t = 1.5, b) t = 3.5, c) t = 6, d) t = 10.

a)

b)

c)

d)

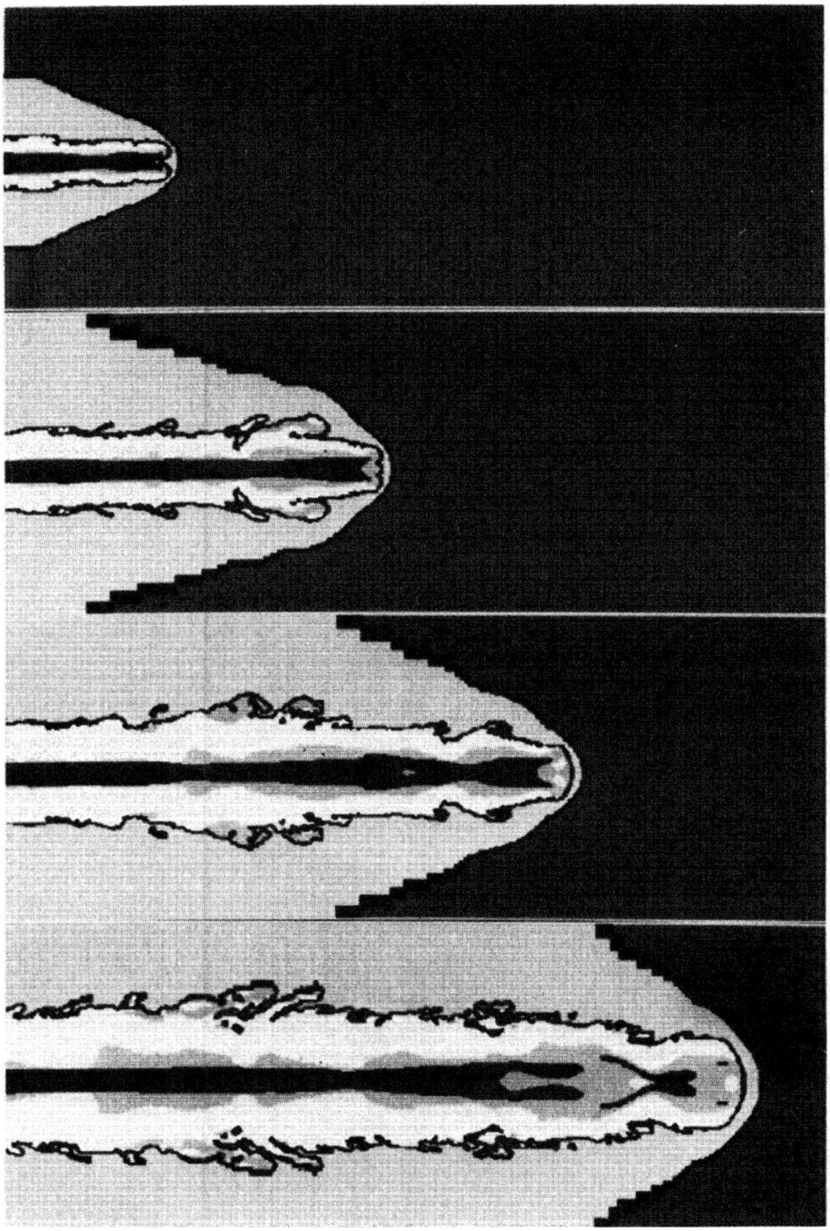

Figure 14. (ρ_b = 0.01, M_b = 12) Kinematic structure at 4 evolutionary times. a) t = 1, b) t = 3, c) t =5 , d) t = 7.6.

Table 1. Propagation velocities and efficiencies*

ρ_b/ρ_m	M_b	V_b	U_{rp}	U_{av}	η_{av}	U_e	η_e	δ
0.01	6	60	5.4	2.77	.51	1.97	.36	-.15
	12	120	10.8	9.47	.88	6.42	.59	-.29
0.1	1.5	4.75	1.14	0.89	.78	0.98	.86	+.08
	3	9.5	2.28	1.13	.50	0.94	.41	-.09
	6	19	4.56	2.70	.59	1.96	.43	-.16
	12	38	9.12	7.65	.84	4.88	.54	-.30
1.0	1.5	1.5	0.75	0.66	.88	0.66	.88	0
	3	3	1.5	0.73	.49	0.56	.37	-.12
	6	6	3	2.18	.73	1.96	.65	-.08
	12	12	6	5.42	.90	4.80	.80	-.10
10.	6	1.9	1.44	1.21	.84	1.17	.81	-.03
	12	3.8	2.88	2.57	.89	2.39	.83	-.06

* Column headings: 1) density ratio, 2) Mach number, 3) beam velocity, 4) ram-pressure estimate for velocity of jet advance, 5) average velocity of jet advance, 6) average propagation efficiency $\equiv U_{av}/U_{rp}$, 7) end velocity of jet advance, 8) end propagation efficiency $\equiv U_e/U_{rp}$, 9) deceleration parameter $\equiv \eta_e - \eta_{av}$.

1) Holding ρ_b/ρ_m fixed, η is high for $M_b = 1.5$, reaches a minimum around $M_b = 3$, and then increases as M_b increases.
2) Holding M_b fixed, η increases as ρ_b increases.
3) δ increases as M_b increases.

Note trends 1) and 2) hold true for both η_{av} and η_e. The most efficient propagators are the dense ($\rho_b/\rho_m \geq 1$), high Mach number ($M_b \geq 6$) jets, and curiously the $M_b = 1.5$ jets; both have $\eta \approx 0.8$. The least efficient propagators are diffuse ($\rho_b/\rho_m < 1$) jets of low to moderate Mach numbers ($3 \lesssim M_b \lesssim 6$), with $\eta \lesssim 0.4$.

Propagation efficiency is closely correlated to the gross morphology of the jet, which is the next topic we wish to consider.

5. FLOW KINEMATICS AND GROSS MORPHOLOGY

Having described how jets of different properties move, we now consider how they appear. We attempt to relate a jet's gross morphology to its underlying flow structure, and examine how these depend on beam density and Mach number as well as evolutionary age.

The most striking morphological dependence is that of jet width on beam density for high Mach number jets ($M_b \geq 6$). This dependence is illustrated in Figure 15 for a sequence of Mach 6 jets. At high densities (Fig. 15a), the beam is in direct contact with the external medium and feeds a working surface that leaves behind dense filaments. At equal densities (Fig. 15b) the beam is surrounded by a thin cocoon of rather uniform thickness, aside from a train of quite regularly spaced surface waves in the process of rolling up. As the beam density is further reduced (Figs. 15c, 15d), the cocoon thickness increases to 5-10 times the beam radius, while surface waves assume a more complex character. The same qualitative behavior is seen to hold for the sequence of Mach 12 jets in Figures 11d-14d. Thus, as jet density decreases we have a trend from naked beams to thick cocoons.

This morphological trend is a consequence of the strong dependence of backflow on beam density. Backflowing gas is shown in white in Figure 15, and can be seen to be absent in dense jets but to dominate the cocoons of diffuse jets. Simple arguments can be made to show why a dense jet should have forward-moving waste matter and a diffuse jet backward-moving waste matter. Note however that forward-flowing gas is also found in the cocoons of diffuse jets (Figs. 15c,d). This derives from large-scale eddies in the cocoon that are associated with surface waves (NSWS, Fig. 7).

At lower Mach numbers the same basic cocoon thickness trend is present (cf. Figs. 3d,4d and Figs. 5d,6d), although more striking morphological differences emerge that are related to interfacial instabilities at the working surface and on the cocoon and beam boundaries. Figure 3 shows pinch instabilities of a naked beam with a small amount of

a)
b)
c)
d)

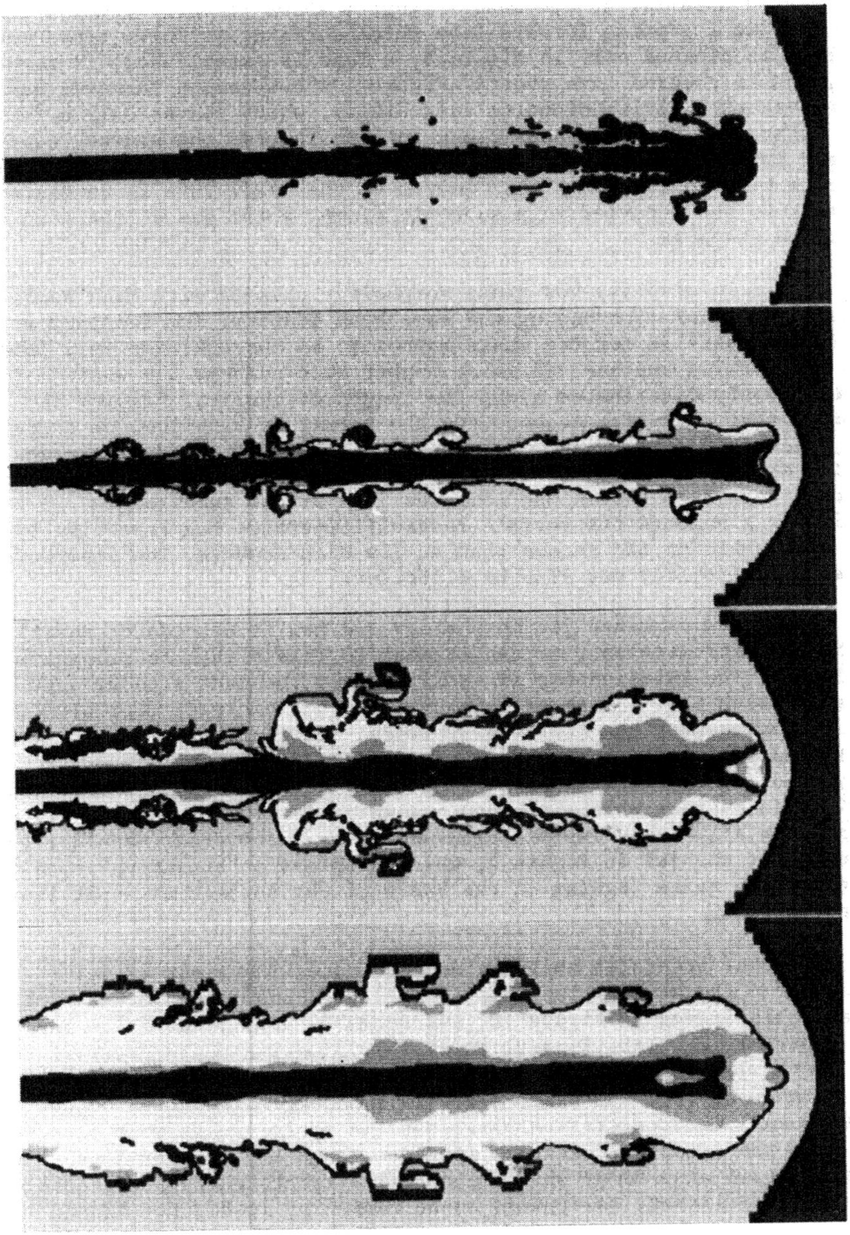

Figure 15. Naked beams and thick cocoons: a sequence of Mach 6 jets having different densities. a) $\rho_b/\rho_m = 10$, b) $\rho_b/\rho_m = 1$, c) $\rho_b/\rho_m = 0.1$, d) $\rho_b/\rho_m = 0.01$. Backflow (white regions) is stronglydependent on density.

accumulated material at the head. Figure 4 shows at early times a stable beam feeding a growing forward lobe which deflates at later times when the beam is pinched off. In Figure 5, a lobe is established, is turned over, and is sheared into sheets. Figure 6 shows cocoon turnover and mixing via the Kelvin-Helmholtz instability, which leaves only a forward lobe intact. As one goes back along the jet towards the source, cocoon gas and confining gas have been more thoroughly mixed because they have had more time to turn over than nearer to the head. This is in sharp contrast to the behavior in a subsonic plume, where gas at the head has mixed the most.

Figure 16 displays how gross morphology changes with Mach number for a sequence of jets having the same beam density. The tendency at low Mach numbers is for the waste matter to be concentrated in a lobe near the working surface, while at higher Mach numbers the waste matter is more evenly distributed along the length of the jet. Figures 16c and 16d show that extensive cocoons are the result of backflow. We argue in NSWS, section VI that backflow strength scales as M_b^2, and determines how far waste matter can be transported away from the head of the jet. An additional factor when one compares jets of the same length, is that at high Mach numbers the surface instabilities have had less time to grow and gobble up the cocoon than at low Mach numbers. The important modes of instability may also be different.

Having now examined jet morphology and how it depends on position in the theoretical plane, we can attempt to relate this to propagation efficiency. The relationship is easily found: the most slender jets propagate with the highest efficiency, and the thickest, most blunt-nosed jets propagate with the lowest efficiency. The minimum in η near Mach 3 is related to the propensity of these jets to develop blunt, or even concave nose geometries which persist in time (cf. Figs.5 and 6). Sometimes it is unclear which is cause and which is effect. Does a low propagation efficiency imply a blunt nose, or vice-versa? Consider the evolution of the jet in Figure 8, which displays a "breathing" mode which is seen to be quite regular on the basis of the surface waves it leaves behind.

Two final trends in the data deserve note. The first is jet "bloating" - the secular increase in jet thickness as it advances. This is best illustrated in Figure 10, and seems to be a property of the most powerful jets (cf. Figs.9, 10,13,14). Bloating would occur if the head of the jet were to decelerate, while all the while receiving the same throughput from the beam. There is a strong correlation between bloating and the deceleration parameters in Table 1. The most powerful jets have the strongest backflow (diffuse, high Mach number jets), and bloating would also occur when the backflow is choked by penetrating cocoon instabilities, as appears to be happening in Figures 9d and 13d.

The second trend concerns where the beam energy is deposited. In Figur 15 we see that the beam terminates further and further from the leading contact surface as beam density decreases. In Figure 16 we see

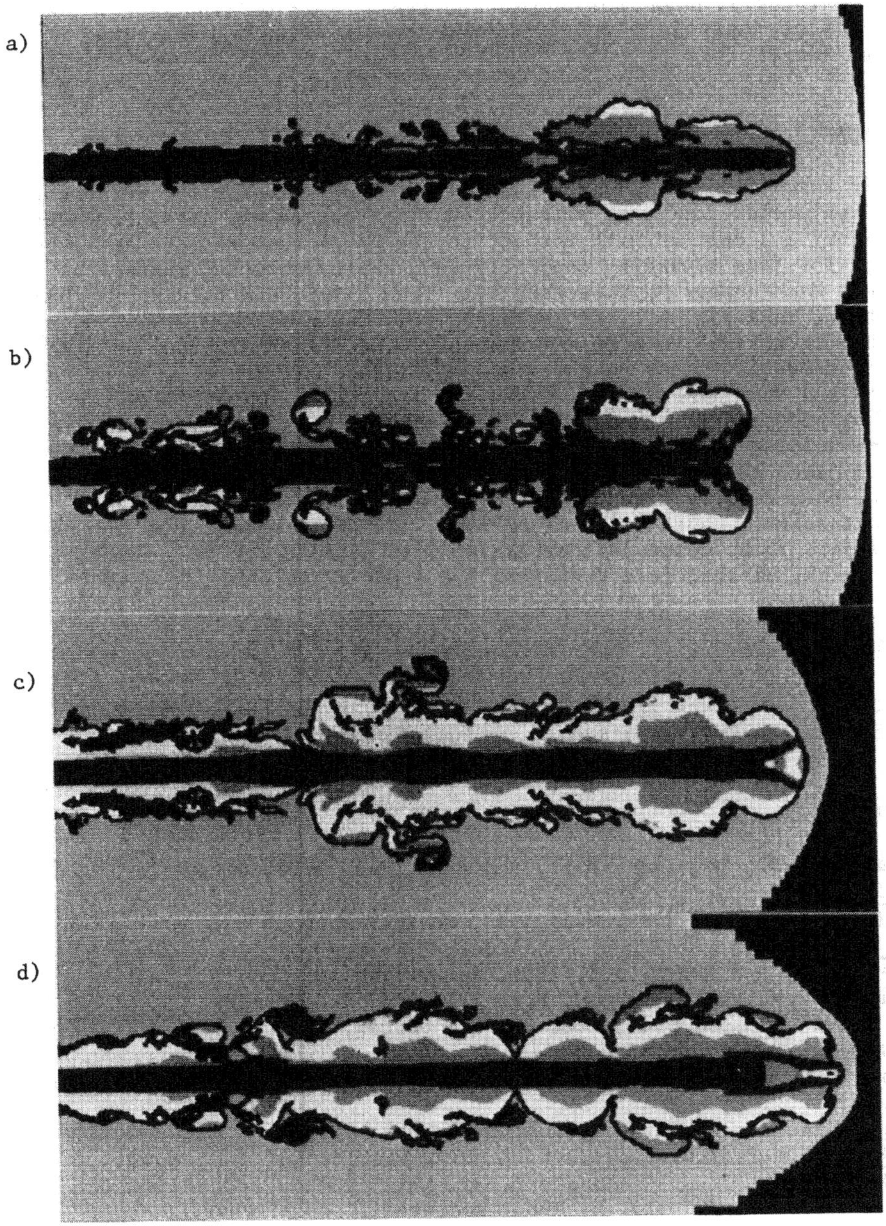

Figure 16. Lobes and unstable cocoons: a sequence of $\rho_b/\rho_m = 0.1$ jets having different Mach numbers. a) $M_b = 1.5$, b) $M_b = 3$, c) $M_b = 6$, d) $M_b = 12$ Backflow (white regions) is strongly dependent on Mach number.

the same trend as beam Mach number increases. Thus we find that high-powered beams dump their energy further from the jet end than low-powered beams.

6. SUMMARY

To summarize, we have studied the propagation and gross morphology of a 2-parameter family of pressure-confined supersonic jets. Propagation efficiency η , defined with respect to the usual ram-pressure estimate, is high for jets advancing sonically into the undisturbed medium. As the beam Mach number M_b increases away from unity, and holding the beam density ρ_b constant, η first decreases reaching a minimum around $M_b = 3$, and then increases. Holding M_b constant, η increases with ρ_b. We also find systemeatic trends in the gross appearance of jets of varying density and Mach number. Dense jets ($\rho_b > \rho_m$) are kinematically the simplest and consist of only a beam in direct contact with the confining medium. Holding Mach number fixed, as beam density is reduced, the trend is from naked beams to jets with thick, backflow-dominated cocoons. Holding beam density fixed, as Mach number is reduced, the trend is from extensive cocoons to lobes of waste matter localized near the head of the jet. Thin, dense jets propagate most efficiently ($\eta \gtrsim 0.8$), while thick, diffuse jets propagate the least efficiently ($\eta \lesssim 0.4$).

ACKNOWLEDGEMENTS

One of us (L.S.) wishes to acknowledge partial financial support from the National Science Foundation under grant no. PHY 80-01496.

REFERENCES

Blandford, R.D., Begelman, M.C., and Rees, M.J. 1982, Scientific American 5, 84.
Blandford, R.D., and Rees, M.J. 1974, Monthly Notices Roy. Astron. Soc. 169, 395.
Harlow, F.H., and Fromm, J.E. 1965, Scientific American 3, 104.
Hjellming, R.M., and Johnston, K.J. 1981, Ap. J. (Letters) 246, L141.
Love, E.S., Grigsby, C.E., Lee, L.P., Woodling, M.J., 1959, NASA Technical Report TR-R-6.
Norman, M.L., Smarr, L., Winkler, K.-H.A., and Smith, M.D. 1982, Astron. Astrophys. 113, 285.
Norman, M.L., and Winkler, K.-H.A. 1983, to appear in Astrophysical Radiation Hydrodynamics, ed. K.-H.A. Winkler and M.L. Norman, D. Reidel.
Woodward, P., and Colella, P. 1983, accepted J. Comp. Phys.

DISCUSSION

M. BEGELMAN: Do any of your lower Mach number jets break up after travelling a certain distance.

M. NORMAN: Yes. Our Mach number = 1.5 jets break up; however, under the assumption of axisymmetry, jet "break-up" proceeds in perhaps an unrealistic fashion. Unstable Kelvin-Helmholtz modes on the surface of the beam grow to the point of chopping off the beam in the form of thin fingers of external gas. The fingers are subsequently shattered by the Rayleigh-Taylor instability as they are accelerated by the beam thrust. External gas is thus entrained into the beam in this form of disruption. Now, since the growth rates of the helical modes are similar to the pinching modes for these parameters, we would expect disruption in the general case to be non-axisymmetric, but nontheless involving significant entrainment of external gas.

M. NEPVEU: Did you perform calculations using half the usual resolution? Motivation: does a Kelvin-Helmholtz instability such as found by you scale down? Is λ independent of the grid used?

M. NORMAN: When computing with half resolution, Kelvin-Helmholtz modes that are still adequately resolved (i.e. $\sim$ 20 zones/wevelength) are quite unaffected. However modes that are marginally resolved at higher resolution are numerically damped. This is the case that the λ of the dominant mode is either independent of the grid, or the mode disappears altogether and the flow is stable.

G. BENFORD: In M87 we know that the X-ray synchrotron implies a substantial magnetic field. How would that affect (qualitatively!) your periodic shock-produced knots?

M. NORMAN: In a fundamental sense, long trains of well-spaced shocks should arise whenever the beam is causally out of contact with its immediate environment. In pure fluid beams this means supersonic flows. So I would expect the same qualitative behaviour in supermagnetosonic flows in $B_\perp$ beams, or superAlfvenic flows in $B_\parallel$ beams.

DETERMINATION OF JET PHYSICAL PARAMETERS FROM OBSERVATIONS

R. Fanti
Istituto di Fisica e Istituto di Radioastronomia
Via Irnerio 46 - 40126 Bologna, Italy

SUMMARY

A review of the methods used to determine physical parameters of jets from the observations is presented. The parameters discussed are: internal energy density, density of thermal plasma, flow velocity, external pressure, energetics. Values of these parameters are given for a number of jet sources.

1. INTRODUCTION

In considering the possible physical situations implied by the characteristics of jets and in attempting to evaluate physical parameters, one faces at least three preliminary problems: a) define what a jet is; b) have a proper representative sample of jets to get unbiased ideas about their general properties; c) be able to extract from observational data informations on physical conditions, as much as possible independent of models.

About the first point there has been a lot of warning in the literature (see Véron 1981; Bridle 1982). To call something a jet (or not) prejudices already the physics that one attempts to make on it. The word itself implies the existence of motion, but motion has never been observed in extragalactic jets except at the m.a.s. level, by VLBI. Nevertheless this is the basic idea in the study of jets and it is assumed all through. Bridle (1982) calls jet any radio feature which i) has a length at least four times its width, ii) is separated, in high resolution observations from other components spatially or by brightness contrast, iii) points from a radio core toward a radio lobe. These three requirements restrict the use of the term to a limited class of radio features. Nevertheless even with these severe restrictions the decision of calling jet a radio feature is sometimes rather subjective.

A. Ferrari and A. G. Pacholczyk (eds.), Astrophysical Jets, 253–270.
Copyright © 1983 by D. Reidel Publishing Company.

Table 1: Large scale jets in double radio sources

Name	red-shift	log P at 1.4 GHz (Watts/Hz)	length of jet (Kpc)	References
Radio galaxies				
Cen A	.0017	24.5	5	1,2
M 87	.003	24.3	1	3,4,5
B2 1321+31	.016	23.7	2x 80	6
3C 31	.017	24.2	2x230	7,8,9,10,11
NGC 315	.017	23.8	250	9,12,13,14
3C 449	.018	24.2	2x 80	15,16,17
3C 66 B	.021	24.6	47	11,18
3C 129	.021	24.6	2x 30	11,19,20
NGC 6251	.023	24.1	140	21,22,23,24,25
B2 0326+39	.024	23.9	2x 63	26
3C 296	.024	24.4	2x 40	16
NGC 1265	.025	24.7	2x 18	27,28
3C 465	.030	24.9	81	29
HB 13	[.034]	[24.0]	[210]	30
NGC 326	.047	24.6	35	31,32
B2 0915+32	.062	24.0	74	33
B2 0844+31	.067	24.8	82	19,34
3C 388	.092	25.7	20	35
4C 74.17	.11	24.9	150	36
4C 53.37(1638+54)	.11	24.8	35	37
3C 219	.17	26.2	35	38,39
3C 33.1	.17	26.0	150	40
3C 133	.28	26.7	14	46
Quasars with double lobes				
4C 32.69	.67	26.7	72	41,42,43
3C 9	2.01	27.9	20	44
3C 280.1	1.66	27.6	35	44
3C 270.1	1.51	27.8	9	45
3C 275.1	.56	27.0	28	45
3C 334	.55	26.9	34	42

REFERENCES TO TABLE 1

1 Schreier et al., 1981, Ap. J. 251, 523
2 Feigelson, 1982, IAU Symp. 97, 107
3 Wilkinson, 1974, Nature, 252, 661
4 Owen et al., 1980, Ap. J., 239, L11
5 Laing, 1980, M.N.R.A.S. 193, 427
6 Fanti et al., 1982, A.A. 110, 169
7 Burch, 1977, M.N.R.A.S., 181, 599
8 Burch, 1979, M.N.R.A.S., 187, 187
9 Fomalont et al., 1980, Ap.J., 241, L145
10 Bridle et al., 1980, Ap. J., 241, L145
11 van Breugel, 1982, A.A., 110, 225
12 Bridle et al., Nature, 262, 179
13 Bridle et al., 1979, Ap. J., 228, L9
14 Willis et al., 1980, A.A., 95, 250
15 Perley et. al., 1979, Nature, 281, 437
16 Birkinshaw et al., 1981, M.N.R.A.S., 197, 253
17 Cornwell, 1982, IAU Symp. 97, 139
18 Northover, 1973, M.N.R.A.S., 165, 369
19 van Breugel and Miley, 1977, Nature, 265, 315
20 Rudnick and Burns, 1981, Ap. J., 246, L72
21 Wagget et al., 1977, M.N.R.A.S., 181, 465
22 Willis et al., 1978, A.A., 66, L1
23 Readhead et al., 1978, Nature, 272, 131
24 Saunders et al., 1981, M.N.R.A.S., 197, 287
25 Willis et al., 1982, IAU Symp. 97, 141
26 Bridle et al. 1982, in prep.
27 Miley et al., 1975, A.A., 38, 381
28 Owen et al., 1978, Ap. J., 226, L119
29 van Breugel, 1980, A.A., 88, 248
30 Masson, 1979, M.N.R.A.S., 187, 253
31 Ekers et al., 1978, Nature 276, 588
32 Ekers et al., 1999, in prep.
33 Ekers et al., 1982, in prep.
34 van Breugel, 1980, A.A., 81, 275
35 Burns et al., 1982, Ap.J., 257, 538
36 van Breugel and Willis, 1981, A.A., 96, 332
37 Burns and Owen, 1980, A.J. 85, 204
38 Turland,1975, M.N.R.A.S., 172, 181
39 Perley et al., 1980, A.J., 85, 499
40 van Breugel and Jagers, 1982, A.A. Suppl. 49, 529
41 Potash and Wardle, 1980, Ap.J., 239, 42
42 Wardle and Potash, 1982, IAU Symp. 97, 129
43 Dreher, 1982, IAU Symp. 97, 135
44 Swarup et al. 1982, preprint
45 Stocke et al., 1982 IAU Symp. 97, 39
46 Laing, 1982, IAU Symp. 97, 161

As a related point, even in cases of generally accepted jets (as for 3C31 and 3C449), it is controversial where the jets end and the broad lobes begin. In the two above mentioned sources, the term jets is generally referred to the inner $\sim \pm 1$ arcmin regions. It seems to me, however, that in those two cases most of the sources can be described as a unique double jet, the outer lobes being virtually absent. It is clear that this alternative interpretation of the morphology may have consequences in any physical interpretation.

As far as point b) is concerned, there are now probably around 100 sources showing presence or possible presence of jets. At the time of the Albuquerque symposium (IAU n. 97), Bridle (1982) reported 69 of them.

These radio features are found in weak luminosity sources as well as in very powerful ones (e.g. 4C32.69 and 3C334). Some of these have been studied in considerable detail (many frequencies and resolutions) and a number of results have been deduced, some of which of considerable relevance, from them. However how representative these sources are for the whole jet population is unclear or uncertain, since they are generally preselected with criteria which may introduce bias in the interpretation. At the moment the only complete sample of sources studied and available from the literature is the one of Ekers et al. (1981).

Several characteristics of the well studied jets are immediately suggestive of qualitative interpretations. The ideal situation would be to be able of evaluating as many physical parameters as possible and use them to construct in a quantitative way the "scenarios" suggested by the morphology.

Unfortunately the physical parameters deducible in a simple way from the observations are rather few and uncertain. Others are based on models so that there is the subtle risk of becoming locked in a loop: model → parameters → model which may be closed so as to become a tautology, or eventually open so as to bring to a non-convergent and self-consistent solution.

Having all of this in mind, here is listed the information available from the observations on the one hand and some questions one would like to answer in order to construct a theory of radio sources.

Observational Information	Questions on jets
- Occurence of jets	- Velocity in jets:
- Symmetry of jets	supersonic?
- Oscillations	relativistic?
- Brightness distribution	constant?
- Expansion rate	- Energy carried by jets:
- Connections to the outer lobes and hot spots	which forms?
	which transformations?

- Ratio of energy radiated by the jets to total source luminosity
- Spectral index along the jets
- Polarization distribution at several wavelengths
- Optical emission: lines and continuum
- X-ray emission in and around the jets
- Social status of the parent optical object

- Mass carried by the jets: where does it come from?
- Jets collimation: free jets? confined? by what?
- Structure of magnetic field: helical? tangled?
- Jets heavy or light?
- Origin of wiggles: gravitational interaction? precession K-H instabilities?
- Why are there one-sided jets?

A systematic discussion of the observational properties of jets can be found e.g. in Miley (1980), Willis (1981), Bridle (1982). Here in the following I will recall only those characteristics directly related to the parameters discussed.

The jets discussed here are those with large linear sizes (L > 1 kpc) available from the literature (see table 1) at the present time and found in radio galaxies and quasars with classical double structure. Jets have been found also in "non classical" double radio sources, often dominated by compact flat spectrum core and asymmetric structure (see e.g. Perley 1981). However they will not be mixed here with the former class owing to the uncertainties about their nature.

Nevertheless a few of them (3C273, Foley and Davis 1982, preprint; 3C138, Geldzhaler et al. 1982, in prep.; 3C147 Readhead and Wilkinson 1980) are, arbitrarily, considered as representative of their class and their parameters compared with those of jets in double sources. Jets at the m.a.s. scale are not considered here.

2. EQUIPARTITION PARAMETERS

From the knowledge of the source geometry and total brightness distribution one can compute the equipartition parameters of the source.

This procedure is well known and accurate formulae for these computations can be found in Pacholczyk (1970). A simple, and accurate enough, expression for the minimum energy density U_{min} is

$$U_{min} \approx 2.16\text{x}10^{-11}[b_{1.4}(\text{mJy/arcsec})/s(\text{kpc})]^{4/7}\ (1+k)^{4/7}(1+z)^{16/7}\ \text{erg/cm}^3$$

where $b_{1.4}$ is the surface brightness at 1.4 GHz, k the ratio of proton to electron energy, s is the source depth along the line of sight.

Besides the assumption of equipartition itself, additional uncertainties are related to the geometry of the emission volume and to the ratio k.

Furthermore, as noted by Willis et al. (1980), in the case of two dimensional expansion (as in the case of a jet) the magnetic and the relativistic particles energy densities have different dependences on the jet radius (r):

$$B^2/8\pi \propto r^{-2} \qquad \text{or} \qquad B^2/8\pi \propto r^{-4}$$

depending on whether the magnetic field is perpendicular or parallel to the flow and

$$u_{part} \propto r^{-8/3}$$

so that, if equipartition holds in some regions, the bidimensional expansion would produce soon departures from it. Finally the run of brightness as a function of transverse size r, as noted by many authors, is often sub-adiabatic, requiring magnetic field amplification or/and particles reacceleration. The details of these processes are largely unknown, and unknown as well is whether they are able to mantain equipartition or not. Since the merit of equipartition is to evaluate the minimum energy required for the observed synchrotron radiation, it is generally assumed as a working hypothesis.

Table 2 lists equipartition energy densities for the considered jets. The values are taken from the references in table 1 and are made uniform adopting the Hubble constant H_o = 100 Km/(sec Mpc) and a value of k = 1.

The values are in the range of 10^{-12} - 10^{-11} ergs/cm^3 for most of the large scale jets in low luminosity radio galaxies, but are much higher for those jets seen in very high luminosity sources (particularly the quasars 4C32.69 and 3C280.1), or in the kpc sized jets (M87, 3C138, 3C147).

3. PARAMETERS FROM POLARIZATION

a) Thermal plasma density

There is now a consistent number of jets whose polarization was studied at two frequencies at least (generally two). In a number of cases depolarization of the radiation is seen at the longer wavelength. This is generally interpreted as an effect of internal differential Faraday rotation. Assuming a slab geometry, Burn (1966) has shown that the complex degree of polarization as a function of the wavelength is given by

$$p(\lambda) = |p(\lambda)| \, e^{i\theta(\lambda)} = p_i \, [1 - e^{-S}] \, /S$$

where p(λ) is the complex intrinsec degree of polarization and

Table 2: Minimum energy densities, thermal plasma densities sound velocities

Name	$U_{min.}$ (ergs/cm^3)	n_{th} (cm^{-3})	v_s (km/sec)
Radio galaxies			
Cen A	1 - 2 10^{-10}		
M 87	7 10^{-9}	< 10^{-1}	2000
B2 1321+31	.6- 4 10^{-12}	2-5 10^{-4}	700
3C 31 North	2-30 10^{-12}	3-15 10^{-4}	1000
3C 31 South	1-20 10^{-12}	7-40 10^{-4}	300
NGC 315	.4-10 10^{-12}	4 10^{-4}	300-1000
3C 449	4-10 10^{-12}	4-20 10^{-3}	150- 300
3C 66B	6-10 10^{-12}	1 10^{-4}	300
3C 129	8 10^{-12}	10 10^{-4}	700
NGC 6251	20-70 10^{-12}	<1-5 10^{-4}	3000
B2 0326+39	.1->1 10^{-11}	35 10^{-4}	400
NGC 1265	10 10^{-12}	4 10^{-4}	500
3C 465	5 10^{-12}	2-4 10^{-3}	400- 800
3C 388	> 20 10^{-12}	10 10^{-4}	> 1000
4C 74.17	4 10^{-12}	40 10^{-4}	200
1638+54	> 60 10^{-12}	10 10^{-4}	> 1700
3C 219	4-10 10^{-11}		
3C 33.1	> 1 10^{-11}	8 10^{-4}	> 800
Quasars with double lobes			
4C 32.69	2-10 10^{-10}	<5 10^{-4}	>14000
3C 280.1	25 10^{-10}	<2 10^{-2}	> 3800
One sided Quasars			
3C 138	1- 4 10^{-6}	< 10^{-3}	∿ 0.5 c
3C 147	.8- 2 10^{-6}		
3C 273	> 4 10^{-9}	10^{-5}	∿ 0.3 c

$$S = (K\, n_{th}\, B_{r\|})^2\, d\, s\, \lambda^4 - i\, 2\, k\, n_{th}\, B_{u\|}\, s\, \lambda^2$$

where s is the source depth along the line of sight, n_{th} the thermal plasma density, $B_{r\|}$ and $B_{u\|}$ the random and the uniform component of the magnetic field along the line of sight respectively, d the scale length of the random magnetic field fluctuations and $K = 2.6\times10^{-17}$ in c.g.s. units. B_r and B_u are related by

$$p_i \,/\, p(\alpha) = B_u^2/(B_r^2 + B_u^2)\ ,\ p(\alpha) = (3\alpha + 3)/(3\alpha + 5)$$

It is easily seen that the imaginary or real part of S will be dominant depending on whether s/d = N (number of fluctuation cells along the line of sight) is greater or smaller than $(B_{r\|}/2\, B_{u\|})$.

The above expression implies both a decrease of polarization percentage and a rotation of the position angle of the electric vector with λ.

It is often assumed that S reduces to the immaginary part and therefore

$$|\, p\ (\lambda)\, | = |\, p_i\, |\ \sin(K\ \lambda^2\ B_{u\|} n_{th}\ s)/\ (K\ \lambda^2\ B_{u\|}\ n_{th}\ s)$$

In a number of cases (e.g. 3C31, NGC315, B2 1321+31) the assumption seems correct in view of the small ratio (B_r/B_u). More sophisticated approaches, both trying to account for more complex geometry (Cioffi and Jones, 1980) and for the tangling of the magnetic field (van Breugel 1980) have also been considered. However these more complex approaches do not give substantially different results from those obtained by the simplified formula and seem not justified in most of the cases where data at only two frequencies are available.

Note that the uniform slab model implies also a rotation of the polarization position angle given by

$$\Delta\ \theta = 0.5\ (K\ B_{u\|}\ n_{th}\ s)\ \lambda^2$$

so that one might expect that the rotation of the angle with λ should be consistent with the estimate of $(B_{u\|}\ n_{th}\ s)$ deduced from depolarization.

However this may well be not the case, for a) additional rotation may occur in a foreground Faraday screen (like in our own galaxy, but also closeby, although outside, to the radio source, and b) any non uniformity in B or deviation from the slab geometry produces a reduction of the linear dependence of θ on λ^2, while it does not change significantly the depolarization dependence (see Burn, 1966). It is therefore preferable to estimate n_{th} from the depolarization ratios more than from the rotation angle.

It has to be noted that depolarization as function of λ could be due also to foreground rotation if the Faraday screen has disomogeneities with sizes smaller than observing beamwidth (side - side depolarization). In general this possibility cannot be ruled out, although it is generally assumed as unlikely (but see the case of NGC 6251, Perley, this Conference). This possibility should be tested by higher resolution polarization observations particularly at low frequencies.

A further complication in interpreting depolarization data arises by the fact that sometimes significant depolarization is seen only at rather long wavelengths (e.g. in B2 1321+31 or 3C66B) where the poor resolution tend to mix the jet emission with other low brightness emission which might be absent at the higher frequencies. Again this problem requires higher resolution to be sorted out.

Table 2 gives values for n_{th} deduced from depolarization ratios, assuming the equipartition magnetic field. Most of the values are in the range of $10^{-4} - 10^{-3}$ cm^{-3}, but note the remarkably low value for the 40 kpc jet of 3C273.

From the equipartition parameters and the above n_{th} one can compute the "internal sound speed" as $(U_{min}/m_p\ n_{th})^{1/2}$. This quantity represents the velocity at which the jet would expand in transverse direction, if let free to do it.

b) Three dimensional structure of the magnetic field.

The second information deducible from multifrequency polarization data is the orientation of the intrinsic polarization vector and therefore of the projected magnetic field. Such results (parallel and perpendicular regimes and transition from one to another) are well known (see Bridle 1982) and I will not discuss them any further. It is stressed however that what one really would like to achieve is the knowledge of the three dimensional magnetic field. Fomalont et al. (1980) have suggested for 3C31 and NGC315 an helical magnetic field structure. However see also the discussion of Laing (1981), who suggests a field structure tangled in a plane perpendicular to the jet axis.

4. JET TRANSVERSE EXPANSION, CONFINEMENT AND AMBIENT CONDITIONS

In several jets associated with low luminosity radio galaxies and mapped with proper resolution (see table 2) there is evidence of increase in transverse size (r) as a function of distance (z) from the core. The ratio "dr/dz", called expansion ratio, is not constant. It often starts with a higher positive value (indicating expansion) and then decreases or even becomes negative (e.g. NGC315), implying some recollimation. Note also that in NGC315 the expansion ratio is quite small at the beginning and increases subsequently in the first arcmin from the core.

The expansion rate represents the ratio of expansion velocity to flow velocity. Its constancy (as thought for a long time for instance for NGC6251) has been considered an evidence for a freely expanding supersonic jet.

The more complex situation, briefly mentioned above, seems to require that these jets, at least to some extent, are not freely expanding, but are subjected to some kind of transverse confinement. It has been noted that confinement could be present even in those regions of constant expansion ratio, while it is certainly required when dr/dz decreases or is zero. A situation in which jets go through different phases of partial or total confinement or no confinement seems to me a more realistic picture (see also Perley, this Conference, for NGC 6251).

The generally first considered confinement process is the static pressure of an external galactic or intergalactic medium. It has been noted (Bridle 1982) that in several jets the recollimation is seen roughly at the same distance from the galaxy on both sides, which suggest that the responsible agent is the thermal plasma pressure in a symmetric galactic atmosphere.

The values of U_{min} in table 2 are immediately converted in estimates for external density n_e required for static confinement by

$$(1/3)\ U_{min} \sim 2\ n_e k\ T_e$$

An typical temperatures of $10^{-7} - 10^{-8}$ K, as indicated in many cases by X-ray emission, one gets

$$n_e \sim 10^{-3} - 10^{-4}\ cm^{-3} \quad \text{for} \quad 10^{-11} > U_{min} > 10^{-12}\ erg/cm^3$$

So for most of the jets associated to low luminosity radio galaxies there appear to be no serious problems in attributing the confinement process to the external hot gas.

However in some other sources this seems far more difficult. One well known case is that of M87. Schreier (1981) discusses this case and concludes that confinement by static gas pressure is quite difficult for knots A and B, but could be possible for the inner lobes.

Even more difficult are the cases of 3C138 and 3C147, which have comparable sizes but higher internal pressures.

Also the quasars 4C32.69 and 3C280.1 (both showing no signs of expansion) are very unlikely to be confined by external gas pressure.

In particular for the former source, the external density required for confinement is incompatible with the X-ray emission. The same situation seems also true for the two other quasars 4C39.27 and 3C334 (Potash and Wardle 1982). However all these jets require to be confined

Table 3: Comparison of internal equipartition pressure and external ambient pressure deduced from X-rays observations

Name	n_e (cm^{-3})	T_e (K)	p_e (c.g.s.)	p_{min} (dyn/cm^2)	Ref.
Cen A	6exp-2	1exp7	9exp-11	1exp-10	1
M 87	5exp-2	2exp7	2exp-10	3exp -9	2
3C 66 B	3exp-3	1exp7-8	.6-2exp-11	1exp-11	3
3C 465	3exp-3	"	.2-2exp-11	5exp-12	4
4C 32.69	<1exp-2	1exp8	2exp-10	2-10exp-10	5
3C 273	<5exp-3	1exp8	6exp-11	>4exp -9	6

REFERENCES TO TABLE 3

1 Schreier et al., 1981, Ap. J. 251, 523
2 Fabricant et al., 1980, Ap.J. 241, 552
3 Maccagni and Tarenghi, 1980, Ap.J. 243, 42
4 Jones et al., 1979, Ap.J. 234, L21
5 Wardle and Potash, 1982, IAU Symp. 97, 129
6 Foley and Davis, 1982, preprint

Table 4: Expansion ratios and "free-jet" flow velocities

Name	dr/dz (arcsec/arcsec)	v_{fl} (free-jet)
M87	∿ .05	40000 km/sec
B2 1321+31	.16	4400
3C 31	.15	6600
NGC 315	.04-.09	10000
3C 449	.17-.05	1800
NGC 6251	.03	100000
B2 0326+39	<.17	> 2400
NGC 1265	<.05	>10000
3C 465	.1	4000-8000
3C 388	<.11	> 9000.
4C 74.17	.04	5000
1658+32	<.05	>34000
3C 219	<.10	
3C 33.1	<.10	> 8000

since their internal sound speed is a significant fraction of the speed of light (unless they are highly relativistic, but see van Groningen et al. 1980). So in these cases a confinement mechanism, at the moment still unknown must occur. Magnetic confinement by currents carried by the jets has been suggested (Benford, this Conference).

The determination of the external density and the comparison with the internal one is important also to evaluate the importance of buoyancy effects, which have been suggested sometimes to explain radio features.

5. VELOCITY FLOW IN JETS

The only cases where motion has been seen on the basis of radio data are at the m.a.s. scale. There it is likely that velocities $v \sim c$ exist.

Velocity determination for jets of much larger extension are largely model dependent. Nevertheless it can be concluded that in the majority of the cases the velocities are not relativistic, being in the range of several hundred to several thousand km/sec. The arguments used are briefly summarized.

a) Free jet expansion.

If the jets are unconfined, they should expand transversally at $V_{exp} \sim (U_{min}/m_p n_{th})^{1/2}$. The jets should have a cone shape with an opening angle 2ψ given by

$$tg(\psi) = v_{exp}/v_{fl} = v_s/v_{fl} = 1 / M$$

Velocities derived in this way for those jets showing expansion are listed in table 4. Most of them are in the range of 1000 to 10000 km/sec. Note however the cases of 3C273 and 3C138 where v_s is a significant fraction of the speed of light, so that if they were free they should be largely relativistic ($M = \gamma$ for relativistic flow, Rees and Blandford 1974). Also the quasar 4C32.69, where confinement is very difficult to achieve, should be highly relativistic if we assume the jet to be free with $M \sim 30$.

Since we do not know at the moment if, or to which extent, jets are free, this method is quite dubious, although the derived values should represent upper limits to v_{fl}, since anyhow $v_{exp} < v_s$.

b) Gravitational interaction.

The parent optical galaxies associated with several sources showing wiggles in their jets have closeby companions at distances of a few kpc (see table 5). The wiggles may be due to orbital motion of the parent galaxy around the baricenter of the system under the effect of the companion (Blandford and Icke 1978). Therefore one expects oscillations in the jets with wavelengths λ (wiggle) $\sim$ T(orbital) x v_{fl}. Assuming reasonable parameters for the galaxies masses and evaluating T(orbital) from the Kepler law, one deduces velocities in the range of several hundred km/sec. Note that, because of projection effects, what is actually deduced is

$$v_{fl} \sin i \,/\, (\sin\phi)^{3/2}$$

where i and ϕ are the inclinations of the jets axis and of the vector radius of the two galaxies respectively along the line of sight. Therefore the real velocities may be significantly lower than the deduced ones, on average by up to a factor 2 (van Breugel and Willis 1981). Note that these values are significantly lower than the "free jet" velocities and sometimes they are also close to v_s. If they represent the real velocities they would necessarily require confinement of the jet.

c) Ram pressure bending.

The existence of jets showing large regular bends of mirror symmetry, allows another method to estimate v_{fl}. The bending is attributed to ram pressure effects produced by motion of the galaxy through an intracluster medium. Two models have been developed (Begelman et al. 1979; Jones and Owen 1980). Both of them express the flow velocity by

$$(n_{th}\, v_{fl}^2) \,/\, R \sim n_e\, v_g^2 \,/\, h$$

where v_g is the velocity of the galaxy, R the radius of curvature of the jets, h a length taken equal to the transverse width of the jets in Begelman et al. and to the size of the galaxy in Owen and Jones. The values deduced for the well known cases of NGC1265 and 1638+53 are in the range of 3000 to 10000 km/sec.

d) Steepening of the spectral index along the jets.

The distribution of the spectral index along the jets has been studied in a number of cases. One might have hoped to detect some steepening of the spectrum along the jet, which would have been interpreted as an effect of radiative energy losses of the relativistic electrons leading therefore to estimates of time scales and consequent flow velocities. However in most of the cases (e.g. 1321+31, NGC315, 3C449, NGC6251, 3C66B) no convincing steepening of the radio spectrum have been found. This would require flow velocities in the range of 1000 to 30000 km/sec. The only case where steepening has been seen is 3C31, although discrepancies exists between the detailed results of the various authors (see Strom et al. 1983). In this source, the most extreme variations seen in the inner part of the jets at higher frequencies would imply flow velocities of the order of 7000 km/sec, while the variations of the spectral index between 610 and 1400 MHz all along the whole source would indicate v $\sim$ 1500 km/sec. This method is unreliable, however, since we know from the sub-adiabatic behaviour of the brightness distribution that mechanism of compensation of the energy losses must exist, which might prevent from observing the

Table 5: Parameters of double galaxies and "Blandford-Icke" flow velocities

Name	M_{pg} R.G.	M_{pg} comp.	projected separation	Wiggle length	v_{fl} (Bl.+Icke)
1321+31	-19.6	-17.1	45 kpc	58 kpc	70 km/s
3C 31	-20.1	-18.3	9	35	600
NGC 315	-21.2	-18.3	68	250 ?	340
3C 449	-19.7	-18.0	11	27	290
NGC 6251	-20.5	-19.5	50	42	70
0326+39	-20.0	-18.0	10	54	770
3C 465	-21.0	-19.0	11	30	580
0915+32	-20.6	-19.0	29	98	370
4C 74.17	-21.3	-20.3	10	52	1100

Table 6: Energetics of the jets

Name	Radio Luminosity (10^7 - 10^{11} Hz) ergs/sec	$v_{fl,8}$ [$F(U_{min})/F(U_{KE})$]	$F(U_{min})$ $v_{fl,8}$ ergs/sec
Radio Galaxies			
M 87	3exp41	8.4	1exp40
B2 1321+31	8exp40	1.0	3exp40
3C 31	3exp41	1.0	5exp40
NGC 315	1exp41	3.0	4exp40
3C 449	3exp41	0.3	8exp39
3C 66B	7exp41	1.0	8exp40
NGC 6251	2exp41	1.5	3exp40
B2 0326+39	1exp41	0.4	3exp40
NGC 1265	9exp41	3.0	1exp39
3C 465	1exp42	0.2	2exp41
3C 388	9exp42	0.2	3exp40
4C 74.17	1exp42	0.1	3exp41
1638+54	1exp42	7.0	1exp41
3C 219	3exp43		1exp41
3C 33.1	2exp43	1.5	4exp41
Quasars with double lobes			
4C 32.69	9exp43	100	1exp42
3C 280.1	7exp44	30	4exp42
One sided quasars			
3C 138	1exp45	1.3 ($v \sim c$)	2exp45 ($v \sim c$)
3C 147	1exp45		2exp45 "
3C 273	5exp43		1exp45 "

$v_{fl,8}$ = 1000 km/sec

steepening in the radio spectrum. The deduced velocities are therefore likely to represent upper limits to the real ones.

e) Gravitational bending.

In the case of 3C31 it is well known that the source is well aligned along the chain of galaxies Arp. Since it is very likely that the potential well is traced by the galaxies chain, we would expect that if the jets are lighter than the external medium, buoyancy forces should tend to remore progressively the jets flow direction away from the chain of galaxies.

On the contrary the jets show several bendings which curve the jets axis toward galaxies of the chain. This seems to indicate that the jets are heavy and, by assuming reasonable parameters for the gravitational field of the galaxies, one is lead to estimate a v_{fl} of few hundred km/sec at $\sim$ 50 kpc from the parent galaxy (Strom et al. 1982 in prep.).

6. ENERGETICS OF THE JETS

Jets carry at least two forms of energy: a) energy of relativistic particles and magnetic field; b) bulk kinetic energy of non relativistic material. Other terms of energy carried by the jets, as thermal energy or turbulence are largely unknown. The two terms of energy flow are

$$F\,(U_{min}) = \pi\, r^2\, U_{min}\, v_{fl}$$

$$F\,(U_{KE}) = (\pi/2)\, r^2\, n_{th}\, m_p\, v_{fl}^3$$

It is easily seen that, with the parameters of table 2, in most of the large scale jets the two terms are comparable for $v_{fl} \sim 1000$ km/sec (but not for 4C32.69 and 3C280.1), so that for even slightly higher values of v_{fl} the first term is likely to become dominant.

If we assume that the radio sources are powered at a stationary level by the energy carried by the jets, the above energy flows should be comparable with the total source luminosities. As seen from table 6, $F(U_{min})$ is about one to two orders of magnitude too low, for $v_{fl} = 1000$ km/sec, requiring therefore velocities correspondingly higher, so that the term $F(U_{KE})$ becomes the dominant one. Thus the bulk kinetic energy could be eventually converted in "useful energy" to produce synchrotron radiation for the whole source. With the values of v_{fl} deduced from the free hypothesis, a moderate efficiency of conversion would be required ($\lesssim$ 10%).

Note also that the velocities deduced from the Blanford and Icke model are too low to have $F(U_{min})$ and $F(U_{KE})$ significant for feeding the whole radio source.

7. VELOCITY DECREASE ALONG THE JETS AND ENERGETIC CONSEQUENCES

It has been noted in at least two sources (NGC315 and B2 1321+31) that the observed depolarization ratios do not behave as expected on basis of simple assumptions in regions of jet expansion. If one assumes a constant v_{fl} and mass flow conservation along the jet one would expect

$$n \propto r^{-2}$$

and for magnetic flux conservation one would have

$$(RM)_i \propto B_{\parallel} \; n_{th} \; r \propto r^{-2}$$

(in the perpendicular regime). In the two mentioned sources on the contrary

$$(RM)_i \sim \text{const.}$$

in regions of jet expansion. This requires that

$n \sim$ const. or

$n \propto r^{-1}$ if we assume $B \sim$ const.

in view of the sub-adiabatic behaviour of the brightness.

These findings imply that v_{fl} decreases along the jets, due to losses of momentum at the boundaries, and/or there is entrainment of matter with consequent non conservation of the mass flow. In the case of B2 1321+31, assuming either conservation of the mass flow ($\dot{M}$), but not of the momentum flow, or conservation of the momentum flow plus entrainment, it was deduced

$$v \propto r^{-\beta} \qquad 0.5 < \beta < 1.0$$

Consequences of velocity decrement can be very important from the energetic point of view. First, even if jets do not carry forms of energy different from the "useful energy" U_{min}, a decrease of v_{fl}, being equivalent to a longitudinal compression, can compensate the adiabatic losses due to transverse expansion and therefore may help in solving the problem of the sub adiabatic behaviour of the brightness distribution (see Fanti et al. 1982).

On the other hand, if the major energy term carried by the jets is bulk kinetic energy of non relativistic material, a decrease of v_{fl} implies a large decrease of $F(U_{KE})$, and the lost kinetic energy must be transferred to some other forms: "heat" or "useful energy" for synchrotron radiation, for instance. However for the range of β given above, the decrease of $F(U_{KE})$ is very large, much larger than what needed to compensate for the adiabatic losses. If the lost energy remains in the jets as "heat" one would expect the jets to quickly expand, owing to the increased internal pressure and reduced v_{fl}. This could well be the case of 3C31 and 3C449, where, after a regime of almost constant opening angle (with possible initial recollimation; see Bridle et al. 1980), the jets rapidly expand to a larger size. In the case of B2 1321+31, on the contrary, where the opening angle remains constant in spite of the presumed strong deceleration, one has to assume that the energy lost goes to build the low brightness broad components surrounding the jets.

8. CONCLUSIONS

a) Although completely model dependent, the more plausible values of the flow velocity in large scale jets showing regimes of expansion are of the order of several thousand km/sec.

b) At these velocities the energy carried by the jets is mainly bulk kinetic energy of non relativistic material and is probably enough to supply the outer lobes, with a moderate efficiency in energy conversion.

c) The behaviour of transverse jet size versus distance from the nucleus in several jets indicates that the expansion velocity decreases and eventually reduces to zero. This strongly indicates the presence of a process exerting transverse confinement. In most cases the parameters of the jets are such as to suggest as responsible agent the thermal pressure of the external gas. There are however cases where this seems not possible.

d) Also for some kpc-sized jets the static pressure of the external gas seems unable to exert the required confinement and furthermore their high degree of collimation together with their very high internal sound speed make it very difficult to believe that they are "free jets". Other confining processes are required, unless one assumes that these jets are "leaky jets".

e) There is at the moment a limited evidence that the flow velocity decreases along the jet. This has important consequences on the energy budget and on the energy transformations within the jet.

REFERENCES

Begelman, M.C., Rees, M.J. and Blandford, R.D., 1979, Nature 279, 770

Blandford, R.D. and Rees, M.J., 1974, M.N.R.A.S. 169, 395

Blandford, R.D. and Icke, V., 1978, M.N.R.A.S. 185, 527

van Breugel, W., 1980, A.A. 88, 248

van Breugel, W. and Willis, A.G., 1981, A.A. 96, 332

Bridle, A.H., 1982, IAU Symp. 97, Extragalactic Radio Sources, ed. by Heeschen and Wade, Reidel, 121

Burn, B.J., 1966, M.N.R.A.S. 133, 67

Cioffi, D. and Jones, T.W., 1980, A.J. 85, 368

Ekers, R.D., Fanti, R., Lari, C., Parma, P., 1981, A.A. 101, 194

Fanti, R., Lari, C., Parma, P., Bridle, A.H., Ekers, R.D., Fomalont, E.B., 1982, A.A. 110, 169

Fomalont, E.B., Bridle, A.H., Willis, A.G., and Perley, R.A., 1980, Ap. J. 237, 418
van Groningen, E., Miley, G.K. and Norman, C.A., 1980, A.A. 90, L7
Jones, T.J. and Owen, F.N., 1979, Ap. J. 234, 818
Miley, G., 1980, Ann.Rev.Astron.Astrophys. 18, 165
Pacholczyk, A.G., 1970, Radio Astrophysics, Freeman and Co.
Perley, R.A., 1981, Proc. Second ESO/ESA Workshop, ESA SP-162,77
Readhead, A.C.S. and Wilkinson, P.N., 1980, Ap. J. 235, 11
Schreier, 1981, Proc. Second ESO/ESA Workshop, ESA SP-162, 109
Véron, P., 1981, Proc. Second ESO/ESA Workshop, ESA SP-162, 7
Wardle, J.F.C. and Potash, R.I., 1982, IAU Symp. 97, Extragalactic Radio Sources, ed. by Heeschen and Wade, Reidel, 129
Willis, A.G., 1981, Proc. Second ESO/ESA Workshop, ESA SP-162, 71

JETS, MAGNETIC FIELDS AND THE CENTRAL ENGINE

Gregory Benford
Physics Department
University of California, Irvine, California 92717

ABSTRACT

Recent observations of jets unconfined by external pressure suggest that self-confinement may be common. This requires current-carrying jets with helical magnetic fields. Such beams occur in the laboratory, in lightning, and in the Crab Nebula, where currents are apparently carried over distances $\geq$ light year. Self-confined jets require a significant torodial magnetic field ($\sim 10^{-3}$-10^{-4} G) emerging from the nozzle. I suggest the ratio $B_{\parallel}/B_{\theta}$ may be the crucial nozzle parameter, causing asymmetries. Helical field configurations have remarkable stability properties and can evolve naturally as synchrotron losses in the jet lead to minimizing $\vec{J} \times \vec{B}$. Current-carrying jets may provide a valuable clue to the physics of the central source.

Every review paper is a form of disguised autobiography, and this shall be no exception. I shall treat aspects of the jet problem as it appears interesting at present, and naturally this depends on what one thinks the important problems are. Our aim is to study processes far down in the parent galaxy, and this bias selects some features of the underlying jet physics.

FREEDOM FOR JETS?

A bent jet is necessarily confined. Free jets must be straight and display a constant opening angle $\theta \sim 1/M$, with M the Mach number. Even during richets from coulds, the jet is confined in the sense that its pressure is dominated from outside. The constant-angle signature is not often seen in jets, probably because external pressure P_e scales with distance z as z^{-x}, with most observations of gas around galaxies yielding $x < 2$. Since a free jet pressure scales as z^{-2} it is eventually overpowered by P_e.

Even worse, some jets we may think are free are in fact confined and $P_e = P_b$. This happens if the jet pressure P_b scales with its

A. Ferrari and A. G. Pacholczyk (eds.), Astrophysical Jets, 271–279.
Copyright © 1983 by D. Reidel Publishing Company.

density ρ as ρ^{β} . For constant flow velocity ρr^2 is constant, where r is the jet radius, and a simple argument shows (Begelman, 1982) that

$$\theta \propto z^{\left(\frac{x}{2\beta} - 1\right)} \quad \text{with} \tag{1}$$

"gas"	β
thermal	5/3
photon	4/3
relativistic	2
$B_\parallel$	2
$B_\perp$	1
rotation	2

Thus $x = 2\beta$ mimics a free jet. The "standard profile" of x=2 gives a constant angle if a $B_\perp$ - gas is the dominant "fluid" in the jet. This $B_\perp$ could be an "internal" field carried with the plasma, or a B_0 field confining the jet.

A jet does not have to be well collimated at the source. Ejection into a cone (steadily or in a precessing pattern) of quite large angle can be confined and refocused in a distance (Eichler, private communication)

$$L_o \cong \left[\frac{E}{(10^{43}\ \text{erg/s})} \quad \frac{(10^{-11}\ \text{dyn/cm}^2)}{P_e} \quad \frac{c}{V}\right]^{1/2} \quad \text{kpc} \tag{2}$$

where E is the energy output and V the jet velocity. When P_e falls in a distance a, collimation fails if $a < L_o$. This seems unlikely and thus most jets spend most of their time confined. However, P_e may not always do the job.

MAGNETIC CONFINEMENT?

A pressure-balanced jet satisfies

$$P_e + \frac{B_0^2}{8\pi} = P_{rel} + \left(\frac{B_\parallel^2 + B_\perp^2}{4\pi}\right) + P_t \tag{3}$$

Equipartition, that universal "leap of faith", sets P_{rel}, the relativistic particle pressure, equal to the magnetic field terms on the right. Note that these field terms represent the field lines which are often said to be "combed out" or docilely carried along by the onrushing jet. They act as energy reservoirs but are usually assumed to passively go along for the ride, exerting an outward pressure on the fluid. Of P_t, the thermal matter pressure, we know little or nothing.

Potash and Wardle (1980, 1982) have found several jets from quasars which cannot be confined by P_e. These, plus M87 and several marginal cases in other sources, lead one to suspect that the azimuthal field, B_θ, may be dominant. When I first proposed this idea in 1976 the Archbishop of Cambridge held it to be heresy! (However, see Rees, 1982, for a Jesuitical assessment).

We know nature does use currents to make beams propagate. Lightning produces fields of $10^3 - 10^4$ G, making a major contribution to the pressure (atmospheric plus magnetic) of $\sim 10^6$ dyne/cm^2. Lightning travels about 10^4 times its radius before it strikes the ground or a cloud, which provides the conducting path for charge balance. The discharge is unstable for hose-like sidewise motions. One does not see lightning strokes wriggling in neat sinusoids because the growth time is comparable to the total stroke time, $\sim$ 1 sec.

The Crab Nebula must return the currents emitted by the pulsar, over a scale > light year (where the "wisps" occur). The Crab current is $\sim 3 \times 10^{14}$ Amp (derived from the Goldreich-Julian corotation currents); this is 10^{10} times the current in a lightning bolt. Confining such quasar jets as 4C32.69 requires a current I $\sim$ 500 times the Crab current, not too difficult a feat to imagine coming from, say, a black hole magnetosphere.

In nature and the laboratory, currents propagate until a conductor affords them a faster path back to the source. In the astrophysical case, the net J leaving a magnetosphere can be (and very probably is) zero. Thus the source black hole, for example, will remain uncharged. As a jet enters a new region, the current transported forward induces, through Lenz's law, an electric field, which drives the return current back to the source. A jacket of return current around the intense jet will have a radius R roughly the size of the flared region at the jet head, and several times the jet radius r_j (Norman, these Proceedings). This jacket will last a resistive diffusion time,

$$t_r \sim 10^{17} \frac{n_{-2}}{B_{-3}} R^2_{kpc} \text{ yrs},$$

where n_{-2} is the cocoon density in units of 10^{-2} cm^{-3}, B_{-3} the magnetic field there in units of 10^{-3} G, and to estimate the resistivity of the cocoon we use a collision frequency $\sim \omega_{ce}$, the electron cyclotron frequency. This is the Bohm value, a worse-case estimate, and still yields very long times, so the cocoon will not be destroyed.

There is some intergalactic magnetic field, too, which will bend the jet through $\vec{J} \times \vec{B}$ forces. A jet carrying current which yields a self-field B, passing through an intergalactic field B* at angle will bend with curvature R* given by

$$R^* \sim 40\pi r_j n_j \left(\frac{V_j}{1000 \text{ km/s}}\right)^2 (B_{-3} B^*_{-6} \sin\theta)^{-1}$$

This yields a value of R^* consistent with the requirements Jack Burns has outlined (these Proceedings) for 3C1919+479 (oppositely directed currents will turn in the same direction, as does a head-tail source). This explanation requires a B* ordered over distances larger than a few Mpcs, which has interesting implications for cluster formation models. He further notes in several curved, head-tail sources a tendency toward helical structure; this hints that the stability of these curved jets is augmented by a magnetic buffer which shields them from the slings and arrows of the intergalactic medium. Certainly something out of the ordinary is necessary to stabilize curved jets (Benford, 1981). The associated cocoon will of course not easily bend, once it is generated, because of its larger mass. If a curving jet leaves its channel, a fresh electric field appears, generating a new cocoon current. This carries the insulating properties of a magnetically dominated cocoon along with the jet -- at an energetic price, of course.

In contrast, it is interesting to explore the notion that energetic particles, blown back from the jet head, could confine the jet for a distance D behind the head. This fresh cocoon will of course lose particles outward in a time R/v_{te}, with v_{te} the thermal velocity. Then $D = UR/v_{te}$ with $U = (\rho_j/\rho_c)^{\frac{1}{2}} v_j$ the jet head velocity. M. Norman finds (these Proceedings) substantial hot blown-back cocoons only when the jets are light, $\rho_j << \rho_c$. I estimate from his results $R < 3r_j$, $v_{te} \geq v_j/3$. For $\rho_j < 0.01\rho_c$ then $D/r_j \sim 1$ and hot cocoon confinement fails not far behind the head. Thus I expect this scheme fails in general. Norman's simulations begin with confinement legislated. It would be interesting to attempt simulation of beams which begin in unconfined modes, and see if they self-confine with a hot cocoon over such a distance.

The magnetic structure in a jet has components

$$\vec{B} \equiv \left[\frac{B_\parallel (r_c)}{x^2} \quad , \quad \frac{B_\perp (r_i)}{x} \right] \tag{4}$$

where $x = r_o/r_i$, with r_i the initial jet radius and r_o the radius further out. It is primarily B_θ, which dominated $B_\perp$, which prevents net charge from escaping radially and "shorting out" the circuit we envision. This is why beams propagate well in laboratory situations, returning net charge to the source through return currents in a plasma, or (finally) through the laboratory chamber itself, when they strike the far end.

If B_θ dominates the jet equilibrium at some distance, we can expect

$$B_\theta^2 \, r_j^{\ 2} \, V_j \sim \text{constant} \tag{5}$$

If this were true all along the jet, we could easily determine the magnetic field conditions near the central engine. Eq. 5 is probably not accurate in most jets, so we would need to study the evolution of MHD jets with several dynamically important "fluids" present -- B_θ, $B_\parallel$ and $\rho_j v_j^2$. This is quite a tall order, but in principle not beyond the present capabilities of plasma physics and numerical methods. Since we suspect v_j constant in at least some jets, this simplifies matters.

It is clear that the high values of $B_{\parallel}$ far from the source must arise from shear enhancement, since (5) predicts rapid decay as the jet expands. If B_{θ} confines the jet at r_o, as implied by pressure arguments such as Wardle and Potash (1982), it must be dynamically important near the nozzle, since it probably scales with z just as does the internal pressure. A confining $B_{\theta} \sim 10^{-3}$ G at r_o yields a confining pressure 10^{-7} dyne/cm^2 at r_i for the Potash-Wardle jet. This is the typical number invoked for nozzles (Norman et al., 1981). Such a high B_{θ} pressure can be overcome by the observed $B_{\parallel}$ far out along the jet, if that $B_{\parallel}$ arises from shearing of velocities. The fact that the implied B_{θ} close to the nozzle is so large means that the torodial geometry there is significant, i.e., a magnetically dominated nozzle may be needed. This points further to magnetic processes of energy extraction, as in Blandford and Payne (1982).

A large B_{θ} can resist incursions of wall material and impede general cloud fragmentation by Kelvin-Helmholtz instabilities. Hoselike motions are also reduced if the B_{θ} field outside the jet is pressed against a larger surrounding conducting sheath (Benford 1978). Pinching modes are suppressed by relativistic fluids since $P \propto r^{-4}$. $B_{\parallel}$, on the other hand, will blow open a nozzle.

This picture suggests magnetic field evolution which can be tested by VLBI polarization studies. As the jet emerges from the nozzle region, $B_{\parallel}$ cannot be too large or it would destroy the nozzle. $B_{\perp}$ should dominate until shear amplification brings $B_{\parallel}$ up. The problem with observing polarization is that projection effects can easily make a helical field show $B_{\perp}$ at the center and $B_{\parallel}$ on the edges, since B_{θ} will point nearly toward us in the jets lying nearly in the plane of the sky (which we select to study because of size etc.). Even allowing for this, the larger pressures at the centers of jets can lead to higher luminosity in the $B_{\parallel}$ region. Only polarization studies with large dynamic range will allow us to see a B_{θ} in the less obvious sheath surrounding the bright jets.

I am proposing here that study of magnetic fields far from the source (alas, usually by indirect pressure arguments) may provide valuable clues about the nozzle environment. Let me sketch one of the implications of strong fields for the ion-supported tori notion discussed in these proceedings by Begelman and by Phinney.

The essential requirement that such tori exist is that the electrons not be heated by the very hot ions. This is fine if classical, Coulomb equilibration times govern the interaction. In as violent an environment as this, however, such an assertion should be studied. One general rule we know from plasma physics is that plasma waves with frequencies $\omega \ll \omega_{ce}$ cannot heat electrons, since the invariant $v_{\perp}^2/B$ keeps the electron temperature steady. This means all instabilities at the ion cyclotron or ion plasma frequency do not heat electrons very well, and thus can occur in an ion-supported torus. To excite modes which <u>do</u> heat electrons, we require $\omega \sim kv_d \geq \omega_{ce}$, where k is the wave number and v_d the net drift velocity of one species with respect to another.

We know $k \leq k_D = \frac{\omega_{pe}}{v_{te}}$, the Debye value, so the condition for suppressing these modes as well is

$$v_d/v_{te} < v_{te}/\beta^{\frac{1}{2}} c < \beta^{-\frac{1}{2}}$$

where β is $8\pi P_t/B^2$, the ratio of thermal and magnetic pressures. For only mildly relativistic electrons, as ion tori require, low β -- i.e., magnetically dominated environment -- insures nearly classical temperature equilibration rates. (Assuming $v_d \gg v_{te}$ is unlikely). Strong magnetic fields thus seem to be implied, once more.

Some theoretical questions we might address in future to resolve these issues are:

Does the central engine prefer $B_\theta \gg B_\parallel$ outflow?
Are B_θ -dominated nozzles plausible?
Does large B_θ in nozzles alleviate instability troubles?

SYMMETRIES

As Fanti has noted (these Proceedings), crude statistics suggest that mirror symmetries occur more often than reflection symmetries. As Ekers (1982) noted, rich clusters tend to convert all morphologies into strong mirror-symmetric distortions. Thus we may be seeing mostly the effect of galaxies coasting through cluster gas. Orbital motion can do this, too. However, if we look far down into the galaxy, where external effects may be reduced, we may get some idea of how often sidewise impacts (which should give mirror symmetries) occur vs. rotational perturbations.

Rather more interesting, though, are the many one-sided jets. This perhaps more telling asymmetry must exist over many distance scales, as in NGC6251 and M87. In such classic sources as Cygnus A (Saikia and Wiita, 1982) the one-sidedness seems unavoidably inherent. Shklovsky (1982) pointed out that if we take the extreme position that there is no momentum flow out the dark side at all, even massive black holes will escape from their galaxies within less than the jet lifetime -- jet propelled! However, rather less exciting models can be invented if we allow the dark side to eject particles in a sloppy, badly collimated way. A valuable hint may lie in Bridle's (1982) noting that $B_\parallel$ -dominated jets are bright and one-sided, while $B_\perp$ -dominated jets are dim and two-sided.

I suggest that:

(a) the bright sources have a larger seed $B_\parallel$ to shear-amplify (and perhaps larger shear), so $I(\nu) \propto B_\parallel^2$ is higher;

(b) the seed $B_\parallel$ is close to the value which would blow open the nozzle, $\sim 10^4$ G;

(c) the other side of the galaxy has a seemingly minor difference in confining gas density, or a slightly higher $B_\parallel$, which takes that nozzle beyond the blowout limit. This jet is then unconfined at small radius and spreads its energy over a region, lowering its surface luminosity below what we can see. (Further out, focusing occurs by (2), to give distant hot spots).

The $B_\perp$ -dominated jets, on the other hand, have smaller seed $B_\parallel$, do not disrupt their nozzles, and because their seed $B_\parallel$ is not close to the blowout field, can tolerate fluctuations in conditions on both sides of the core.

This demands that asymmetry deep within the engine persist for many dynamic time scales. By relating the failure to focus to one parameter, $B_\parallel/B_\perp$, which governs confinement at small radii, it connects several observed systematics. The obvious way to test this idea is through VLBI polarization, if we can see a $B_\perp$ -dominated region closer in than the $B_\parallel$ -strong regime.

WHY ARE THEY SO (RELATIVELY) STRAIGHT?

The well-aligned classical doubles imply beams travel $\sim 10^3$ times their radius, aiming to within a few degrees -- better than lightning! To keep sources such as Cygnus A and 3C236 from wriggling too much we search for stabilizers of sidewise motion.

A simple method is raising the Mach number or the beam energy factor, γ. However, it is difficult to see how high M jets can be steadily produced, and they are not necessarily slow-growing configurations (Ferrari, Trussoni & Zaninetti, 1981; Hardee, 1979, 1981). High γ jets over > 10 kpc scales seem difficult to keep free of entrainment.

If magnetic fields from jets thread a bulky cocoon plasma, the growth scale for hoselike motions increases by a factor of M_c/m_{jet}, with M_c the threaded cocoon mass (Benford, 1981). This, together with other stabilising properties of helical magnetic fields, may do the job. Filling the cocoon, though, may imply high rates of mass loss from the core, telling us something about the past evolution of sources. The stabilising agent must be able to fail, though, in order to allow for the many wiggling jets. It is comforting that instability onset analysis seems to show mild Mach numbers $M \sim 5 - 10$ and dense beams, since these indeed should be unstable (Trussoni et al., these Proceedings).

Similarly, instability probably destroys jets that try to bend when they encounter dense clouds. Simple analysis of fluid equations (M. Smith, private communication) shows that beams should turn around. Beam instability work suggests that such wrenched-about jets should go to pieces very rapidly, in a variety of modes, however (Norman and Smith, 1981). The non-observation of such reversals is thus consistent with what we believe.

ARE WE STUDYING "WEATHER"?

Some aspects of jets are certainly sensitive to cluster motions, cross-winds, clouds in the way or carried out, etc. This "weather" can obscure connections to the core. Other features can be -- theory must separate out the cases -- equally sensitive to basic jet parameters, however. For example, radio surface brightness is proportional to (relativistic electron density) x B^2 x (path length through jet), and for pinching motions this scales as r^{-3} for $B_\perp$ - dominant jets, r^{-5} for $B_\parallel$ -dominant jets. Thus brightness is sensitive to field structure, which in turn may be dependent on fields carried out from the central engine.

I have stressed the possible roles of magnetic fields -- B_θ, particularly -- in order to underline how we might be able to trace jet characteristics back into the central engine. Even much-maligned linear stability analysis plays a role, for numerical simulations of flows often cannot include crucial physics, such as three dimensions. Thus I think we should keep our minds open about what jets can use of standard physics (carrying currents, etc.) to make their way. Arguments based on our own lack of imagination should be treated warily. Canonical assumptions, such as MHD pictures and equipartition, we should keep in mind as mere orthodoxies.

This work was supported by the Air Force Office of Scientific Research. Grant ≠ 82-0233.

REFERENCES

Begelman, M., 1982, Proceedings of the I.A.U. Symposium n. 97, Reidel
Benford, G., 1978, MNRAS, 183, 29
Benford, G., 1981, Ap. J., 247, 458
Blandford, R. and Payne, D., 1982, MNRAS, 199, 883
Bridle, A., 1982, Proceedings of the I.A.U. Symposium n. 97, Reidel
Ekers, R., 1982, Proceedings of the I.A.U. Symposium n. 97, Reidel
Ferrari, A., Trussoni, E. and Zaninetti, L., 1981, MNRAS, 196, 1051
Hardee, P., 1979, Ap. J., 234, 47
Norman, M.L., Smarr, L., Wilson, J. and Smith, M., 1981, Ap. J., 247, 58
Potash, R.L. and Wardle, J.F.C., 1979, Astron. J., 84, 707
Rees, M., 1982, Proceedings of the I.A.U. Symposium n. 97, Reidel
Saikia, D.J. and Wiita, P.J., 1982, MNRAS, 200, 83
Shklovsky, I.S., 1982, Proceedings of the I.A.U. Symposium n. 97, Reidel
Smith, M.D. and Norman, C.A., 1981, MNRAS, 194, 771

Wardle, J.F.C. and Potash, R.I., 1982, Proceedings of the I.A.U. Symposium n. 97, Reidel

DISCUSSION

P. Biermann: There is information on the magnetic field at compact scales from our sample of flat spectrum radio-sources (Biermann et al., 1981 Ap.J. 247, L53 and 1982 Ap.J. 252, L1). Optical polarization is typically parallel to the VLBI structure and intrinsic radio polarization angle perpendicular to the VLBI structure (radio polarization unpublished).

G. Benford: This is the kind of data we need, especially with time dependence.

WIGGLES AND KNOTS IN RADIO JETS

E. Trussoni
Istituto di Cosmo-geofisica del C.N.R., Torino, Italy
and
Max-Planck Institut fuer Extraterrestrische Physik,
Garching, FRG

A. Ferrari
Istituto di Cosmo-geofisica del C.N.R.,
and
Istituto di Fisica Generale, Università di Torino, Italy

L. Zaninetti
Istituto di Fisica Generale, Università di Torino, Italy

ABSTRACT

Dynamical effects in binary nuclei inside parent galactic cores, gravitational interactions with companion galaxies and Kelvin-Helmholtz instabilities in the flow propagation have been proposed as mechanism responsible for the formation of the low amplitude morphologies, wiggles and knots, observed in radio jets. Here we discuss the basic characteristics and implications of these models, and confront them with the limited sample of existing data. In conclusion we shall show that present observations cannot discriminate definitely these theories; conversely different mechanism may be at work together in any jet at the same time.

I. INTRODUCTION

Extragalactic Radio Jets have attracted the interest of the astronomical community (both observers and theorists) due to their tight relation with the physics of extended radio sources and nuclear activity in galaxies and QSOs. The recent collection of VLA data provide a preliminary overview of the main properties of these objects. From the point of view of the radiation emission most known jets (> 70) are associated with low-luminosity sources, and their emissivity decreases with increasing radius along the jet at a rate slower than consistent with purely adiabatic expansion (→ relativistic electron acceleration). The magnetic field is usually parallel to the flow in

A. Ferrari and A. G. Pacholczyk (eds.), Astrophysical Jets, 281–290.
Copyright © 1983 by D. Reidel Publishing Company.

regions close to the nucleus and then flips to perpendicular farther out. Concerning structural morphologies, very rarely jets appear to be perfectly straight: most of those so far observed show more or less peculiar structures, like head-tail bendings, wiggles, knots, etc. Theoretical interpretation of these features and, in general, our understanding of the physics and evolution of jets is still rather controversial, also due to the rough and incomplete knowledge of basic parameters, e.g. density and temperature of the jet plasma, flow velocity, magnetic flux, etc.

In this review we shall discuss the theoretical models proposed for peculiar morphologies, in particular wiggles and knots. We want to outline the consequences of various theories and to define, as clearly as possible, the constraints obtained from the comparison with available observational data.

II. GENERAL MORPHOLOGIES OF JETS

Two classes of morphological structures can be defined:

1. Large amplitude structures: this class characterizes head-tail sources (e.g. NGC1265, 3C129), the "Z"-type (inverse) symmetry source (e.g. 3C315) and "C"-type (mirror) symmetry sources (e.g. NGC326).

2. Small amplitude structures, also nicknamed wiggles and knots: they do not affect the overall directivity of the flow, but only introduce oscillations in the globally straight pattern and modulate the brightness distribution (e.g. NGC6251, M87).

Of course the distinction between the two classes is loose. Actually both kind of structures are often found in the same object. For instance, small amplitude wiggles are observed in 3C129, which is a "Z"-type, and in the initial part of 3C315, which is a "C"-type. At the opposite limit, one can take 3C449, in which a weak "C"-type symmetry is superimposed on a clear series of wiggles. However no obvious correlation has been found so far between large and small amplitude features. Here we discuss specifically the models proposed for wiggles and knots, i.e. for the second class of morphologies.

III. MODELS FOR WIGGLES

The proposed theories for the formation of wiggles in jets can be classified under two groups essentially: jet oscillations are either (i) due to orbital or precessional motions of parent galactic cores, excited intrinsically or by the external gravitational pull of a companion galaxy (gravitational models), or (ii) they arise from intrinsic instabilities of the flow due to the interaction with the external medium through which the beam propagates (plasma instability models).

III.1 GRAVITATIONAL MODELS

Many jets are associated with galaxies in clusters or associations; therefore it is reasonable to consider effects of gravitational interactions among nearby members. Obviously interactions can either induce orbital motions of the galaxy and/or precessional motions of the nucleus associated with the jet, or act directly on the jet perturbing its directed propagation.

1. Two orbiting galaxies separated by 1 to 10 kpc (then well outside dynamical friction effects) have typical periods of revolution of some units of 10^5 years. As discussed by Lupton et al. (1982), the shape of ballistic jets coming out from one of the two orbiting galaxies is fixed by the orbital velocity V_{orb} and beam velocity V_b. If $V_{orb} < V_b$, symmetric oscillations are expected (with decreasing amplitude for $V_{orb} \rightarrow c$); while for $V_{orb} > V_b$ irregular and complex morphologies arise. A similar approach was followed by Blandford and Icke (1978), but assuming that the jet's parent galaxy had been perturbed by a companion following an open parabolic orbit.

2. The galactic halo and/or potential well, inside which the jet propagates after breaking out of the core, can be perturbed by an interacting galaxy. As shown by Wirth et al. (1982), the jet morphology is fixed by the geometry of the interaction. For high-ellipticity orbits or casual encounters (impulsive interaction), the deformed halo or potential well leads to more or less evident inverse "Z"-type symmetries. For orbital motion conversely a mirror symmetry is expected, in particular a "C" shape, if the beam velocity is smaller.

Model 1) and 2), even though starting from different physical processes, produce similar morphological effects of both large and small amplitudes. We remember that large scale symmetries (like head-tail sources) can be affected by the motion of the parent galaxy through the external medium. This effect can overlap with the processes discussed above (e.g. in 3C449 the smooth "C" shape may have such an origin); however this interaction is not apt to produce any wiggling structure (cfr. Burns, these Proceedings).

An alternative to (induced) precession beam models has been proposed by Ekers et al. (1978) and investigated by Begelman et al. (1980); it is independent of the presence of a companion galaxy, because it assumes that during the collapse of an active nucleus, two compact objects (one of which emits jets) are formed in the core, and are tight together, without coalescing because of their relatively large angular momentum. Comparison of different time scales of dynamical friction (during the collapse phase of the double nucleus) and gravitational wave emission, shows that the two compact objects can rotate without coalescing for a time of the order 10^{7-8} years on orbits with radii in the range 10^{16-19} cm. Therefore they have a revolution period of the order of

$$\tau = 0.6 \times 10^{2} \, r_{16}^{5/2} \left\{ \left(\frac{M}{m}\right) M_8 \right\}^{-3/2} \quad \text{years} \qquad (1)$$

($r_{16} = r/10^{16}$ cm), $M_8 = M/(10^8 \, M_\odot)$, (M/m) = mass ratio of the two orbiting galaxies) and can undergo a large precession, provided their spin axes are non-aligned. If the double nucleus separation is wide, a "Z"-shaped structure is generated; if it is small, we get a spiral structure winding up on a cone of constant angle. Small amplitude wiggles can also occur if the precessional motion is modulated by harmonics of the fundamental period. The specific case of relativistic flow velocity in the jet has been investigated by Gower et al. (1982), who showed that, only for a suitable choice of the precession angle, line-of-sight angle and velocity, "C"-type symmetries can arise from precessing beams.

III.2 PLASMA INSTABILITY MODELS

Extragalactic jets are beams of plasma moving through the intergalactic medium and therefore are subject to Kelvin-Helmholtz instability. Several authors (Hardee 1979, 1982a, Ferrari et al. 1981, 1982, 1983, Ray 1981, 1982) have studied the linear instability problem and showed that wiggles and knots may be associated in cylindrical beams with unstable helical and pinching modes of long wavelength.

Details of such study are presented in the paper by Zaninetti and Trussoni in these Proceedings.

For our discussion here we recall that the instability excites both pinching and helical modes for any value of the Mach number; in particular, in the case of supersonic flows the two types of modes have similar growth rates. The typical scales of the fastest growing perturbations are larger than the beam radius, and the most unstable helical modes have wavelengths 2 or 3 times larger than pinching. Note, however, that growth rates of short wavelength modes are not much lower, so that the instability may in fact release its energy into several channels: only very short wavelengths are stabilized by the presence of boundary layers at the contact discountinuity between jets and the outside medium. Eventually the whole spectrum of unstable modes is determined by mode coupling, producing a turbulent cascade from large wavelengths. This cascade does not involve the fastest growing modes, because the largest turbulent eddies must have sizes smaller than the geometrical extension of the plasma, i.e. the jet radius. In this sense, the fastest growing modes determine the morphology of the flow, the shorter wavelengths create a turbulent cascade and, if the flow is magnetized, support charged particle acceleration and jet scattering (Ferrari et al. 1979, Benford et al. 1980).

III.3 A TEST CASE: 3C449

Almost all authors have tested their suggestions on the source 3C449. We present a rapid overview of the conclusions of each model when confronted with observations:

1. The source is known to be associated with a cD galaxy and its jets have wiggles symmetric with respect of the core. This type of morphology requires that projection effects are invoked, if we want to interpret them in terms of precessional motions of the parent core. Apart from this difficulty, one can define the measured oscillation wavelength in terms of the jet's flow propagation time scale and precession period (section III.1) applying simple geometrical arguments, and obtain that a relativistic velocity of the flow is required, namely 0.7c (Gower and Hutchings 1982).

2. If we want to explain the wiggles in terms of an orbital motion of the parent galaxy around a companion galaxy (Lupton and Gott 1982), the separation of the two objects should be $\sim$ 1 kpc ($\sim$3") and the beam velocity $\sim$ 2000 km/sec, i.e. > 6 V_{orb}. At present, we have only weak evidence for the existence of such a companion.

3. Kelvin-Helmholtz instabilities predict wavelengths for helical modes (Hardee 1981, Ferrari et al. 1983) with Mach numbers $10 < M < 60$ and relative densities (inside/outside) $1 < \rho < 10$ (such large ranges are due to uncertainties in defining the e-folding length of unstable modes). If the beam is confined, such values constrain the jet velocity, $2000 \lesssim V_b \lesssim 10000$ km/sec, and external density, $3 \times 10^{-4} \lesssim n \lesssim 5 \times 10^{-3}$ cm.

From a general point of view, we conclude that the requirements of models 2) and 3) appear less demanding than assuming a relativistic jet velocity, for which there is no other evidence. On the other hand the instability model implies jets heavier than the outside medium; there are no measurements for the environment of this specific source, but in other sources indirect arguments would suggest light jets. Then, at least for 3C449, one may conclude that the most promising model for explaining the origin of wiggles is the one based on orbital motions.

IV. MODELS FOR KNOTS

Knots are much less known features of jets: so far such morphology has been detected in only three objects, i.e. M87, CenA and, possibly, NGC6251 (Perley, these Proceedings). But this appears to be simply due to instrumental limitations, to be overcome by new devices as the VLA and MERLIN.

Rees (1978) has associated these knots with a periodic activity of the central engine, that produces changes of velocity of the jet flow and, correspondingly, leads to the formation of shocks, where enhanced particle reacceleration and nonthermal emission occur. Alternatively, Blandford and Königl (1979) assumed that heavy, slow clouds of gas, periodically emitted by the nucleus, are entrained by the fast moving jet plasma which impinges upon them. A shock is then formed on the upstream side of the clouds, where again particle acceleration mechanisms are favoured.

Knots can be produced directly by the development of pinching modes in Kelvin-Helmholtz instabilities (Ferrari et al. 1983, Hardee 1982b): as already recalled above, such perturbations have similar growth rates and typical scales two or three times shorter than helical modes. The possible evolution of pinching modes yields again the formation of compressions (clouds) and shocks, with local particle acceleration.

A last model applies to magnetically confined jets and is related to the occurrence of resistive (tearing) instabilities: preliminary results have been presented by Ferrari and Trussoni (1981b) and are at present under study. The mechanism follows from experiments in laboratory plasmas: the cylindrical flow wrapped by helical fields, breaks up into regularly spaced, wiggling magnetic islands, originated by resonant superposition of reconnecting modes of different helicity and longitudinal wavelength. These islands are zones of turbulent magnetic field, providing conditions for local stochastic particle acceleration along the beam.

V. CRITICAL EVALUATION OF MODELS

We must now discuss the theoretical models presented above in terms of the observational constraints, that we summarize here:

1. There is no evidence of correlation between large and small amplitudes morphologies in jets.

2. The general trend suggested by the VLA sample (Bridle 1982) is that 2 or 3 oscillations only develop before the beam looses its strict collimation (in this context it is worth recalling that the term "jets" is here used not to indicate the extended pattern from core to lobes, but only the initial, strongly collimated portion, up to arcsecond resolution, as displayed by VLA and MERLIN maps).

3. Typical wiggle wavelengths are in the range 5 to 20 times the jet radius; however, more than one harmonic can be found in the pattern of a single object (e.g., see the case of NGC6251, Perley, these Proceedings).

4. Wiggles are possibly more frequently formed than knots and have scalelengths 2 or 3 times larger.

5. As for the whole extended radio source, also specifically for the jet region, spectral data are represented by a flat power law, which require local acceleration of relativistic electrons for synchrotron emission.

6. Most jets are associated with cD galaxies situated in groups or clusters of galaxies, i.e. regions of relatively large mass concentration. However evidence for physical associations with companion galaxies is only marginal.

Dynamic models based on orbital and (externally) induced precessional motions are based on point 6); long orbital periods fit well with the few oscillations observed (point 2). In addition they can easily lead both to "Z"- and "C"-type symmetries, depending on the kind of interaction. On the other hand this model would predict one wavelength only for wiggles (point 3). No direct relation is expected between gravitational interaction and knot formation.

The intrinsic precession beam model is subject also to some difficulties. The periodic motion could lead, depending on the parameters of the system, to formation of wiggles of several scalelengths on the large amplitude structure. On the other hand, the intrinsic precession period predicted from Eq.(1) is short, and we should see several oscillations, which is in contrast with point 2).

A limitation of all precessing beam models is that they are able to produce "Z" symmetries only, unless relativistic velocities and projection effects are considered; conversely it appears that jets are non relativistic, except, possibly, in their very initial propagation phases close to the parent nucleus (superluminal sources). In addition these models do not provide any correlation with the formation of knots.

Kelvin-Helmholtz instability can work independently from the presence of companion galaxies, although pressure equilibrium conditions require the interaction with an external medium as found in clusters (point 6); in addition they can develop in jets with large amplitude structures already active.

Most unstable helical modes have wavelengths which fit those of wiggles; in addition perturbations with different wavelengths are predicted by the theory (point 3). The few observed oscillations may agree with the fact that the nonlinear growth of modes makes the jet to flare after a few e-folding lengths. Unstable MHD modes can support efficient acceleration of charged particles (point 5) both through resonant interaction with small wavelength modes (Ferrari et al. 1979, Eilek 1979) and stochastic interaction with large scale modes (Ferrari and Trussoni 1981a). An important element in support of the instability theory is its prediction of the formation of both pinching and helical modes at the same time (point 4): the same process would lead to the two different morphological features. In fact they could be associated with periodic structures of knots recently detected by Nieto and Lelièvre (1982). The regularity of such structures cannot be interpreted by the heavy cloud (Blandford and Konigl 1978) or shock (Rees 1978) theories.

However, even though promising, the instability process has some drawbacks also. First of all, in order to fit observed wavelengths and growth rates, jets should possess densities higher than, or at least comparable to that of the external medium: so far there is no evidence of such "heavy jets" (see Ferrari et al. 1983, Zaninetti and Trussoni, these Proceedings). A way out from this difficulty would be to consider the presence of a light cocoon around the jet, as shown in some of the numerical results presented by Norman et al. in this Workshop; then the jet might be denser than the cocoon, although lighter than the medium outside to which X-ray data refer.

A second uncertainty is that knots might actually be less frequent than wiggles, while the instability predicts that both perturbations have similar growth rates. The possible explanation is that, due to their shorter wavelength, a selection bias affects observations; this difficulty could be overcome by higher resolution and dynamic range observations.

A more fundamental problem for the instability theory is that most results come from a linear analysis of the process, and it is very difficult to complete nonlinear calculations. Results so far presented (Nepveu 1981, Norman et al. 1982, and these Proceedings) indicate a rough agreement with linear models. On the other hand all nonlinear computations are based on 2-D numerical codes, excluding "a priori" the intervention of wiggles. Even more uncertain is the "numerical" viscosity introduced in codes by the computational method, which seems to be much higher than the real physical viscosity.

Finally we must remark that Kelvin-Helmholtz instability arises in pressure confined jets, while it is still highly controversial whether all jets are actually confined by an external medium. An interesting check has been suggested to distinguish between instabilities and other mechanisms, i.e. that of following the evolution of wiggle amplitude: an exponential increase would support the instability theory. Leaving aside all technical difficulties of using uncertain instrumental resolution and applying them to very few wiggles, such test might be non conclusive after all. In effect even in a linear regime, an expanding jet displays a linear growth (Hardee 1982a), and finally, the observed wiggles might already follow a nonlinear regime, where saturation effects limit the amplitude growth to a flatter evolution.

VI. CONCLUSIONS

The present analysis is far from conclusive and, in a way, disappointing. In effect for every theoretical model difficulties seem to balance supports. In addition it is very unlikely that one model can apply to all jets and, in a single jet, to all morphologies. In particular, it seems reasonable to associate the large scale (mirror, inverse, head-tail, etc.) structures with gravitational interactions and dynamical effects (plus, possibly, an interaction with the external medium). On the other hand, the requirement of acceleration of particles means that the turbulent modes coming from instability should be a common ingredient in these objects. In other words, gravitational interaction and Kelvin-Helmholtz instabilities are probably present (although not necessarily correlated) in all beams. Thence they give rise to different morphologies, depending on local conditions (jets inside and outside clusters, confinement of the jet, velocity with respect to external medium, etc.).

In conclusion we draw the following unifying scenario:

1. Jets with small or any large scale structures (straight jets like NGC6251) are not affected by gravitational or dynamic interaction. Wiggles (and knots when present) are likely to be due to the developments of large scale perturbations. The amplitude of wiggles is fixed by the efficiency of dissipation processes and the condition of confinement.

2. In largely structured beams instability and precession produce similar small amplitude morphologies (see e.g. Icke 1981, for 3C129). It looks then difficult to distinguish between these two processes by observations.

These general conclusions could be modified by forthcoming, more detailed and suitable theories (e.g. the formation of knots from magnetic reconnection is worth to be analized in detail). However it seems unavoidable that jets are common features in high-energy astrophysics, but every object has its own story.

ACKNOWLEDGMENTS

We wish to thank A. Bridle, E. Fomalont, J. Burns and G. Tanzella-Nitti for useful discussions and comments.

REFERENCES

Begelman, M.C., Blandford, R.D., Rees, M.J.: 1980, Nature 287, 307
Benford, G., Ferrari, A., Trussoni, E.: 1980, Ap. J. 241, 98
Blandford, R.D., Icke, V.: 1978, MNRAS 185, 527
Blandford, R.D., Königl, A.: 1979, Astrophys. Lett. 20, 15

Eilek, J.A.: 1979, Ap. J. 230, 373
Ekers, R.D., Fanti, R., Lari, C., Parma, P.: 1978, Nature 276, 588
Ferrari, A., Trussoni, E., Zaninetti, L.: 1979, Astron. Astrophys. 79, 190
Ferrari, A., Trussoni, E., Zaninetti, L.: 1981, MNRAS 196, 1051
Ferrari, A., Trussoni, E.: 1981a, Journal of Geophys. 50, 70
Ferrari, A., Trussoni, E.: 1981b, Space Science Rev. 30, 135
Ferrari, A., Massaglia, S., Trussoni, E.: 1982, MNRAS 198, 1065
Ferrari, A., Trussoni, E., Zaninetti, L.: 1983, submitted to Astron. Astrophys.
Gower, A.N., Hutchings, J.B.: 1982, Ap.J. Lett. 258, L63
Hardee, P.E.: 1979, Ap. J. 234, 47
Hardee, P.E.: 1981, Ap.J.Lett. 250, L9
Hardee, P.E.: 1982a, Ap.J. 257, 509
Icke, V.: 1981, Ap.J.Lett. 246, L65
Lupton, R.N., Gott. III, J.R.: 1982, Ap.J. 255, 408
Nepveu, M.: 1980, Astron. Astrophys. 84, 14
Nieto, J.L., Lelièvre, G.: 1982, Astron. Astrophys. 109, 95
Norman, M.L., Smarr, L., Winkler, K.H.A., Smith, M.D.: 1982, Astron. Astrophys. 113, 285
Ray, T.R.: 1981, MNRAS 196, 195
Ray, T.R.: 1982, MNRAS 198, 617
Rees, M.J.: 1978, MNRAS 184, 618
Wirth, A., Smarr, L., Gallagher, J.S.: 1982, Astron. J. 87, 602

REFLECTION JETS AND COLLIMATION OF RADIO SOURCES

A. G. Pacholczyk
Specola Vaticana, Castel Gandolfo
Vatican City State

1. INTRODUCTION

It has become increasingly clear to me as I listened to the papers presented at this workshop that in discussing astrophysical jets we are talking not only about apples and oranges, but also about pears and peaches (and perhaps an occasional lemon), as the large observational material presented was not pointing towards one single phenomenon. We cannot, therefore, reasonably expect to explain the observations in terms of one single picture. With this in mind I would like to add here one more hypothesis aimed at a description of only a certain class of jets in extended radio sources by discussing hydrodynamics of jets formed by discrete portions of material ejected from the parent galaxy through a channel and reflected back into it as a result of an encounter with the material accumulated at the end of the channel.

The picture presented here is that of Christiansen, Pacholczyk, and Scott (1977, 1981, 1982a, 1982b); it combines some older ideas with the recent ones. The older ideas consist of modeling of extended radio sources in terms of multiple ejection of plasmons (plasmoids) through a channel ploughed by the first few plasmons in the ambient medium (Christiansen 1973, Christiansen, Pacholczyk, and Scott, 1977) with a resupply of energy in plasmons through the conversion of bulk kinetic energy into relativistic electron energy through instability driven turbulence (Pacholczyk and Scott 1976, Pacholczyk and Scott 1977; for a summary see e.g. Pacholczyk 1977, Chapter 8). The recent ideas concern the formation of retro-jets as the result of interaction of a plasmon with the dense relic material at the end of a channel (Christiansen, Pacholczyk, and Scott 1981, 1982a) and the collimation of plasmon material in channels (Christiansen, Pacholczyk, and Scott 1982b).

2. EXISTENCE AND STABILITY OF CHANNELS

The crucial elements of the multiple plasmon model are the existence and stability of channels in the circumgalactic medium through which the ejecta are traveling outwards from the active nucleus of a parent galaxy. This is discussed in detail in Christiansen, Pacholczyk, and Scott (1982a). Hypersonic passage of

A. Ferrari and A. G. Pacholczyk (eds.), Astrophysical Jets, 291–302.
Copyright © 1983 by D. Reidel Publishing Company.

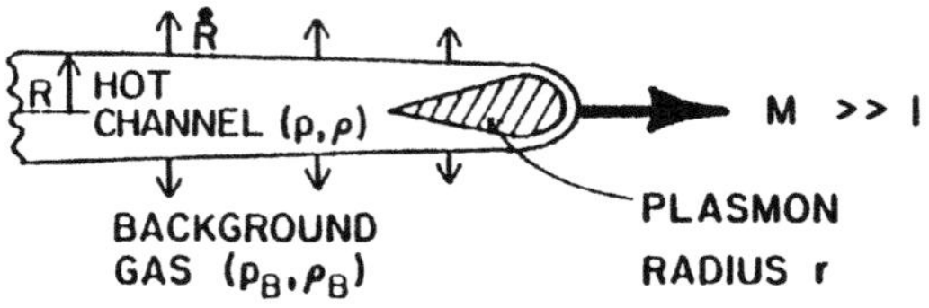

Figure 1. Geometry of a hot channel behind a hypersonic plasmon.

plasmon of radius r (of the order of 10 kpc), velocity v and Mach number M (of the order of 10-100) "instantaneously" heats up gas behind the shock to temperatures $\sim M^2T$ (where T is the background temperature of the order of 10^8 K) releasing energy $\mathcal{E}$ per unit length proportional to the average drag force on plasmon

$$\mathcal{E} \sim \langle F \rangle \sim \pi r^2 \cdot \rho_B v^2$$

resulting in very high line pressure and therefore in an almost cylindrical shock expansion (described in terms of a two dimensional Sedov solution). The geometry is illustrated in Figure 1. Numerical calculations by Plooster (1970) for a cylindrical shock (thunder) following an "instantaneous" injection of energy along a line (lightning) are representative of the situation considered here and are illustrated in Figure 2. These numerical solutions can be approximated by an exponential behavior of density and pressure inside the channel (behind the shock), as indicated in Figure 2. This figure represents the cylindrical shock at a time short (0.03t) compared with the relaxation time t of the shock which is of the order of sonic crossing time at background temperature multiplied by Mach number of the plasmon

$$t \sim \frac{2}{3}\frac{r}{c} M$$

(c is the background sound speed). Figure 3, based on Plooster's (1970) weak shock limit calculations, illustrates a situation at much later times (1.4t and 6t) when the channel is practically pressure relaxed after a pressure balance is gradually restored. An important result represented on this figure is the presence of a deep axial hole in the density distribution which persists for times at least up to 6t approaching a limiting size of the order of 1/4 Mr. Therefore, up to times t of the order of 10^{9+} years a low density channel of the size (1/4)Mr continues to exist in the ambient medium following the passage of a hypersonic plasmon of radius r and Mach number M. It is therefore likely that secondary plasmons will be ejected into those low density tunnels.

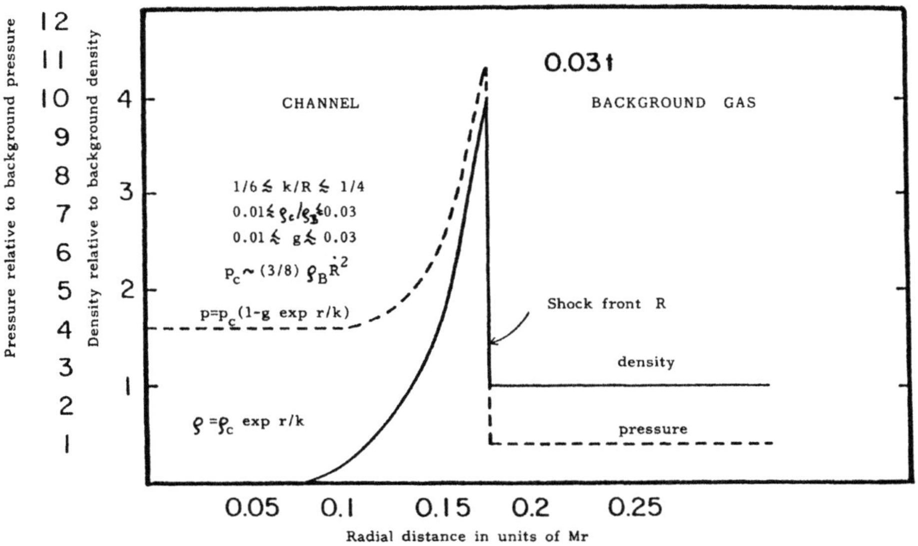

Figure 2. Radial distributions of pressure and density in a channel according to Plooster's solutions.

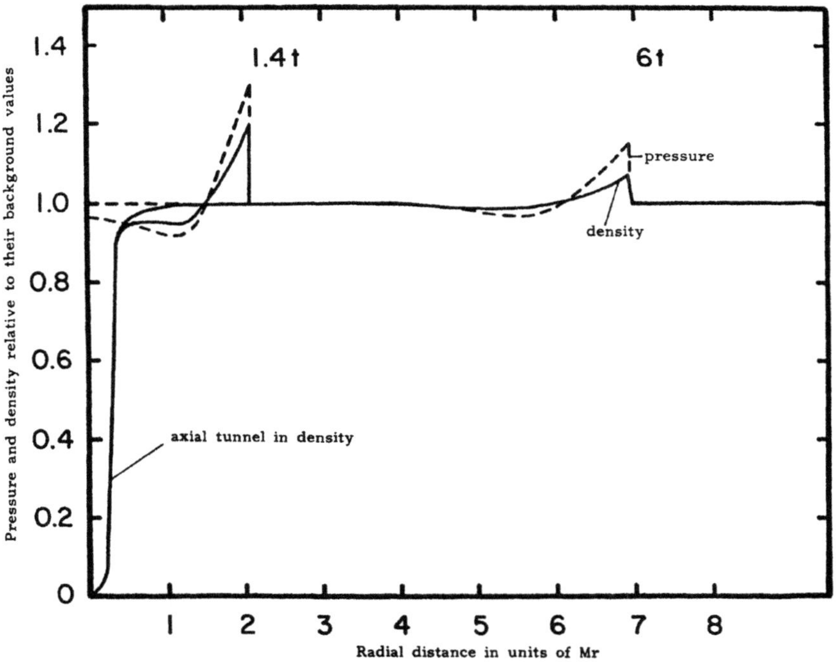

Figure 3. Radial distributions of pressure and density in a channel at times larger than the relaxation time.

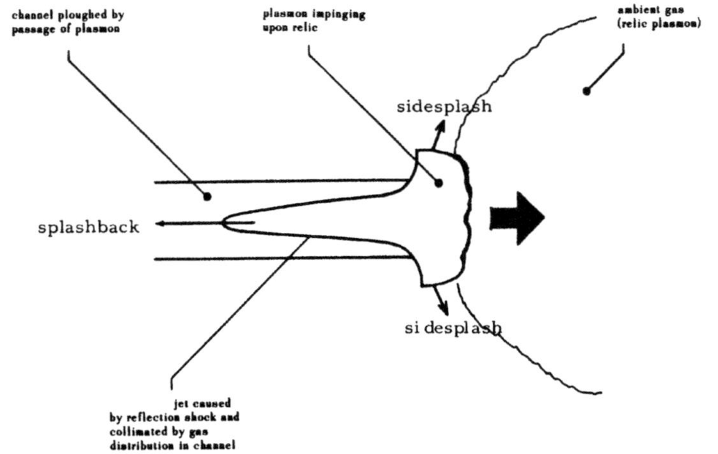

Figure 4. Interaction of the plasmon with the relic at the end of channel: first phase (transient splashes).

The stability of the channel depends on the efficiency of the closing of the channel by refilling it from the sides or by a reexpansion of the plasmon backwards into the channel.

Refilling from sides will occur in sonic crossing time (at the background temperature) multiplied by Mach number of plasmon; for T 10^8 K, that is, in $10^7 \cdot$M years. For Mach numbers from 10 to 100 the side-refilling time is same as the interval between ejections of plasmons estimated from the statistics of bridges.

Reexpansion of plasmon backwards will occur when plasmon velocity is smaller than its internal sound speed. The backfilling time is larger than the distance travelled by the plasmon (channel length) divided by the internal sound speed (as we have neglected the pressure in the channel). For a typical channel length of 100 kpc, the plasmon to background density ratio of 10 and the plasmon velocity of 10^9-10^{10} cm/sec, backfilling time is much larger than 10^7 years. The backfilling process is negligible if radiative losses are dominant over shock heating in a plasmon.

3. FIRST PHASE: TRANSIENT SPLASHES.

We will discuss now the phenomena occuring when the n-th plasmon traveling through a channel encounters a relic plasmon (n-1)-th at the end of the channel (a "rigid wall"). During such a collision the material of plasmon is compressed, thermalized and eventually re-expanded, both sideways (sidesplash) and backwards (splashback) into the channel. This stage is illustrated in Figure 4.

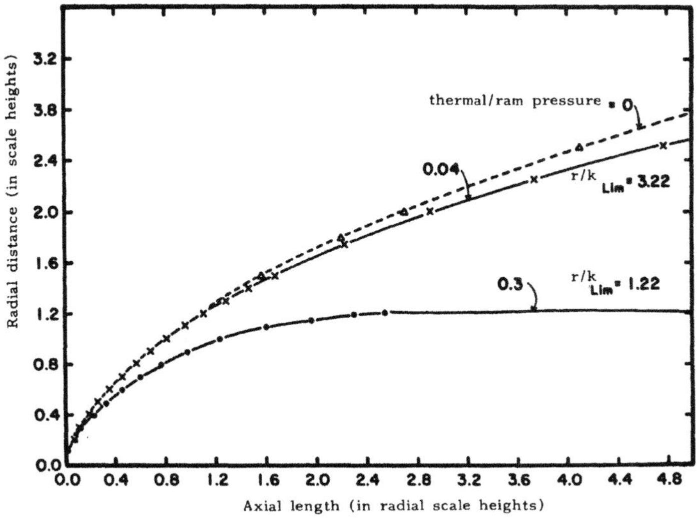

Figure 5. Shape of a splashback jet in a channel uniform along the axis.

Initially sidesplash is the dominant process: as long as trailing portions of plasmon continue to fall on the emerging reflection shock, its velocity will be much less than plasmon's velocity (Landau and Lifshitz 1959). For details see Christiansen, Pacholczyk, and Scott (1982b). The time t at which shock will break through the channel side (radius of channel $R \sim 5k$) and sidesplash is

$$t \sim 5 \frac{R}{v} \left(\rho_c / \rho_p \right)^{1/2}$$

after the initial impact, ρ_c / ρ_p is the ratio of the channel central density to the plasmon density at the stagnation point.

The shape of the axial reflection front is determined by a pressure balance; in the Kompaneets (1960) approximation the average thermalized pressure inside the shocked volume is equal to the sum of the external ram pressure and of the external thermal pressure

$$p = \rho_c v^2 \cos^2\alpha \exp(r/k) + p_c + g p_c \exp(r/k).$$

(See Figure 2 notation). For r=0 the above equation gives

$$p = \rho_c v^2 + p_c (1+g).$$

Introducing

$$\cos^2\alpha = \frac{(dr/dz)^2}{1+(dr/dz)^2}$$

one can solve this equation for dr/dz and integrate it to get the shape of the splashback jet. In a channel with an exponential radial density distribution (with a scale height k) but independent on the axial coordinate z the shape of the splashback jet is illustrated in Figure 5. From Figure 5 we see that the resulting jet is narrowly confined. For example, for the ratio of thermal to ram pressures for the shocked channel equal about 0.3 the radial extent of the jet is about one scale height. Since the channel radius is between four and six scale heights, we see that the jet's radius constitutes roughly only a fifth of the channel radius indicating an efficient collimation of the jet.

In the preceding we discussed the hydrodynamics of the splashback jet disregarding the (presumably small) effects of the magnetic field and relativistic particles. But the appearance of the jet will depend critically on these hitherto neglected components of the plasmon. Assuming the magnetic field of the order of 10^{-5} gauss we see that the synchrotron lifetime of the jet is short: it is of the order of 10^{7} yrs, that is, the jet is visible for a short time only. The consequences of the short synchrotron lifetime in the jet are the following:

(1) jets will be seen in a small fraction of sources,
(2) jets will be seen predominantly on one side of double source as the synchrotron lifetime is smaller than the difference in arrival times,
(3) in a splashback jet we will not see the superposition of spectra with high frequency cut offs like in the head since there is no acceleration of electrons in the jet. The jet's spectrum (as compared to that of the head) is therefore either/or:
    a. flatter - like the original pre-losses spectrum
    b. invisible - if we look as a frequency beyond the cut-off.

4. SECOND PHASE: QUASISTATIC CONICAL TRAIL.

After the transient features described above cease to be visible, quasistatic ram deceleration phase begins and a pressure balance is established between dense compressed head material, characterized by an exponential axial distribution of pressure and a hot tenuous channel gas, characterized by an exponential (Sedov) radial distribution of pressure. As a result, a conical shaped lobe is formed behind the radio source's head, as illustrated in Figure 6. Matching pressure in plasmon with that of gas in the channel one gets the boundary of plasmon in channel with the shape given in Figure 7. Observationally the typical the ratio of plasmon to channel scale heights is 4 to 8.

The properties of the quasistatic trail region are as follows.

(1) Plasmon density at the end of trail (reduced by 3e foldings from head density) is larger than the central channel density (which is a factor 30-100 below background density):

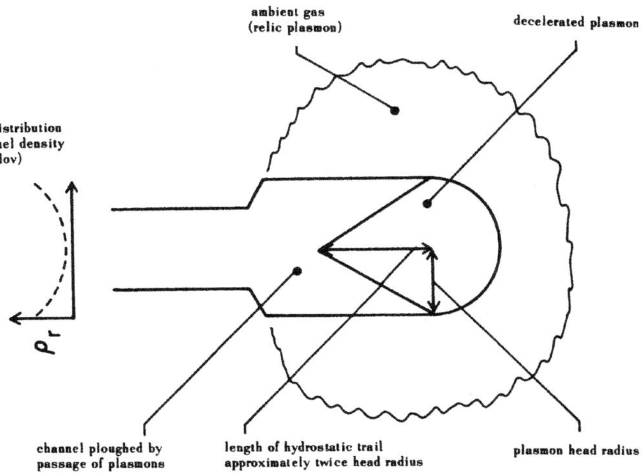

Figure 6. Interaction of the plasmon with the relic at the end of the channel: second phase (quasistatic conical trail).

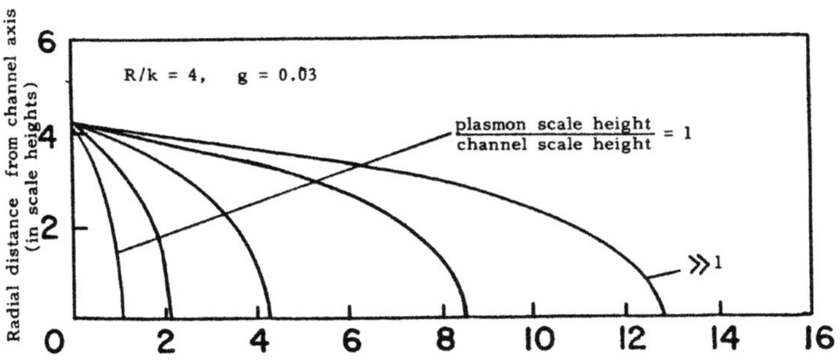

Figure 7. The shape of the plasmon boundary in a channel

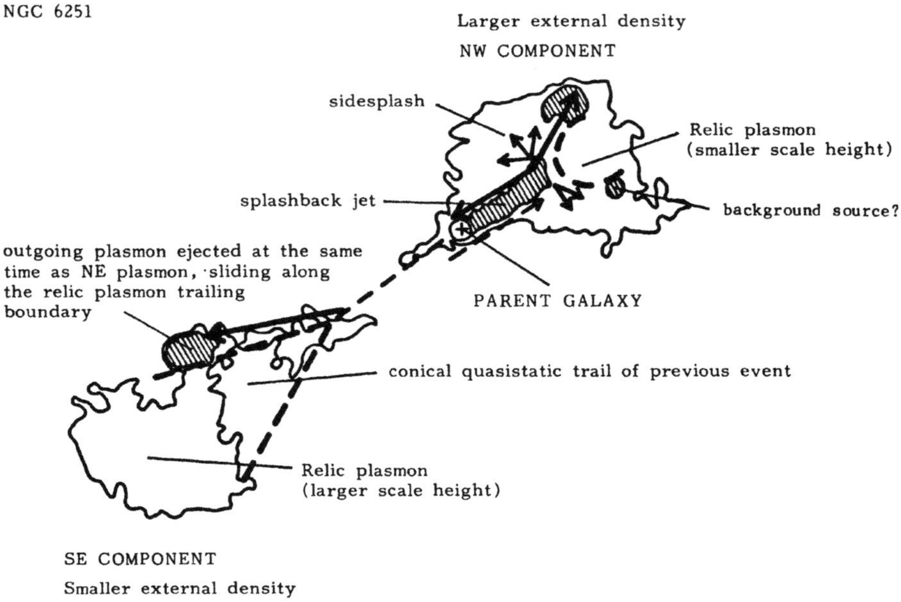

Figure 8. An illustration of the features discussed in this paper (NGC 6251).

therefore trail is always denser than channel.

(2) From the above it follows that the boundary is Rayleigh-Taylor stable.

(3) Trail material is the same as head's inter-spot medium. Being removed farther from the regions of instability in the head (hot spots) it will have less accelerated electrons by field amplification and turbulence in the instability regions. Therefore trail spectra will show more high frequency curvature than head spectra and the two frequency spectral index is higher in the trail than in the head.

(4) All trail material is about of the same age; therefore its spectrum is practically independent on distance from the head: the trail length may be only weakly frequency dependent.

(5) The trail region is relatively stable; there is little instability driven turbulence. Therefore the trail's polarization is higher than the head's.

5. AN ILLUSTRATION OF A SPLASH BACK JET.

The picture of the radio source NGC 6251 (Willis et al 1982), Figure 8, could serve as an illustration of the features discussed in the preceding sections, in particular, of the one-sided splashback jet.

6. COLLIMATION OF RADIO SOURCES.

The last (but not least) remark I would like to make is to refer to Figure 5, illustrating the calculated shape of a narrow retro-jet splashed back into the channel and to point out to the channel distribution of density as to the a factor responsible for an efficient collimation process that could operate in radio sources (Christiansen, Pacholczyk, and Scott 1982b). An outgoing n-th plasmon ejected into a channel can be described similarly to a splashback jet by the use of Kompaneets' approximation (the driving energy in this case is released in the parent nucleus), resulting in a similar efficient collimation by the density tunnel formed by the preceding (n-1-th) outgoing plasmon and its returning splashback. The initial collimation of the first event needs to be only very moderate, such as, for example, that resulting from a blast wave collimated by the flattened density distribution surrounding the parent nucleus, discussed by Saunders (1976). Our self-collimation process is a progressive effect increasing initially with the number of ejected plasmons, every plasmon being confined more narrowly by the tunnel resulting from the preceding event, thus itself producing a still narrower channel for the subsequent plasmon. In reality, however, this narrowing process is self limited by the multiplicative heating of the channel material making subsequent ejecta less hypersonic and eventually limiting the collimation process when the Sedov type approximation breaks down.

REFERENCES

Christiansen, W.A.: 1973 MNRAS 104, 211

Christiansen, W.A., Pacholczyk, A.G., and Scott, J.S.: 1977 Nature 264, 543

Christiansen, W.A., Pacholczyk, A.G., and Scott, J.S.: 1981 IAU Symposium #97, 51

Christiansen, W.A., Pacholczyk, A.G., and Scott, J.S.: 1982a Steward Preprint #384

Christiansen, W.A., Pacholczyk, A.G., and Scott, J.S.: 1982b Steward Preprint #399

Kompaneets, A.S.: 1960 Doklady 5, 46

Landau L.D. and Lifshitz E.M.: 1959 Fluid Mechanics, Oxford: Pergamon Press

Pacholczyk, A.G.: 1970 Radio Astrophysics: Non thermal processes in galactic and extragalactic sources, San Francisco: Freeman

Pacholczyk, A.G. 1977 Radio Galaxies: Radiation transfer, dynamics, stability and evolution of a synchrotron plasmon, Oxford: Pergamon Press

Pacholczyk, A.G., and Scott, J.S.: 1976 Ap.J., 203, 313

Pacholczyk, A.G., and Scott, J.S.: 1977 Ap.J., 210, 311

Plooster, M. N.: 1970 Physics of Fluids 13, 2665

Saunders, R.H.: 1976 Ap.J., 205, 335

Willis, A., Strom, R., Perley, R. and Bridle, A.: 1981 IAU Symposium #97, 141

DISCUSSION

BURNS: WRST observations by Jaffe and Perola (1974, Astron. Astrophys. 32, 223) revealed that something strange had occurred during the evolution of 3C338. First of all, the integrated spectral index of 3C338 above 1.4 GHz is greater than 2.0 (defined $S_\nu \sim \nu^{-\alpha}$). This makes 3C338 one of the steepest spectrum radio sources known. Second, the flat-spectrum nuclear core was found to be asymmetrically displaced from the centroid of the radio emission in the north-south direction. However, the core was found to lie midway between two diffuse lobes in the East-West direction. Unfortunately, the resolution was too low to investigate the nature of the north-south asymmetry and to find the detailed distribution of spectral index.

New VLA observations at 6 and 20 cm have revealed further details of the structure of 3C338. The origins of these substructures and their spectra are puzzling. Figure 1 shows the 20 cm VLA map at a resolution of 2" along with listings of the 6 to 20 cm spectral indices and percentage polarization (m). Note that the apparent cause of the north-south asymmetry with respect to the core seen on lower resolution maps is a southern ridge of emission which may resemble a jet.

Is this elongated structure displaced 6 kpc from the core a jet? In some ways, the answer seems to be yes. It is narrowly collimated and appears to connect two diffuse lobes of emission much like the jets in 3C31. But, the nuclear core is not within the jet or at the beginning of theas is always seen. In addition, the spectrum of the "jet", like the rest of the diffuse emission, is very steep ($\alpha > 1.5$); typically jets have $\alpha \approx 0.6$. If it is a jet, then something peculiar must have occurred during its radiative/transport history.

One possible clue lies in the nature of the optical identification. 3C338 is associated with the cD galaxy NGC6166 at the center of the rich cluster of galaxies Abell 2199 (Z=.031). The galaxy actually consists of 4 separate nuclei, the brightest of which coincides with the core of 3C338 and the others appear to be radio-quiet (the three brightest nuclei are labeled A, B, and C in Fig. 1). Jaffe and Perola first suggested that motion of the radio nucleus about the barycenter of the cD system could be responsible for the distorted appearance of 3C338. For this scenario to be plausible, the radio nucleus must have previously existed to the south of its current location. During this time it ejected dual beams of radio plasma to form jets. The engine must have turned off while the nucleus travelled to the north. The radio plasma in the jets and lobes then must have aged via synchrotron losses (being well confined by a dense, x-ray emitting gas; Feigelson and Berg 1982, preprint) so that the spectrum steepened. Both the curvature of the integrated spectrum and the minimum energy assumption produce timescales of $\sim 10^7$ yrs since the last injection of particles in the jet/lobes. Interestingly, this is on the same order as the time for nucleus A to move to its current position given the radial velocities of the nuclei and an assumption of circular orbits.

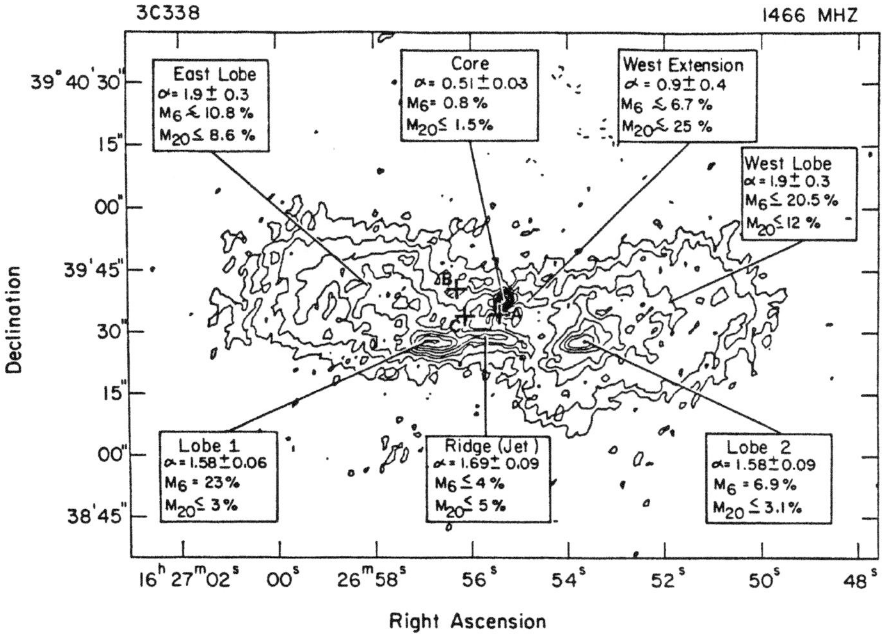

Figure 1: Contour map of 3C338 at 20-cm. The contour intervals are -2, 2, 4, 7, 10, 15, 20, 25, 30, 40, 60, 80, and 100 percent of the map peak which is 153 mJy/clean beam. The shaded ellipse is the clean beam whose parameters are 2.64" x 1.95" (FWHM) p.a. = -48°. The crosses represent the location of the optical components within NGC6166 as measured by Griffin (1963). The boxes contain values of the 6 to 20 cm spectral index (α_6^{20}), and the percentage polarization at 20 cm (m_{20}) and 6 cm (m_6).

The presence of the jet is still curious, however, within the framework of this model.One would expect that the channel should have become evacuated and collapsed after the last bit of radio plasma passed through to the lobes. Possibly a "splash-back" of plasma from the lobe back into the jet as suggested by Pacholczyk (this workshop) has refilled the jet with radio emitting material. This would also be necessary to maintain the pressure balance of the channel with the surrounding medium.

There are other problems, however with this model. In particular, the orbital model requires nuclei masses which are 16 times larger than that suggested by the optical data in order to gravitationally bind the system. Possibly, an oscillation of the nuclei about the center of the cluster which contains a massive ($>10^{13}$ $M_\odot$) "black pit" (Blandford and Smarr 1982, preprint) of dark matter can bind and move the nuclei.

An alternative hypothesis is that the radio ridge is not really a jet at all. A2199 is known to possess a "cooling core" of gas accreting onto NGC6166. X-ray spectroscopy suggest that 100 $M_{\odot}\ yr^{-1}$ is infalling onto the cD (Mushotzky 1982, private communication). If this infall is an asymmetric flow, the resulting ram pressure could compress and confine diffuse radio plasma to produce the southern radio ridge near the core. In fact, we find that the minimum particle plus field pressure in the jet is on the same order as implied ram pressure of the flow. It should be emphasized that this flow must be strongest to the south of the core. But no signs of this asymmetry are seen in the x-ray continuum emission or the Hα distribution (Cowie et.al. 1982 preprint). This explanation, also, is not totally satisfying.

Further details of this source and possible interpretations will appear in Burns, Schwendeman, and White (1983 submitted to Ap.J).

PACHOLCZYK: Indeed it is clear that it is very difficult to interpret the feature you just described in terms of an outgoing jet.

PORCAS: We have heard in this workshop that many (maybe all!) of the known superluminal sources have classical double structure on the arc-second scale if one looks hard enough. They also have arc-second scale jets, separated from the cores by gaps in some cases, and presumably easily explained by your plasmon model. In at least three cases (3C273, 3C345, 3C120) one apparently sees repeated ejections of blobs from the nucleus, with intervals of a few years. Should we identify these blobs with individual plasmons?

PACHOLCZYK: My answer is yes, we should, or at least, I would. But, as I mentioned at the beginning of my talk, I did not intend to apply the retro-jet model to compact super-luminal sources at the present stage; the blobs you are talking about would be the forward traveling plasmons.

PERLEY: A quick calculation shows that the minimum energy contained in the jet of NGC6251 is $\sim 1.5 \times 10^{56}$ erg. Presumably the splashback jet contains less energy than its creator plasmon. Thus if you wish to interpret the entire NGC6251 jet as the reflection of a single event, isn't the amount of energy involved a problem? The collimation of such an event would seemingly be difficult, especially within the 3° overall cone angle of this source. The entire reason for the plasmon was to reduce the event energy to manageable bounds. It appears to me that this jet, in particular, is not a good model for a splash-back jet.

PACHOLCZYK: The presence of extended relics on both sides in NGC 6251 points towards many plasmon events in its history. True, NGC 6251 would belong in the upper range of energies and therefore should be described in terms of more energetic than the average plasmons, but still quite manageable.

INSTABILITIES IN ASTROPHYSICAL JETS

M. Nepveu
Astronomische Institut der Universität Bonn,
SFB Radioastronomie
Auf dem Hügel 71, D-5300 Bonn, 1, FRG

ABSTRACT

Instabilities in astrophysical jets are studied in the nonlinear regime by performing 2D numerical classical gasdynamical calculations. Two kinds of instabilities are studied:

(1) Those which arise from unsteadiness in output from the central engine feeding the jets, and
(2) Those which arise from a beam in a turbulent surrounding.

An extra power output an order of magnitude higher than is normally delivered by the engine over a time equal to

(nozzle length)/(sound velocity at center)

causes a nonlinear Kelvin-Helmholtz instability in the jet walls. Constrictions move outwards, but the jet structure is left untouched. A beam in turbulent surroundings produces internal shocks over distances of a few beam widths. If viscosity is present the throughput of material is hampered on time scales of a few beam radius sound travel times. The implication are discussed.

1. INTRODUCTION

The fluid dynamical instability of astrophysical jets connected with double radio sources and possibly head tail sources is a longstanding problem. Many workers have already attacked the linear problem (see references in Nepveu 1980); the follow up of the nonlinear part is considerably harder and essentially requires numerical analysis (for an exception, see Nagano 1979). A further problem lies in the host of possible instabilities, that is in the host of possible perturbations of the "beam/surrounding" system. One should therefore look to those that seem to arise naturally from these systems.

In the following I describe results of numerical computations on two systems.

A. Ferrari and A. G. Pacholczyk (eds.), Astrophysical Jets, 303–308.
Copyright © 1983 by D. Reidel Publishing Company.

1) A ϕ- independent beam with a nozzle, consisting of light, hot plasma that is suddenly injected with hot, denser plasma.
2) An infinite slab consisting of light, hot plasma moving through denser, cooler surroundings that are turbulent.

In both cases initially pressure balance exists between beam/slab and surrounding material. Case 1) mimicks the tortures inflicted upon the beam walls through unsteady feeding by the central monster; case 2) offers the possibility to study which unstable modes the beam itself picks out. The velocity fields inside the beam and outside were endowed with turbulent viscosity in case 2).

2. RESULTS

I'll give the results sketchily: methods etc. are given in Nepveu 1982a, b.

Case 1) At t = 0 a fluid beam with a nozzle was constructed (Landau, Liftshitz 1974 Ch. IX) in pressure equilibrium with cooler, denser surroundings. In these surroundings pressure chopped like

$$p(R) = p_{origin} \exp(-R^2/K^2) \qquad (1)$$

$$K^2 = \frac{3k_B T}{Gm_{pr}\rho_s}$$

where k_B is Boltzmann's constant, m_{pr} is a proton mass, T a temperature, ρ_s the mass density near the origin, which coincides with the beams central region. Inside the beam the temperature at the origin was taken ten times higher than outside.

In the region between the origin and the nozzle throat extra mass and extra internal energy was added:

$$p_{extra} \lesssim 10\, p_{equil.}$$

The beam was no longer in equilibrium and the walls were inflated.

In fig. 1 it is shown what happens. A constriction forms at the beam wall W, behind a shock front which races ahead inside the beam and moves sideward into the cool medium. The inflation becomes increasingly serious ((a) → (c)): the constriction goes up to 90% in volume and moves steadily outward with the sound speed of the cool external medium. The beam is not totally wrecked! If a magnetic field B is applied along the beam the extra power output leads to wavy disturbances in the beam wall. The constrictions described before do not show up if

$$B^2 > 4\pi\rho\, v^2_{beam} \qquad (2)$$

Even for fields a factor two or three smaller, constrictions cannot nearly reach 90% level. This 90% constriction level for non magnetic cases is reached in a time

$$t_{90\%} \cong \frac{\text{throat circumference}}{\text{beam sound speed at origin}} \qquad (3)$$

Case 2) Turbulent surroundings induce "irregularities", shocks in the slab boundaries with spatial periods of a few beam widths. In the inviscid case these shocks are outspoken; in the cases with viscosity (Re $\leq$ 150) they are not (fig. 2, see the crosses). More important is that the slab broadens and that turbulence is induced inside. This broadening is an overheating effect as is easily established by looking at the pressure equation in presence of viscosity. After a time of order of

$$t_c \cong 5 \times \frac{\text{slab width}}{\text{sound speed inside slab}}$$

the neat flow pattern is completely gone: material inside is not transported unhampered anymore. The slab (beam) lost its coherency.

3. DISCUSSION

Case 1) At the place where constrictions rise through the beam the magnetic field (if sufficiently small) will be strongly tortured; relatively big current densities are to be expected. Micro instabilities might trigger relativistic particle acceleration (Norman and Smith 1978) and these places would show up in radio maps as brightness enhancements between center and hot spots. The distances between two such enhancements would be of order

$$\Delta L \simeq \underset{(c)}{(\text{external sound speed})} \times \underset{(\tau)}{(\text{memory loss time of central engine})}. \qquad (4)$$

This would be to order of magnitude:

$$\Delta L \simeq \frac{c}{(1000 \text{ Km s}^{-1})} \times \frac{\tau}{(10^6 \text{ yrs})} \text{ kpc} \qquad (4a)$$

Because of the diffusion speed for relativistic electrons $V_{\text{Alfvèn}}$ (Kulsrud and Pearce 1969) distinct blobs would "grow together" in radio maps after

$$t_{blurr} \simeq \frac{c_e}{V_A} \tau . \qquad (5)$$

In the supersonic parts of the beam it is well possible that $t_{blurr} \gg \tau$. Unfortunately the rising speeds of constrictions are so small

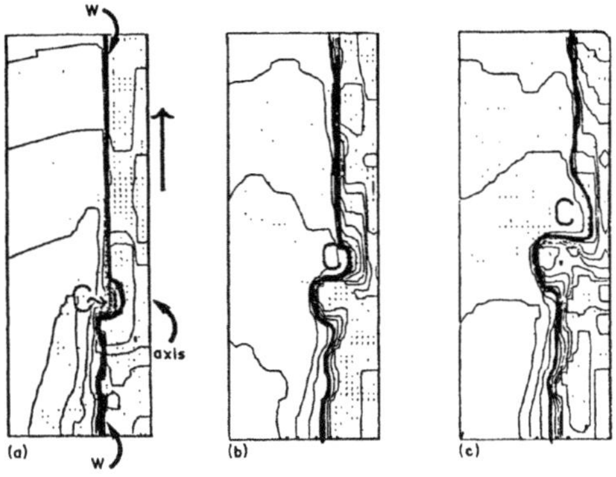

Fig. 1 - Evolution of a fluid beam with a disturbance. The arrow gives the flow direction in the beam. Disturbance energy density = 3 x thermal energy density at place of application at t = 0. Constriction at C.
The bundle of lines W representing density contours gives the boundary between beam and external medium. At b) the constriction at C has reached 90% level.

a) : time = 1.6 x (R_*/c)

b) : time = 5.6 x (R_*/c)

c) : time = 9.7 x (R_*/c)

R_* = throat radius
c = sound speed at center

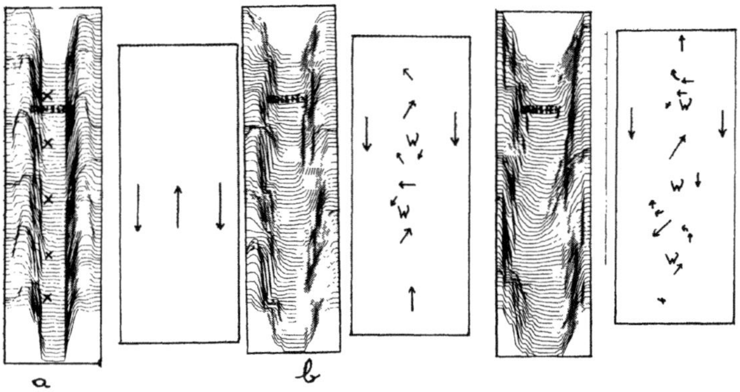

Fig. 2 - Ruled surface plots of plasma densities at 1.3 resp. 2.5 resp. 3.8 slab sound crossing times (a,b,c). Next to each ruled surface plot the vector fields are globally displayed. Turbulent eddies are seen at the places marked W. At the crosses (a) shocks can be seen. They are not very outspoken due to viscosity.

($< 10^3$ km s^{-1}) that there is no hope of determining them by observation. The explanation for the fact that the beam does not explode is connected with the shock system: shocks in the cool medium and the hot beam plasma are connected in such a way that the pressure balance between inside and outside beam is not heavily broken.

Case 2) Assuming that our unrealistic geometry nevertheless teaches us applicable physics we may assert following. The shocks along the beam boundaries are not unexpected and have been found earlier (Nepveu 1980, Tajima and Leboeuf, 1980). These shocks may well have the nice property of transferring bulk kinetic energy into other energy modes (Ferrari and Trussoni, 1981) thereby damping the large scale fluid instability. After all, quite a few straight jets are known (e.g. NGC6251; Blandford 1981).
If such energy conversion does not happen the distance over which the beam loses its coherency must be equal to

$$L_{blurr} \simeq O[\text{ Machnumber x beamwidth }] \quad . \qquad (6)$$

If the jet radius is 10^{1-2} pc (Blandford and Rees 1974), the flow Mach number 10^{1-2} (Spangler 1979) a blurring distance of $10^{\pm 1}$ kpc is to be expected. This, then, explain why a head tail like NGC1265 (Miley et al. 1975) blurrs over a short distance within this range.

A few last comments would be in order. Firstly, it is amazing that some relations which are supposed to hold only in the linear regime extend their validity into the nonlinear one. It was found, for instance, that the rms width of an infinitesimal slice of the slab boundary case 2 grew exponentially, fast up to the time that the internal velocity field becomes incoherent and fully turbulent.
Secondly, simple criteria for stability or instability do not work. If instability sets in certain relations might be linear longer than expected; whether instability sets in is not well determined by relations like the one by Miles (1958), concerning the values of Mach numbers.

4. CONCLUSIONS

1. Beam integrity is determined by the fate of imposed disturbances - not necessarily infinitesimal - when followed in the nonlinear regime. Those instabilities should be studied which are likely to occur in the system, e.g. those arising from external turbulence or internal beam unsteadiness.

2. Turbulent viscosity may hamper beam plasma flow (for Re = 150 it does so within a few beam width sound crossing times).
   Sudden changes in radiation patterns may arise over $10^{\pm 1}$ kpc (NGC 1265).

3. Brightness enhancements between center and hot spots may well be caused by unsteadiness of the central feeding monster. In that case these enhancements should move down the pipeline at speeds $\sim 10^3$ km s^{-1} or less; the central unsteadiness does not kill the beam structure if the power outputs are an order of magnitude higher over times necessary to fill the whole central region with extra, hot (dense) plasma.

ACKNOWLEDGEMENTS

I thank all those many people discussing with me problems related to this work.

Sonderforschungsbereich Radioastronomie supported this project financially.

REFERENCES

Blandford, R.D., 1981, in Plasma Astrophysics: Proceedings Varenna Workshop 27. Aug. - 7 Sept. 1981, ESA, p. 207
Ferrari, A. and Trussoni, E., above ref., p. 353
Kulsrud, R. and Pearce, W.P., 1969, Astroph. J. 156, 446
Landau, L.D. and Liftshitz, E.M., 1974, Hydrodynamik, Akademie Verlag Ch. IX.
Miley, G.K., Wellington, K.J. and v.d. Laan, H., 1975, Astron. & Astrophys. 38, 381
Miles, J.W., 1958, J. Fluid Mech. 4, 358
Nagano, H., 1979, J. Plasma Phys. 22, 27
Nepveu, M., 1980, Astron. & Astrophys. 84, 14
Nepveu, M., 1982a, Astron. & Astrophys. 105, 15
Nepveu, M., 1982b, Astron. & Astrophys. 112, 223
Norman, C.A. and Smith, R.A., 1978, Astron. Astrophys. 68, 145
Tajima, T. and Leboeuf, J.N., 1980, Phys. Fluids 23, 884

MHD KELVIN-HELMHOLTZ INSTABILITIES AND LARGE SCALE PHENOMENA IN JETS

L. Zaninetti (1), E. Trussoni (2, 3)

(1) Istituto di Fisica Generale dell'Università, Torino, Italy
(2) Istituto di Cosmogeofisica del C.N.R., Torino, Italy
(3) Max-Plank Institut für Extraterrestrische Physik, Garching b München, Germany

I. INTRODUCTION

Recent VLA data have shown that wiggles might be a common feature of extragalactic radio-jets; knots instead have been so far detected only in M87, Cen A, and possibly NGC6251. Several theoretical models have been proposed to explain these observational morphologies (see Trussoni et al., these Proceedings), as gravitational interaction (orbital motions, precessing beam, etc.) or plasma instabilities intrinsic to jet propagation. In this Note we want to investigate in detail this second possibility, i.e. that large scale phenomena are ruled by Kelvin-Helmholtz magnetohydrodynamical instabilities due to the beam interaction with an external (intergalactic or intracluster) medium. In particular from the comparison of observed scalelengths of wiggles and jet's lengths with theoretical wavelengths, λ_{max}, and e-folding lengths, l_e, of the most unstable helical modes, the physical parameters of the flow (Mach number and density) can be obtained. In Table I we have reported a selected sample of jets for which these properties of wiggles and lengths have been well studied.

II. INSTABILITY ANALYSIS AND COMPARISON WITH OBSERVATIONS

Assuming an axisymmetric flow confined by an external medium and with magnetic field parallel to the velocity, standard MHD equations can be linearised assuming for the perturbations a behaviour like

$$g(r) \exp \{i (k z + n \theta - \omega t)\} \qquad (1)$$

with ω complex frequency and n azymuthal number (Ferrari et al. 1981). Unstable phincing (n=0) and helical (n=1) modes can be directly related with observed knots and wiggles. The main results of the stability analysis are summarised as follows:

A. Ferrari and A. G. Pacholczyk (eds.), Astrophysical Jets, 309–313.
Copyright © 1983 by D. Reidel Publishing Company.

Table I

SOURCE	λ_w /R	D/R
NGC6251	17.5	55
3C449	14	45
NGC315	6	41
M87	22	53
4CT74.17.1	7	20
0326+396W	12	18
3C31	6	16

Observational parameters of 7 VLA Jets

column 1: Source name
column 2: Ratio between λ_w, wavelength of the observed oscillations, and R, jet radius taken as the mean HWHM of the jet over the region of wiggling
column 3: Jet's length taken as the linear distance from the core after which the flow looses directivity.

1. Instability is present for all Mach numbers, M, and growth rates are similar for n=0 and n=1 in the supersonic regime.

2. The fastest growing pinching and helical modes have wavelengths λ_{max} of the order or larger than the beam radius (namely $\lambda_{hel} \sim 2\div3\, \lambda_{pinch}$). In addition several other modes with shorter wavelengths can grow with sligthly smaller growth rates ("overreflected" perturbations).

3. An aligned magnetic field does not critically affect these results as far as the Alfvén velocity is lower than half of the jet velocity.

The fastest growing wavelengths λ_{max} and the corresponding e-folding lengths l_e (= phase velocity/growth rate) are interpolated from the numerical solutions of the dispersion relation. One obtains the following analytical relations between λ_{max} and l_e in terms of the Mach number M and jet density $\bar{\rho}$ (in units of the density of the external medium) (see Ferrari et al. 1983):

$$\lambda_{max}/R = M\,(a_\lambda + b_\lambda \ln\bar{\rho}\,)$$
$$l_e/R = M\,(a_{ef} + b_{ef} \ln\bar{\rho}\,) \qquad (2)$$

which are valid for $\bar{\rho} \leq 100$ and $M \geq 10$ (the coefficients a, b are given in Table II). Identification of λ_{max} in the first of 2) with the

Table II				
n=1				
*	a_λ	b_λ	a_{ef}	b_{ef}
$V_a/V_s=1$	1.3	-0.21	0.63	0.03
$V_a/V_s=0$	1.8	-0.31	0.93	0.04

Values of the coefficients defined in equations (2) for unmagnetized and MHD flows. In this last case Alfvēn and sound speed are equal, and the magnetic field ten times higher in the jet than in the external medium.

observed wavelength of wiggles or knots is obvious. Concerning the e-folding length, this provides an estimate of the length over which nonlinear effect can arise. In particular we can roughly say that the beam length corresponds to the distance after which the instability destroys the collimated flow. We then measure the length D of the beams in units of l_e's:

$$D = f\, l_e \qquad f \geq 1 \tag{3}$$

From Eqs. (2) and (3) density and Mach number of jets are easily obtained; the results for our sample are in Table III. From this we conclude that, if the wiggles are due to the development of unstable K-H modes, jets must be dense and supersonic (unless $f \geq 5$).

III. SUMMARY AND SUGGESTIONS TO OBSERVERS

The occurence of Kelvin-Helmholtz instabilities can be reasonably related with the features of extragalactic jets as follows:

1. Their oscillatory behaviour is associated with the most unstable wavelength of helical modes;

2. "Harmonics" of the largest scale oscillations (as suggested for NGC6251 by Perley in these Proceedings) are due to the effect of overreflected modes;

3. A decrease of the wiggles wavelength along the jet (e.g. in 3C273) is connected with a decrease of the Mach number;

Table III						
	f=1		f=3		f=5	
SOURCE	M	$\bar{\rho}$	M	$\bar{\rho}$	M	$\bar{\rho}$
NGC6251	70	160	25	19	17	3.3
3C449	58	160	21	21	13	3.6
NGC315	51	290	18	100	10	27
M87	69	110	25	8.4	17	1.0
4CT 74.17	26	140	9.4	15	6.1	2.2
0 326	24	48	9.4	1.2	6.5	.08
3C31	20	200	7.2	37	4.6	8.3

Physical parameters of jets derived from the comparison of our model with observations: M, Mach number, $\bar{\rho}$ density contrast. These quantities are evaluated for different f's the number of e-folding lengths after which we observe the occurence of nonlinear phenomena.

4. Knots, when observed, are due to the same instability process, but correspond to pinching modes; the interblob distance in M87 is roughly 3 times shorter than the wiggle lengths, in accordance with the theory. The fact that knots are a less common feature is likely to be due to observational limitations.

5. The abrupt termination of the jet after a few oscillations could be related to a nonlinear development of instability.

On other hand a difficulty is that our model predicts dense jets: this could lead to heavy energy and mass requirements in the phase of origin and acceleration.

Finally we indicate which measurements on well resolved jets might provide tests for our suggestions:

1. Our intrinsic model implies that both λ and D must be correlated with the radius of the jet, taken as the mean HWHM of the emissivity over the region of wiggling.

2. The lateral displacement of a jet from a straight line must be studied as a function of the distance from the core in order to discriminate between exponential and linear growths.

3. The transverse and longitudinal dimensions of classical doubles must be correlated to test the hypothesis that hot spots are generated by the disruption of invible beams.

4. Higher resolution data could detect knots in other jets allowing for a comparison with wiggle wavelengths.

ACKNOWLEDGMENTS

L. Z. thanks the "Fondazione Angelo della Riccia" and "Commissione per gli Scambi Culturali fra l'Italia e gli Stati Uniti" for financial support during a visit to the National Radio Astronomy Observatory. We thank A. Bridle, E. Fomalont, J. Burns and R. Perley for valuable discussions.

REFERENCES

Ferrari, A., Trussoni, E., Zaninetti, L., M.N.R.A.S. 196, 1051
Ferrari, A., Trussoni, E., Zaninetti, L., submitted to Astr. Astrophys.

DISCUSSION

G. BENFORD: The problem with applying linear analysis is that, by the time you can see a wiggle, it exceeds R (beam radius)/M (Mach Number) for $M > 4$, and is thus nonlinear, since it excites shocks in either the beam or the cocoon. Can you deal with this.

L. ZANINETTI: Although it is clear that nonlinear effects provide formation of shocks in unstable beams, we do not know yet whether wiggles are formed in a linear or nonlinear regime. This point will perhaps become more clear with more detailed observations of the wiggle amplitude evolution along the jet. At this stage we assumed that nonlinear effect are responsible for the flaring of the jet, while for about 5 growth lengths we may still apply linear results. This seems to be confirmed by numerical results (see Akira & Miura, Phys. Rev. Lett. 49, 779 , 1982).

CONCLUDING REMARKS: A PROGRESS REPORT ON OUR UNDERSTANDING OF JETS

G.C. Perola
Istituto di Astronomia, Università di Roma, Italy
A. Ferrari
Istituto di Fisica Generale, Università di Torino, Italy

I. INTRODUCTION

This Workshop has been so full of interesting contributions that the time devoted to open discussions has progressively shrinken to coffee and lunch breaks. Therefore we think useful to attempt a summary of which we gathered the audience's feelings were about the basic problems of the physics of jets.

We shall follow the list of comments presented by one of us (GCP) in his concluding talk at the Workshop; we have rewritten them in the perspective of the floor's discussions which followed. Obviously our presentation will be somewhat personal in the choice of the highlights. However we hope to be able to give a fair report on the status of the art by the end of 1982.

II. OBSERVATIONS

The observational situation has been further enriched after the IAU Albuquerque Symposium. As discussed in their reviews by Browne, Preuss, Burns and Fomalont, we have a collection of radio maps of jets at different resolutions, from milliseconds to several seconds of arc, showing a large variety of morphologies at all scales: knots, wiggles, large bendings, etc., often with contrasting symmetries which do not allow straightforward kinematic interpretations valid for all cases. This means that, at present, we are not yet in the condition to define statistically the "standard properties" of jets.

Concerning nuclear jets, the situation is still very uncertain. Although in many cases observers claim a definite alignment of nuclear jets with extended collimated structures, this is not always the case, also due to the presence of strong bends. However, as it was recalled by Browne, projection effects are likely to be responsible for a conspicuous enhancement of rather small (a few degrees) intrinsic misalignments.

A. Ferrari and A. G. Pacholczyk (eds.), Astrophysical Jets, 315–320.
Copyright © 1983 by D. Reidel Publishing Company.

In large scale jets, instead, we begin to find systematic patterns in the variety of parameters and morphologies. A preliminary classification has been presented by Fomalont, who, generally speaking, has confirmed most of the ideas put forward by Bridle in Albuquerque: in particular the correlations between total power and linear size, ratio $B_{\parallel}/B_{\perp}$ and distance from core, wiggle and knot scale and jet radius, average local emissivity and distance from core, etc.

Given this situation, it was naturally the large scale jet to attract much of the attention during the Workshop. Fanti has presented a rigorous collection of the physical parameters which can be derived from observations in terms of standard arguments (equipartition, pressure confinement, free jet expansion, etc.). In particular large scale jets do not appear to be relativistic in the flow speed, their typical plasma densities are not much different from, in fact possibly smaller than, that of the external medium, in many cases they may well be pressure confined by hot gas. The picture may be quite different for nuclear jets, whose flow velocity is definitely relativistic in a few cases, and where the impact of the interaction with the environment on their dynamics is yet very unclear.

Obviously all these intrinsic physical parameters could vary substantially in the future, if phenomenological arguments should be found too approximate. At present, however, they agree in giving us a picture in which jets are externally confined, supersonic (but not relativistic) and lighter than the external medium.

III. DEFINITION OF MODEL PARAMETERS

What kind of picture of a jet arises from these preliminary data? Namely, what is the general consensus on basic parameters to which theoreticians should apply their models?

III.1 Propagation and confinement

Most of the theoretical papers presented at the Workshop have been devoted to explain the propagation mode with special interest for (a) jet confinement, (b) morphological features related to external effects (from large scale bendings as in NGC1265, to wiggles as in M87, NGC6251, 3C449, etc.).

With regard to jet confinement, observational data are rarely clear-cut. The external pressure by hot gas is derived from measurements of X-ray diffuse emission, which are available only for galaxy clusters with strong emission. Internal pressures are derived from the usual equipartition, which might be wrong; in particular the thermal plasma might provide the dominant pressure term (we have no measurements of the jet's plasma temperatures). In fact the only clear-cut evidence would be a negative one, where the external thermal pressure terms come out to be significantly less than the "minimum" internal pressure compatible with the observed radio parameters. In only a few cases such negative evidence exists; the jet in M87 is one of them, with the external pressure being at least a factor 10 smaller than the minimum internal one. In general specialists seem to be inclined to think that jets are confined, either by fluid pressure or by magnetic fields. The idea of free jets has lost most of its appeal, because VLA data have clearly shown regions of recollimation also. Of course there must be phases of free (or almost free) propagation, in order to allow a jet to expand from a core of size 10^{-3}pc to a radius often larger than 50 pc, but on the average a confining agent has to be present, if we have to avoid flow speeds in large excess over predictions from stopping distances and wiggle amplitudes. Benford has given arguments for magnetic confinement, and we shall return to his ideas. Even if we may not want to use magnetic fields in all cases (ordered currents would be required, whose origin should be explained in terms of origin models), these seem to be the only explanation for the strict collimation of jets like the one in M87.

If one agrees upon the necessity of confinement, the next step is to realize that a confined flow is subject to several kinds of fluid instabilities. These may destroy completely the condition for confinement, mixing the jet plasma with the external medium: exactly what happens in the case of smoke coming out of a chimney. In particular we know from these studies that the instability responsible for beam diffusion is the so-called Kelvin-Helmholtz instability. In fact several authors, since 1976, have published papers on K-H instability in supersonic and magnetized flows, and considering also the specific geometry of cylindrical flows. Today we have a clear picture showing that these beams are actually unstable, with growth time scales fast as compared with the flow convective timescales. This result originated the mode of free jets to avoid any interaction with external perturbing agents; this simplistic view is however incorrect, because instability can always arise in the surface layers of beams due to the presence of shears. In addition it has been realized that instabilities, provided they are saturated at low amplitudes, can be an efficient way to extract energy from the flow to support local jet emissivity. This mechanism has been referred to by several speakers during the Workshop (for instance Perley, Benford, Zaninetti).

Trussoni has discussed the consequences of these instabilities in terms of morphologies, referring to observed small amplitude structures as wiggles and knots; in this way one can also derive limits on the flow parameters. For instance one can discuss whether jets are denser or lighter than their immediate environment. For most jets the indication is that the flow Mach number is ~ 3-10 and the ratio of internal to external density is of the order of, or larger than unity. While the values of Mach numbers are in the generally accepted range, the estimate on density does not agree with others obtained from free-jet models or from confinement considerations, which predict lighter beams.

Norman has discussed a series of very interesting numerical calculations on jet propagation. They seem to confirm the picture presented above, especially in what refers to the global saturation of instabilities. It is not clear if these calculations have an immediate application to the interpretation of astrophysical jets, since they use a viscosity whose value cannot be defined "a priori" in terms of observations. Furthermore, note that the numerical plots represent the thermal component of the plasma, while our direct informations come from synchrotron emission, i.e. are related to the nonthermal component.

III.2 The role of magnetic fields

Benford has discussed in detail the role of magnetic fields in confining jets. An azimuthal field is required, which can be generated by a current along the jet itself, possibly due to a small charge separation left by electrodynamic processes of acceleration inside the deep galactic core. For what it concerns observations, this model does not substantially differ from those presented above. In this case, however, resistive instabilities arise, whose large scale topology easily applies to knots and wiggles (Ferrari and Trussoni, Varenna 1981). In addition they provide a continuous generation of MHD turbulent modes.

Unfortunately polarization measurements, which allow to determine directly magnetic field structures, do not display any clear evidence of azimuthal fields; even worse, the field at the beam edges is always parallel to the flow direction. Of course one can always imagine that the confining field is invisible because, due to its effect, relativistic particles are concentrated on the axis, so that the emissivity from circumferential regions is too low to be observed.

An important advantage of magnetic confinement is that of being a self-collimating process; this is an interesting fact to explain jets in which the transverse dimension displays many regimes, and also recollimations, as for instance in NGC315. In this connection we would like to recall also that the presence of an electromagnetic component in the flow, as for instance a large amplitude coherent wave, would provide also a self-focusing effect, independently of any "ad hoc" magnetic field. Models of this type, originally presented by Martin Rees for a many-pulsar galactic core, are today out of fashion because it seems very difficult to find observational evidence in their support. Nevertheless they did not encounter so far any basic objection.

III.3 Observational tests

Which possible observational tests of proposed models were suggested during the Workshop? Actually all speakers were very careful in discussing this point, mainly because it looks very difficult to find clear-cut arguments. We believe that one interesting possibility would be to measure, when feasible, proper motions of jet's features, such as in the knots of M87. This measurement has been already attempted (M. Read, private communication), but results are still under analysis. If proper motions were observed, one could directly determine the flow velocity (optical lines do not seem to be a good tracer, as discussed by G. Miley); in addition one could discuss whether these features are convected by the flow, which is perfectly compatible with Kelvin-Helmholtz instabilities, and not with resistive instabilities, related to an ordered (steady) magnetic field.

Another interesting point put forward by Fomalont (Section 2) and Fanti (Table 1 and 4) is that in many cases the jet's opening angle is anticorrelated with the total luminosity. Correlations of opening angles with other physical parameters are in fact expected in self-consistent models. In these models the jet emissivity is due to some dissipation mechanism of the flow kinetic energy; obviously, energy dissipation entails momentum dissipation also, and therefore the flow morphology (namely, opening angle), is directly connected to the jet emissivity, with specific patterns for each model. However, apart from very few attempts the observational implications have still to be fully investigated. In addition, from the observational side, such a comparison requires once more the definition of a statistically complete sample of jets.

An important aspect of the interpretation of jet physics that did not receive appropriate attention during the Workshop was the existence of one-sided jets, associated with double radio lobes (as for instance M87, again). This has been suggested as an evidence for the occurence of relativistic flow velocities on kpc scales (or smaller) and, correspondingly, Doppler enhancement of brightness of the jet pointing to the observer. Other independent observations on the same sources, however, do no support the existence of relativistic velocities; nevertheless, we believe that this point should be investigated by large dynamic range instruments, as they might provide definite informations on the flow origin and typical acceleration phases.

IV. ORIGIN AND TERMINATION OF JETS

Some discussions were devoted to the physics of termination of jets with formation of lobes, and plasma heating for optical emission. Pelletier suggested the formation of standing shocks at the head of beamed flows and discussed how a cascade of turbulent MHD modes can be formed leading to particle acceleration. Zaninetti indicated also the possibility that the nonlinear development of fluid instabilities can produce flarings of the collimated flows and, eventually, halt the directional motion. This global instability might actually develop into the shock structure proposed by Pelletier. In this connection it is important to recall that, in the case of the radio jets associated with the galactic X-ray source, ScoX-1, proper motion measurements have allowed us to determine an upper limit on the velocity of advancement of the lobes. This velocity is much smaller than what is required by travel time or energy flow arguments. This fact implies that the process responsible for the beam termination must be very efficient in depositing and isotropizing momentum. If this is the case for extragalactic jets also, light jets would be favoured, and electromagnetic beams also which display a large variety of fast instabilities on short scales.

Concerning the origin of jets, Calvani and Nobili, Phinney and Begelman have presented new developments of the ideas based on accretion disks; the crucial point is still that of finding a way to produce supersonic (perhaps relativistic) flows from cores which are not particularly luminous, and therefore corresponding to subcritical accretion rates. Electromagnetic processes of acceleration appear to be favoured at the moment, although they appear very complicated. From the theoretical point of view wind-type solutions are also under investigation (Tsynganos, Ferrari, Rosner, in preparation), which might satisfy observational requirements without too exotic assumptions as in electromagnetic models. Tests for these mechanisms will possibly come from further observations of VLBI jets.

Another point that would require more attention is the physics of the local acceleration of relativistic electrons and, correspondingly, nonthermal radiation emission and transport, especially in the deep core regions. At the Workshop no new developments were reported, although the requirement for a better understanding of the problem were expressed in many comments.

V. CONCLUSIONS

We finally express our feeling that the Workshop was very useful for all participants in providing an opportunity to state once more the difficulties in the study of this very peculiar objects of our modern astronomy. Although not many answers were obtained, nevertheless new insights were discussed, which will guide us all in the next year or so, before gathering again.

SUBJECT INDEX

OBJECT INDEX

9 789027 716279

GPSR Compliance

The European Union's (EU) General Product Safety Regulation (GPSR) is a set of rules that requires consumer products to be safe and our obligations to ensure this.

If you have any concerns about our products, you can contact us on ProductSafety@springernature.com

In case Publisher is established outside the EU, the EU authorized representative is:

Springer Nature Customer Service Center GmbH
Europaplatz 3
69115 Heidelberg, Germany

Batch number: 10816188

Printed by Printforce, the Netherlands